The Business of Bees

An Integrated Approach to Bee Decline and Corporate Responsibility

THE BUSINESS OF BEES

An Integrated Approach to Bee Decline and Corporate Responsibility

EDITED BY JILL ATKINS AND BARRY ATKINS

LONDON AND NEW YORK

First published 2016 by Greenleaf Publishing Limited

Published 2017 by Routledge
2 Park Square, Milton Park, Abingdon, Oxon O14 4RN
711 Third Avenue, New ork, N 10017, USA

Routledge is an imprint of the Taylor & Francis Group, an informa business

Cover by Sadie Gornall-Jones.
Illustrations by Jolyon Hartin.

British Library Cataloguing in Publication Data:
A catalogue record for this book is available from the British Library.

ISBN-13: 978-1-78353-522-4 [hbk]
ISBN-13: 978-1-78353-435-7 [pbk]

To all of our beloved family

Contents

Preface

> Bees are the most excellent of all insects whatsoever,
> and expresse both worth and wonder in all their waies
> (opening to Samuel Purchas's *A Theatre of Politicall Flying-Insects*, 1657).

Reading and relaxing in the idyllic, peaceful bliss of our Welsh mountains garden, soaking up the warm, embracing rays of the early summer sun, it is almost impossible to imagine that the world is in the grips of climatic catastrophe, mass species extinction and the potential end to all life on Planet Earth. The bumblebees wend their clumsy way from our lavender bush to the pot of fat, purple chives, bumping through the breeze to pots of herbs and assorted flowers. A bee crisis? A catastrophe in global bee populations? Threats to the world's food supplies? Surely not. Certainly not in our garden, anyway. We are fortunate enough to live in a country which is free from mass poverty, war, famine, disease. We, personally, live in a little piece of heaven on earth and are grateful every day for our good fortune. Others, many millions of others, are far less fortunate. Vast areas of our planet are inhabited by people and cultures facing severe and life-threatening conditions on a daily basis. There is no "us and them". We all share one planet and ultimately what threatens some threatens all, if not now, in the future. As scientists become increasingly aware of the, possibly irreversible, effects of global warming and climate change, societies worldwide are starting to appreciate the danger to themselves and to humanity arising from these effects. The climate change debate and the urgent need for countries, politicians, governments, local councils and religious leaders to take action is not the subject of this book. Instead, we take one tiny element of this global and immense problem: the role of the bee. At first, this may seem like a fanciful diversion from the elephant in the room and the impending problems associated with climate change. However, we feel that it is in the detail, the small elements of the broader picture, where change and action are crucial. As we discuss below, if a systems and ecosystems approach is applied to global problems then the destruction or disappearance of one tiny cog in the wheel of Planet Earth can affect the workings of the whole system. The devil is in the detail. If bees disappear this could, ultimately, have catastrophic effects on nature and certainly on humankind. But this story is not simply about the survival and comfort of humanity but also about Earth, nature, flora and fauna and the survival and continuance of all life on Earth.

In writing this book, we have read hundreds of books, articles, reports, scientific studies and websites only to be confronted with the stark reality that not only are bees in serious decline but that this fall in numbers is affecting food production, stock markets, investors' pockets and corporate value, as well as destroying the long and cherished relationship between bees and the human race. Bee decline appears to be a result of not one isolated cause but instead, a great many factors including (possibly) pesticide use, habitat loss, reduction in floral diversity and natural pests such as mites, wax moths and "foul brood". The relationship between people and bees can be traced back to the earliest times of human activity and is a vital part of the exploration and discussion within this book. We are immensely grateful to the many writers, academics and practitioners who have kindly contributed chapters ranging from discussions of bee science to financial studies and accounting research.

This book begins with a leisurely meander along the age-old path of bees and the way the relationship between bees and humans has evolved over many centuries. Then we start a purposeful trek through the "Business of Bees" and the financial and financially material repercussions of bee decline. Specifically, the authors examine the nature of corporate accountability for bee decline and what companies are doing to counteract the fall in bee numbers, as well as the impact on the financial markets of bee decline. Ultimately this book aims to raise awareness of the crucial importance bees play in our lives and our future, not just within the academic community and the business world but among people generally. The financial and accounting aspects represent the ways in which bee decline is affecting world business and corporate governance. However, there are so many more ways in which the decline in bee populations worldwide is a threat to our culture, our heritage and our way of life. These issues are also core to the book and its discussions.

Bees are starting to take centre stage on the global agenda and media platforms. Governments worldwide are introducing regulation and initiating policies to tackle the rapid decrease in bee populations and indeed the decline in pollinators generally. Governments, companies, local communities, investors, accountants and NGOs are mobilizing to save bees worldwide. Bee decline affects and will continue to affect people at all levels and in all places around the world: rich and poor alike. Bee decline is a true "leveller". Bees are a natural asset which people have used as a free good without considering the consequences of their exploitation. Unless urgent action is taken, future generations will not have these natural assets available to them.

In this book, we explore the multifaceted aspects of bee decline where human history, science and scientific discovery, philosophy, religion, music, art, technology, accounting, finance and business are all stitched together in a global patchwork quilt of interdependent factors, problems and concepts. The primary focus of *The Business of Bees* is to explore the current and growing impact of bee decline on the corporate sector, on company performance, on financial return, on institutional investment, on the world's stock markets and on business generally. However, by adopting a "systems" approach to our analysis and discussion, we acknowledge and indeed seek to promote a broad and integrated appreciation of the relevance of bees to every aspect of human and non-human life and, indeed, existence, on Planet Earth and certainly the continuance of humanity.

Acknowledgements

In writing, or rather, "constructing", this book, we are immensely grateful to our contributors who come from all corners of our planet. Coming from academia, from the investment industry, from NGOs, they provide a mixture of scholarly and practitioner/professional perspectives which ensure that the final product addresses bee decline from all sides. In alphabetical order, we acknowledge the contribution of the following people, and thank them for their hard work and commitment to this project. Without them, it would have been impossible to produce this book:

- Elisabetta Barone, Brunel University
- Christoph Biehl, Beijing Institute of Technology and Henley Business School
- Jack Christian, Manchester Metropolitan University, UK
- Margaret Clappison, University of Athabasca, Canada
- Samuel Discua
- Koichi Goka, National Institute for Environmental Studies, Japan
- Abigail Herron, Aviva Investors
- Joël Houdet, African Centre for Technology Studies (ACTS); Albert Luthuli Centre for Responsible Leadership (ALCRL), University of Pretoria; Synergiz
- Kristina Jonäll, University of Gothenburg, Sweden
- Scott Longing, Texas Tech University, USA
- Warren Maroun, University of the Witwatersrand, South Africa
- Martina Macpherson
- Masahiro Mitsuhata, Arysta LifeScience, Japan
- Carol Reade, San José State University, USA

- Gunnar Rimmel, University of Jonkoping, Sweden
- Andrea Romi, Texas Tech University, USA
- Aris Solomon, University of Athabasca, Canada
- Rick Stathers, Schroders Investments Ltd (at the time of writing)
- Olivia Stewart, Preventable Surprises, UK
- Raj Thamotheram, Preventable Surprises
- Robbin Thorp, University of California, Davis, USA
- Ruan Veldtman, South African National Biodiversity Institute (SANBI)
- Marius Wasbauer, University of California, Davis, USA

We would like to thank a number people who have commented at presentations and discussions on the working drafts of the book including, Michelle Aucock (University of the Witwatersrand), David Coldwell (University of the Witwatersrand), Robert Garnett (University of the Witwatersrand), Rob Gray (University of St Andrews), Philip de Jager (University of Cape Town), Mervyn E. King (Chair, International Integrated Reporting Council, IIRC) and Nirupa Padua (University of the Witwatersrand). We would actually like to mention a certain Rob Gray again, as he read early drafts and helped to guide us in orientating our book theoretically and stylistically. We are also hugely indebted to our publishers for their encouragement, support and guidance throughout the production of this book, especially to Rebecca Macklin, Rhian Williams and Cathie Manis. Thanks also to Jolyon Hartin for providing sketches of bees to illustrate the text: you are a star.

Jill and Barry Atkins
April 2016

Part I

The historical, cultural, philosophical and scientific context of bee decline

1

Bee decline

An integrated approach

Jill Atkins
University of Sheffield, UK

Barry Atkins
University of South Wales, UK

This is a book about bees. We were inspired to put the book together because of the critical situation bees find themselves in. Bees are under serious threat, with their numbers declining worldwide. As this book bears the title, *The Business of Bees*, there is clearly a focus on the business-related aspects of bee decline. Throughout the book, the contributors discuss bees in relation to business, finance and accounting. There is also a focus on bee science and discussions around the causes of bee decline, from a scientific perspective. However, business and science are not enough to understand and appreciate the historical, cultural and philosophical importance of bees to the human race. Chapter 2 is therefore devoted to discussing bees from a wide range of different perspectives. Furthermore, it is important to consider bees in terms of their role in Nature and their intrinsic value. By intrinsic value, we mean the value of bees as a species within Nature without any consideration of their relationship to the human species. In order to think about bees in this wider, universal context, we discuss various frameworks, from theory and philosophy, which can be used to place bees at the centre of our analysis, rather than focus on a human-centred approach.

Bees have been identified as a critical species within our ecosystem, providing an ecosystem service through pollination, as well as honey production, without which world food production would become extremely problematic. A book which reflects on bees from a societal perspective stated that,

> ...human industry has not become so big that it can do without bees. Where bees are scarce, plants fail to pollinate, fruits fail to grow, species die out, and the air is less oxygen-rich for us to breathe. Even in these days of agribusiness, bees remain a more efficient way to pollinate an orchard than any other, and farmers will pay beekeepers to set up apiaries on their land.[1]

The decline in the population of *Apis mellifera* (the European honey bee) has caused considerable alarm, since it is one of the most common "commercial" bees (we will return to the concept and meaning of commercial bees later). It is widely recognized among the scientific community that, "Pollination services strongly depend on the total of all individuals in a community, the so-called aggregate abundance of flower visitors".[2]

Bee decline, bee statistics and financial risk

The United Nations' Food and Agriculture Organization has highlighted reports of pollinator decline in at least one region or country on every continent (except Antarctica).[3] Reasons for this loss of pollinators are thought to include habitat decline, pesticide use and the spread of disease. Bees are among a wide range of pollinators, which also include humming birds and bats. The situation for bees is critical, with some species listed as endangered and at risk of extinction. Colony collapse disorder (CCD), where bee colonies disappear suddenly, has become a serious problem in countries across the world.

Ultimately, no pollinators mean no pollination, which means no crops and no food. Not all food is produced via insect pollination but a large amount is, as we shall see in discussions throughout this book. Certain species would be particularly affected by declining pollinator populations, namely sensitive plants including: apples, apricots, avocados, cashew apple, cherries, cranberries, kiwi fruit, peaches,

1 This quotation is taken from Bee Wilson, *The Hive: The Story of the Honeybee and Us* (London: John Murray, 2004), which traces the relationship between bees and humans through history.

2 This quotation was taken from S. Naeem *et al.*, *Biodiversity, Ecosystem Functioning and Human Wellbeing: An Ecological and Economic Perspective* (Oxford, UK: Oxford University Press, 2009).

3 FAO, *Rapid Assessment of Pollinators' Status: A Contribution to the International Initiative for the Conservation and Sustainable Use of Pollinators* (Rome: FAO, 2008).

nectarines, pears, plums, sloes, raspberries, vanilla, cocoa beans, coffee, almonds, Brazil nuts, cashew nuts, melons, pumpkins, squash, gourds and watermelons.[4] Losing these types of fruits and nuts is now thought to pose a serious risk to human health, given the potential deterioration in our varied diets. It would also be an unpleasant prospect for most to lose something as dearly loved as chocolate and coffee!

As mentioned above, a massive financial risk is associated with decline in bee populations and other pollinators. There are many estimates of the financial impact of losing pollinators, going back several decades. As early as 1968, it was estimated in South Africa that cross-pollinated fruits realized South African R48,000,000 but that without honey bees this amount would be R16,000,000.[5] In the UK, in 2007 the National Audit Office (NAO) concluded from collating research that the value of bees' pollinating "services" to the UK economy was approximately £200 million a year.[6] The potential financial impact of bee decline has been estimated more recently at $15 billion in the US alone.[7] For 2005, scientists from France and Germany estimated that the worldwide economic value of pollination "services" "provided" primarily by bees was $153 billion for the main crops that feed the world.[8] In 2013, Greenpeace quoted research which estimated the global economic benefit of pollination as around $265 billion, based on the value of crops dependent on natural pollination.[9] However, this estimate, and all the other approximations discussed here, does not take into account the fact that it would be practically impossible to replace natural pollination and the cost would be unimaginably high. Whichever study we look at, the figures differ but the fact remains that they are very large. They also grossly underestimate the costs of trying to "replace" natural pollination. Furthermore, they do not make the slightest attempt to account for what we would term "intrinsic value", i.e. the value of bees *per se*, as creatures on the planet, and their role in the wider ecological system. How can we possibly "value" in monetary terms, the loss of a species? From a business point of view, it is important to attempt estimates of the value of bee decline in order to appreciate why loss of bees is important for food production, for food retailing, for food security, for business

4 These species are summarized in R. Stathers, *The Bee and the Stock Market: An Overview of Pollinator Decline and its Economic and Corporate Significance* (Research Paper, Schroders Investment Ltd, 2014).

5 This estimate was given for the South African context in A.F. May, *Beekeeping* (Cape Town, South Africa: Haum, 1969).

6 See N. Holland, "The Economic Value of Honeybees", *BBC World News Online*, 23 April 2009.

7 See M.B. Albright, "Could Robot Bees Help Save Our Crops?", *National Geographic*, 21 August 2014.

8 Helmholtz Association of German Research Centres, "Economic Value of Insect Pollination Worldwide Estimated at U.S. $217 Billion", *ScienceDaily*, 15 September 2008.

9 Greenpeace, *Bees in Decline: A Review of Factors that put Pollinators and Agriculture in Europe at Risk* (Technical Report (Review), Greenpeace Research Laboratories, January 2013).

profits and for investors. However, the business chain does not even begin to assess the impact of bee loss on people's diet and ultimately survival as well as on the ecosystem as a whole.

The impending financial disaster relating to bee decline predicted by scientists around the world has spurred the White House into issuing a presidential memorandum involving the creation of a Pollinator Health Task Force, led by the President. The Task Force aims to explore causes of CCD, covering a range of possible causal factors from insecticide usage to bee diseases and issues arising from different beekeeping methods. The global economic value of pollination was estimated at US$217 billion in 2008.[10] American scientists are even exploring mechanical replacements for bees. One possibility, being investigated by scientists at Harvard University in the USA, is to create a "robo bee", a mechanical bee, which can pollinate crops instead of live bees.[11] Another "quick fix" under consideration is the creation of a virus-resistant "superbee".[12]

The "business of bees" has become big business for humans over the last few centuries. Honey production on a commercial basis has burgeoned into a massive business worldwide. In the late 1950s, it was estimated that the USA produced 242 million pounds (110,000 tonnes) of honey in a year as well as 4,476,000 pounds (2,030 tonnes) of beeswax.[13] At the same time, Russia (then the USSR) was producing a similar amount of honey to the USA, and Australia was producing 40.5 million pounds (18,400 tonnes) of honey per year.

So, what will happen when, or if, we lose bees altogether? Well, this has actually already happened which means we have a real live example to take note from. In the Chinese province of Sichuan, pear trees now have to be pollinated by hand, as all honey bees died off, possibly due to pesticide use, in the 1980s. This situation provides a micro image of how the world could look without bees. The hand pollination process is slow and labour-intensive. Writers suggest that if this form of pollination had to be used in the USA it would cost approximately $90 billion per year.[14] Something as seemingly natural and apparently simple as bees pollinating our crops is becoming something unnatural where people are having to consider pollinating crops themselves. How can we have come to this point? A natural part of the ecosystem, which humans have effectively taken for granted, is under severe threat.

In the high octane world of finance, large-scale investment institutions have become increasingly engaged in dialogue with their investee companies on environmental, social and governance (ESG) issues. By engagement and dialogue we

10 Mark L. Winston, *Bee Time: Lessons from the Hive* (Cambridge, MA: Harvard University Press, 2014).

11 See Albright, "Could Robot Bees Help Save Our Crops?"

12 This is mentioned in A. Benjamin and B. McCallum, *A World Without Bees: The Mysterious Decline of the Honeybee—and what it means for us* (London: Guardian Books, 2009).

13 These figures are provided in Francis G. Smith, *Beekeeping in the Tropics* (London: Longmans, 1960).

14 These figures are provided in Benjamin and McCullum, *A World Without Bees.*

are referring to a practice, increasingly prevalent within the investment industry, whereby investors have face-to-face conversations with representatives from companies in which they hold shares, often directors, on issues such as climate change, in order to alter the companies' behaviour. This practice is commonly known as "responsible investment" in the business community. The reason investors are taking action in this way is that they increasingly perceive ESG factors to be financially "material". Material basically means the issue represents a risk which could significantly affect their financial return from investment. Bees are one such risk and investors are starting to engage with investee companies on how they are attempting to mitigate the risk of bee decline. Institutional investors are demanding more and more ESG disclosures and also requesting regular meetings with investees to discuss ESG concerns. A recent report by Schroders Investment Management explored whether the institutional investment community should be concerned about bees, concluding that there should be greater engagement with investee company management on their bee-related activities,

> ...It is too early to reflect a consideration of pollinator decline in valuations but it is not too early to engage with companies throughout the investment universe on this issue in order to develop a response to halt and reverse this decline in a valuable part of our economy before its consequences become more significant.[15]

The report also identified certain industries as having significant exposure to declining bee populations including food retailers, agrochemical companies and cosmetic firms. These are companies which are the focus of research in later chapters of the book.

So, what is killing bees?

There seems to be an endless list of culprits behind bee decline. By means of a brief introduction, we take a look here at simple bee science and the problems bees face, both from their natural environment and from human intervention. Bee science is covered in far greater detail in Chapter 3 but we feel a brief summary is useful at this point, to outline the issues. Bees as pollinators fall into two groups: social bees which live in hives, and solitary bees which live alone but are also efficient pollinators. Bees appear to be threatened in a multitude of ways. In fact when we researched the diseases and natural predators as well as the current concerns over chemical pollutants and pesticides we thought it was amazing that they have survived at all! From what we have read, bees apparently do not have the best immune systems and are sensitive to disease, climatic changes and many other environmental factors.

15 Stathers, *The Bee and the Stock Market*, 12.

Scientific research has revealed a wide array of threats to bee populations especially bees in hives (social bees). An interesting book on beekeeping, dating back to the 1960s, described a whole host of diseases which affect bees.[16] American foul brood is deemed the most serious bee disease which affects bees in hives. This disease has affected bees in the USA, England and Wales. It is only called *American* foul brood because it was named in the US. It is a contagious disease, caused by *Bacillus larvae,* a spore-forming organism which attacks bees' larva. A related disease, European foul brood, also kills bees. This highly infectious disease is caused by the organism *Bacillus pluton.* Another bee disease, not as serious as the above diseases, is sac brood which affects bee larvae and is related to chalk brood which is caused by a fungus, *Pericystis apis* Maasen, again only affecting larva. Addled brood is connected but not infectious. Dysentery affects adult bees and is similar in presentation to dysentery in humans. Another bee disease, acarapisosis, emerged in 1904 and became known as the Isle of Wight disease as it destroyed nearly the whole bee population on this island found off the south coast of England. The disease is caused by a parasitic mite, *Acarapis woodi,* and kills adult bees. A further disease, nosema, occurred in South Africa and was discovered by Dr S.H. Skaife.[17] This disease affects bees' digestive systems and derives from a single-celled protozoan parasite, known as *Nosema apis.*

As well as diseases, there are also many insect pests and parasites which affect bees. For example the small ant (*Iridomyrmex humilis*) has been viewed a significant threat to honey bee populations.[18] The ants bite bees to death. South Africa also suffers from robber wasps which paralyse bees and lay their eggs inside them. Other similar threats include: wax moths (in warm climates) whose larvae develop in combs; small hive beetles (*Aethina tumida*) which lay their eggs in beehives; the death's head moth (*Acherontia atropos*) prevalent in South Africa, which robs beehives; the tachinid parasite (*Rondanioestrus apivorus*) or Senotainia fly, prevalent in Russia, South Africa and Europe, which lay eggs inside the bees' bodies and as the egg hatches the creature eats the bees from the inside; the conopid parasite, which lays eggs on honey bees as they are visiting flowers, again feeding on the bee as the larvae grow; the bee louse (*Braula coeca*) which clings to fluff on bees' backs and eats the food from the bees' mouths (mainly in the USA); and chelifers which cling to bees by their pincers and steal nectar. Larger bee predators include honey badgers which rob hives, as well as toads and lizards which eat bees. Furthermore, spiders, dragon flies, mice and rats eat bees or enter hives and cause harm.[19]

As mentioned above, scientific research has shown that bees are extremely sensitive to diseases and threats to their health. They do not have a strong immune system. As well as natural bee diseases and parasites, there is the extremely significant human dimension of the problem. For decades it has become clear that corporate

16 See May, *Beekeeping.*
17 *Ibid.*
18 *Ibid.*
19 *Ibid.*

activities, resulting in the use of pesticides and insect poisons, are thought by some scientists to have had a devastating effect on bee populations, and especially on commercial beehives grown for agricultural pollination purposes. There has been research into the effect of agrochemicals on bee health dating back to the 1950s.[20] We feel this an important point to make as current concerns about pesticide use and the possible effect on bees, which preoccupies many chapters of this book, is by no means a new worry but one which has been studied and written about for over 60 years. There were reports of South African apiarists losing thousands of bees from careless spraying of pesticides during the 1960s. There is also of course, the well-known work by Rachel Carson, *Silent Spring*, which is referred to in many places throughout this book and which raised the issue of chemicals and their impact on birds and insects. She has been described as, "the quintessential defender of the environment, a champion of human health and ecological wellness".[21] In the 1960s, three groups of poisons were identified as potentially posing a serious threat to bees, namely: arsenic-based poisons; dichlorodiphenyltrichloroethane (DDT); and poisonous gases such as carbon bisulfide and methyl bromide.[22] It was DDT and the catastrophic effects of this family of chemicals on nature, which prompted Rachel Carson to write her globally influential book, which contributed to the creation of the US Environmental Protection Agency (EPA) and consequently to its banning of DDT in 1972.[23] A book which promotes more organic and natural farming methods highlights the need to temper pesticide and insecticide use, with the author commenting, "Our war against pests has exacted a heavy toll...not only do we annihilate pests, but we also take our plants, insect, fish, birds, and myriad other organisms when we spray".[24]

As we can see from this discussion, the generally acknowledged negative impact of insecticides on bee populations is not a 21st century phenomenon but one that was recognized decades ago. As early as the 1950s and 1960s, the toxic effects of chemical pesticides on bees became evident. Beekeepers were deeply concerned about the toxicity of agricultural chemicals on honey bees and insisted that DDT should not be sprayed directly on bees visiting blossoms but only during morning or evening when bees are not flying.[25] May also explains that other agrochemicals, including Metasystox, dieldrin, lead arsenate, malathion and parathion are very toxic to bees and should never be applied when trees are in blossom. May commented that,

20 T. Palmer-Jones and I.W. Foster, "Agricultural Chemicals and the Beekeeping Industry", *New Zealand Journal of Agriculture*, October 1958.

21 M.L. Winston, *Nature Wars: People vs. Pests* (Cambridge, MA: Harvard University Press, 1997), 9.

22 May, *Beekeeping*.

23 Winston, *Nature Wars*

24 *Ibid.* 186.

25 May, *Beekeeping*.

> An apiarist in a large fruit-growing district complains that he loses tens of thousands of bees by careless spraying... An official of the Western Province Fruit Research Station at Stellenbosch suggests that apiarists shut up their bees for a day while spraying is being done.[26]

As we mentioned at the beginning of this chapter, the latest serious threat to bees appears to be colony collapse disorder (CCD). At the present time there is no absolute consensus as to the causes of CCD, although there are a whole host of studies which attempt to pinpoint the culprit and lots of conflicting evidence. One of the first places where CCD was identified was in France in 1994, when worker bees flew away from their hives and never returned, resulting in the loss of many colonies. CCD wiped out over 10% of the 2.4 million hives in the USA in 2006/2007. Concerns began to be raised about relatively new pesticides used on commercial crops. The insecticide Gaucho® was suspected to be harmful to bees. It had begun to be used on sunflowers and flowering food crops. Gaucho contains an ingredient called imidacloprid (IMD), which is a chlorinated nicotine-based insecticide known as a neonicotinoid. Like DDT, this chemical seems to be a neurotoxin similar to nerve gas. Bayer CropScience created IMD in 1985 and the company declared Gaucho® was safe for bees, as the dose would not be lethal for them in nectar. However, French researchers tested sublethal doses in the 1990s and suggested that even a few parts of IMD per billion in the nectar could make bees groggy and block bees' short-term memory. Although IMD would not kill bees there were concerns that it could disrupt their ability and desire to feed and forage, essentially intoxicating them. It is this effect which is thought to be a possible causal factor for CCD. However, there are other contemporary explanations, as we shall see throughout this book.

Current concerns over neonicotinoids (or neonics, for short) are similar to worries over DDT some 50 years ago and unfortunately echo the fears of Rachel Carson and others at that time. There is a deep and uncomfortable resonance between concerns over DDT and current worries about neonics. Indeed, some of the latest scientific research has suggested that the widespread use of neonics could be killing bees all over the world. We are not even going to begin a comprehensive summary of these studies, partly because it is beyond the scope of this chapter but also because we are not scientists! However, one interesting recent study found that where bees were presented with a choice of collecting honey from crops grown from seeds treated with neonics or those that were not, they tended to favour the ones infused with insecticide. As the taste does not differ, the scientists concluded that perhaps the bees were getting a "high" from the nicotine in the nectar.[27] To us, this seemed an important finding, because if neonics are as harmful to bees as some fear, then their automatic preference for contaminated nectar puts them in even worse danger. We leave it to the expert entomologists in Chapter 3 to explore these scientific issues in far more detail, from a scientific perspective. We now move

26 *Ibid.* 11.

27 These findings are reported in S.C. Kessler *et al.*, "Bees Prefer Foods Containing Neonicotinoids", *Nature* 521 (2015): 74–6.

on to exploring overarching frameworks for analysing bee decline in an integrated manner, as a backdrop to the rest of this book.

Towards a bee-centric framework?

Before we explore the historical, cultural and philosophical basis of the relationship between bees and the human race in Chapter 2, we first consider a number of ways in which we can interpret nature, species and the relationships between living creatures, flora and fauna across our planet. These different ways of looking at nature and interpreting linkages between the millions of species inhabiting the planet may be referred to as "theoretical frameworks". There is nothing magical or complicated about "theories" and theoretical frameworks: they are simply ways of looking at the world, ways of interpreting what is happening around us. A theory is not "right" or "wrong" but is simply an approach to understanding our world. For the purposes of this book, there are a number of approaches, or theories, which we feel are useful to understanding the problems associated with bee decline and also the import and impact of these problems for humanity and ecosystems.

General systems theory has been applied across many disciplines and seems a good place to start for addressing bee decline using an integrated approach. The pioneer of general systems theory was Ludwig von Bertalanffy, who described the theory as a "general science of wholeness".[28] Essentially, general systems theory is about interconnectedness, wholeness, it involves adopting an integrated and holistic approach.[29] The intention was to break down barriers between systems of knowledge and prevent the entrenchment of knowledge in "silos", as we often discuss in terms of university research today. There has been a long-standing tendency in natural science but also across academic disciplines, towards what we call "reductionist reasoning", whereby researchers tend to focus on an issue or a particular phenomenon by isolating it from everything else.[30] General systems theory approaches problems and issues from the perspective of trying to understand the complete context within which the issue exists. Although general systems theory was derived for natural science, "systems thinking" has been applied across a wide

28 Ludwig von Bertalanffy, *General System Theory* (Harmondsworth, UK: Penguin, 1971).

29 General systems theory has been applied in the area of archaeology and even in the area of ancient earth mysteries because of the multi-faceted nature of these disciplines. See P. Devereux, *The Illustrated Encyclopedia of Ancient Earth Mysteries* (London: Blandford, 2000) for a discussion of systems theory in relation to ancient earth mysteries, an interesting application.

30 There is an excellent discussion of general systems theory in:
R. Gray *et al.*, *Accounting & Accountability: Changes and Challenges in Corporate Social and Environmental Reporting* (Harlow, UK: Pearson Education Limited, 1996) and similarly in R. Gray *et al.*, *Accountability, Social Responsibility and Sustainability: Accounting for Society and the Environment* (Harlow, UK: Pearson Education Limited, 2014).

variety of intellectual and practical areas, including natural science, social science and the business domain. Social science provides an ideal crucible for introducing systems theory, since humans interact with all other species, the land, water, the air, even outer space and therefore have an impact on, and are affected by, everything. A system has been defined as,

> ...a conception of a part of the world that recognizes explicitly that the part is (a) one element of a larger whole with which it interacts (influences and is itself influenced by); and (b) also contains other parts which are intrinsic to it.[31]

Although we, as humans with limited intellectual capabilities, can only ever see the part rather than the whole, as we are not omniscient, adopting a systems theory approach allows us to at least attempt to understand problems and issues within the broader, and indeed broadest possible, context. A reality exists "out there" in which everything is related to everything else, and although we may accept this and wish to bring it into writing and research, we have to find ways of building interpretations of "systems" which are as all-encompassing and as holistic as is possible within the constraints of the current state of human knowledge, understanding and the limits of the human mind and ability to take such a universal perspective.

At the expense of being labelled "New Age tree huggers", we now delve into some other theoretical frameworks which have been developed to try to capture the immensity of nature and the interrelationships between species of flora and fauna, and Planet Earth. These frameworks, although competing to some extent, succeed in our view, in different ways, in grasping the enormity and endless complexity of life on Earth and its essential vulnerability. This combination of complexity and vulnerability has to, in our view, be core to any discussion of one particular species and threats it may encounter in the 21st century. As we mentioned earlier, Rachel Carson, in her seminal work *Silent Spring*, focused mainly on birds but she was also concerned about the impact of agrochemicals on bees. New chemicals since the 1960s may pose an even greater threat to bees than those around in Carson's time.

One theory, dating back to the 1970s, provides further insights into the interconnectedness of all life on Earth and into the complexity and sensitivity of integrated systems. Gaia Theory, developed by James Lovelock, views the Earth as a self-regulating system consisting of the totality of organisms, surface rocks, oceans and the atmosphere, tightly coupled in constant evolution. The system is perceived as having a goal, specifically that of, "...the regulation of surface conditions so as always to be as favourable for contemporary life as possible".[32] In relation to the (very significant) effects of human activity on nature, Lovelock says, "Keep in mind that it is hubris to think that we know how to save the Earth: our planet looks after itself. All that we can try and do is try to save ourselves".[33] He makes a clear statement about

31 This definition comes from Gray *et al.*, *Accounting & Accountability*, 13.
32 J. Lovelock, *The Vanishing Face of Gaia: A Final Warning* (London: Allen Lane, 2009), 166.
33 *Ibid.* 9.

the interconnectedness of everything in relation to the threats of climate change in his 2009 book, *The Vanishing Face of Gaia*, by stating that humans could not survive for an instant on a dead planet. Lovelock highlights the need for humanity to realize that a holistic and, effectively, system-based approach is crucial to the continuing existence of all life on Earth:

> The ideas that stem from Gaia theory put us in our proper place as part of the Earth system—not the owners, managers, commissars or people in charge... This way of thinking makes clear that we have no special human rights; we are merely one of the partner species in the great enterprise of Gaia.[34]

On reflection, it seems unfair, to us at least, that deep ecology perspectives, theories of the Earth as a living, breathing system and anti-anthropocentric models are often branded as "loony" or "deep green" in a negative way. It is absolutely certain that as we move further into the 21st century, the intrinsic value of all species and the desperate need to maintain and conserve our biodiversity and the natural world is becoming more and more recognized. Despite efforts to deride theoretical frameworks which prioritize nature and conservation, as well as the people who have proposed them, there has to remain hope that people and planet can start to work together, before it is too late. Rachel Carson herself faced much derision following the publication of *Silent Spring*. It makes one wonder whether such opposition to ecological perspectives, which can be quite aggressive at times, may arise from fear at the threat of having to change the way we live?

One of the most fascinating aspects of the development of Gaia theory is the use of the "Daisyworld" model. This computer model essentially "proved" some of the aspects of the Gaia theory. It showed, through computer simulation, that if there is a tiny biodiversity (two species) which have restrictions on their living conditions and also affect their environment through the reflection of light from their white petals, that their environment (world) will actually turn itself into an environment most suitable to their survival. This model supports the idea that our planet will design its conditions around life on Earth and its continued survival.

The message and essence of Gaia may be viewed as a theory of a living Earth in the sense of an Earth Goddess, or Mother Nature, or a living organism. However, our interpretation is one which is quite in keeping with the notion of systems and systems theory: that of an interconnected Earth, where ecosystems, biodiversity and all living things are inextricably linked, their fates being identified and inseparable from each other.

Thinking about systems and about bees led us to think also of chaos theory, a strand of mathematics and science which seeks to explain apparently completely

34 *Ibid.* 6.

random behaviour.[35] The most obvious example is the weather. It is almost impossible to forecast weather exactly. One of the well-known concepts arising within chaos theory is the "butterfly effect". Scholars discussed the possibility that a butterfly flapping its wings in one place could have a ripple effect on events elsewhere, even at the other side of the world; not a direct effect but an effect as part of a change to the system as a whole.[36] The butterfly effect basically seeks to explain how small initial differences or events may lead to significant and relatively much greater unforeseen consequences and effects over time. As we would not know about these tiny events we could not possibly be in a position to factor them into any mathematical or logical model. According to chaos theory, a butterfly flapping its wings can affect a whole system (defined in maths as a non-linear system). As with general systems theory, it seems that chaos theory assumes that simple, reductionist models do not work in practice and that in reality, models are complex, unpredictable and apparently random (although not if we can understand the system in its entirety). Indeed, efforts by the United Nations and leading climate scientists around the world to estimate the impacts of climate change and global warming are, at best, vague and at worst, possibly inaccurate. However many variables and factors scientists include in their models, they can never represent the whole picture, the whole "system" and will always fall short of providing a detailed forecast. Chaos theory and the butterfly effect help here, as it is impossible to know, and therefore to factor in, every tiny flap of a butterfly's wings.

If we apply the idea of the butterfly effect to the notion of bees and bee decline, a bee buzzing its wings as it dies in one part of the world could have repercussions through the whole planetary system and could potentially lead to massive changes on the working of this system, whether it is the food system, the climate system, the human population system, climate change, the ecosystem or any other "system" we want to identify. If we consider the deaths of millions of bees as bee populations decline, how much greater could these repercussions be on the planet and on human and non-human life?

35 Please see the following for a more detailed exposition of chaos theory and the butterfly effect:
R.L. Devaney, *Introduction to Chaotic Dynamical Systems* (Boulder, CO: Westview Press, 2003).
J. Gleick, *Chaos: Making a New Science* (New York: Viking, 1987).
R.C. Hilborn, "Sea Gulls, Butterflies, and Grasshoppers: A Brief History of the Butterfly Effect in Nonlinear Dynamics", *American Journal of Physics* 72 (2004): 425–27.

36 Edward Lorenz came up with the notion of the butterfly effect in his studies of weather systems: a butterfly flapping its wings can cause, in association with other factors, a hurricane somewhere else on earth.

In a lovely work of modern natural history, *A Buzz in the Meadow*, Dave Goulson describes a genuinely systems-based and holistic approach to nature, which appears to come from personal observation and experience of nature, rather than theoretical analysis,

> Plants compete for space, water and light, are food for herbivores, hosts to parasites and diseases. They use diverse strategies to tempt pollinators to visit them, and in turn their pollinators have evolved numerous tricks so that they can learn which flowers are most rewarding and can gather those rewards quickly... Every species is linked, one way or another, to hundreds of others, in a web of interactions that are at present far beyond our ability to comprehend.[37]

From this whistle-stop tour of theories which relate loosely to a systems-type approach to the world and its inhabitants, it appears, at least to us, that it is not possible to think of bee decline in reductionist, or isolated, terms. It is highly likely, adopting a systems approach, that the decline in bee populations worldwide is having and will continue to have substantial, if not catastrophic, effects on human life, on ecosystems, on ecology, on nature and on the planet.

Perhaps one way of approaching the ideas within this book, both in terms of bees and bee decline as well as in relation to bees within a "system" is to take a "bee-centric" viewpoint. Instead of the anthropocentric, or human-focused, approach to reading this book, let's experiment by considering the world as a web of interconnected and interrelated factors which are all related to the bee. The spiral image attempts to present a bee-centric system where humans are removed from the "top position", so common to human attitudes, and replaced by the bee. This "experiment" may serve as a means of introducing a different way of thinking, a way of thinking "out of the box" as it were.

If we were to apply reductionist reasoning to the problem of bee decline, this book would focus on a very narrow set of issues, probably solely on bees and what the cause of the decline may be, with possibly a discussion of the repercussions on food security and the very direct effects of bee decline on pollination. However, adopting a general systems theory approach is radically different. Instead of focusing on the closest resolution around bee decline, we discuss bee decline in the context of the historical relationship between bees and the human race, the impact of bees on all aspects of human culture, and ultimately the implications of bee decline for food production, business more broadly, accounting, finance, stock markets, the ecosystem, people and the planet.

An integrated approach is especially pertinent for a book focusing on the business, financial and accounting aspects of bee decline, given the relatively recent emergence of integrated reporting, a new form of accounting which relies on businesses adopting "integrated thinking". Integrated thinking essentially means that the business must take account of the interconnectedness of all risk factors and all

37 Dave Goulson, *A Buzz in the Meadow* (London: Vintage Books, 2014).

issues affecting the business and avoid dealing with issues in a reductionist manner. In a way, integrated reporting provides a systems approach to accounting. We will return to discussions of integrated reporting and the ways in which bees "fit" into the integrated reporting framework in later chapters.

The structure and content of *The Business of Bees*

Part I of this book adopts a broad and interdisciplinary perspective on bees and bee decline, considering the ways bees have influenced our world history and their ecological value within the planet's ecosystem. The first few chapters bring together bee science with cultural bees, economic bees, philosophical bees, artistic and musical bees. Chapter 1 has provided a framework for discussion grounded in a range of theoretical perspectives. In Chapter 2, we explore the historical, cultural and philosophical aspects of bees and their age-old relationship with the human race. In Chapter 3, Longing and Discua provide an overview of bee science, examining the various species of bees, the threats they are faced with and a scientific discussion on the causes of bee decline and colony collapse disorder. Reade *et al.* discuss the bumblebee trade from a Japanese markets perspective in Chapter 4 and the reasons for, and challenges of, using bumblebees for pollination "services". This ecosystem services approach to the bumble is contrasted sharply by a discussion in the following chapter of the bumblebee from a deep ecology perspective. In Chapter 5, Christian presents a personal ecological account of bumblebees, drawing on Taoism and a deep ecology framework for understanding the intrinsic value of the bumblebee.

In Part II, the book focuses on the financial impacts of bee decline, examining the financial materiality of bee decline from the perspective of the institutional investment community: how bee decline is affecting stock markets and the ways in which investors can alter corporate attitudes towards bees through active engagement policies. Stathers, in Chapter 6, builds on his earlier report into bees and the stock market, explaining why the institutional investment industry is concerned about bee decline: the ways in which falling bee populations worldwide are affecting company values and consequently investors' pockets. The financial materiality of bee decline is explored further by Herron in Chapter 7, in which she considers bee decline as an investment portfolio risk and provides a case for investor action: namely engagement and dialogue with companies around the need to protect and enhance bee populations. In Chapter 8, Clappison and Solomon discuss the issues arising in relation to a court case being fought in Canada involving a pesticide company and local apiarists and agricultural businesses. This chapter provides some context and an in-depth case study, which demonstrates the potential financial and ecological problems associated with the commercial bee industry and the use of agrochemicals. Chapter 9 takes a similar approach; Thamotheram and Stewart provide a detailed guide and road map for investors to tackle bee decline. Overall, this part of the book demonstrates the crucial role being played by the world's financial industry in saving bees, for economic reasons admittedly, but a role that is critically important in making a difference to bee populations and the future of pollinators more generally.

Part III approaches bee decline from a corporate perspective and essentially from a corporate accountability point of view. Do companies have any accountability for bee decline? If so, why and how should they discharge this accountability? Houdet and Veldtman, in Chapter 10 offer a framework for bee accounting for companies' internal reporting purposes. The other chapters in Part III gauge the current state of bee accounting around the world, analysing bee-related disclosures by leading listed companies from the US to the UK, across Europe, to South Africa. In Chapter 11, Atkins *et al.* investigate corporate reporting by companies listed on the London Stock Exchange (LSE) on bee-related issues. Romi and Longing in Chapter 12 assess accounting and accountability for bees by US companies and focus on initiatives at government and NGO level to address bee decline. In Chapter 13, Maroun provides evidence that some large multinationals listed on the Johannesburg Stock Exchange (JSE) are reporting their initiatives aimed at saving bees. Jonäll and Rimmel investigate bee accounting and accountability in the Swedish context in Chapter 14 and, in Chapter 15, Biehl and MacPherson provide a German perspective.

Bibliography

Albright, M.B. "Could Robot Bees Help Save Our Crops?", *National Geographic,* 21 August 2014.

Atkins, J., Gräbsch, C. and Jones, M. J. "Biodiversity Reporting: Exploring its Anthropocentric Nature". In chapter in Jones (eds.) *Accounting for Biodiversity,* edited by Michael Jones. Abingdon, UK: Routledge, 2014.

Atkins, J. and Thomson, I. "Accounting for Nature in 19th Century Britain: William Morris and the Defence of the Fairness of the Earth". *Accounting for Biodiversity,* edited by Michael Jones. Abingdon, UK: Routledge, 2014.

Benjamin, A. and McCallum, B. *A World without Bees: The Mysterious Decline of the Honey-bee—and what it means for us.* London: Guardian Books, 2009.

von Bertalanffy, L. "General Systems Theory", *General Systems Yearbook* 1 (1956): 1–10.

von Bertalanffy, L. *General Systems Theory: Foundations, Development, Applications.* Harmondsworth, UK: Penguin, 1971.

Carvalheiro, L. G. "Species Richness Declines and Biotic Homogenisation Have Slowed Down for NW European Pollinators and Plants", *Ecology Letters* 16 (2013): 870–8.

Constantino, M. *Bees and Beekeeping.* Worcestershire, UK: King Books, 2011.

Devaney, R. L. *Introduction to Chaotic Dynamical Systems.* Boulder, CO: Westview Press, 2003.

Devereux, P. *The Illustrated Encyclopedia of Ancient Earth Mysteries.* London: Blandford, 2000.

FAO. *Rapid Assessment of Pollinators' Status: A Contribution to the International Initiative for the Conservation and Sustainable Use of Pollinators.* Rome: FAO, 2008.

Gleick, J. *Chaos: Making a New Science.* New York: Viking, 1987.

Goulson, D. *A Buzz in the Meadow.* London: Vintage Books, 2014.

Gray, R., D. Owen and C. Adams. *Accounting & Accountability: Changes and Challenges in Corporate Social and Environmental Reporting.* Harlow, UK: Pearson Education Limited, 1996.

Gray, R., D. Owen and C. Adams. *Accountability, Social Responsibility and Sustainability: Accounting for Society and the Environment.* Harlow, UK: Pearson, 2014.

Greenpeace (2013) *Bees in Decline: A Review of Factors that put Pollinators and Agriculture in Europe at Risk.* Greenpeace Research Laboratories Technical Report (Review), January 2013.

Helmholtz Association of German Research Centres. "Economic Value Of Insect Pollination Worldwide Estimated At U.S. $217 Billion". *ScienceDaily,* 15 September 2008.

Hilborn, R. C. "Sea Gulls, Butterflies, and Grasshoppers: A Brief History of the Butterfly Effect in Nonlinear Dynamics". *American Journal of Physics* 72 (2004): 425–27.

Holland, N. "The Economic Value of Honeybees". *BBC World News Online,* 23 April 2009.

Jones, M. J., ed. *Accounting for Biodiversity* Abingdon, UK: Routledge, 2014.

Jones, M. J. and J.F. Solomon. "Problematising Accounting for Biodiversity". *Accounting, Auditing & Accountability Journal* 26 (2013).

Kessler, S. C., E.J. Tiedeken, K.L. Simcock, S. Dervear, J. Mitchell, S. Softley, J.C. Stout, and G.A. Wrigth. "Bees Prefer Foods Containing Neonicotinoids". *Nature* 521 (2015): 74–6.

Lovelock, J. *GAIA: A New Look at Life on Earth.* Oxford, UK: Oxford University Press, 1979.

Lovelock, J. *The Vanishing Face of Gaia: A Final Warning.* London: Allen Lane, 2009.

Lovelock, J. *A Rough Guide to the Future.* London: Allen Lane, Penguin Books, 2014.

May, A. F. *Beekeeping.* Cape Town, South Africa: Haum, 1969.

Naeem, S., D.E. Bunker, A. Hector, M. Loreau, and C. Perrings. *Biodiversity, Ecosystem Functioning and Human Wellbeing: An Ecological and Economic Perspective.* Oxford, UK: Oxford University Press, 2009.

Palmer-Jones, T. and I.W. Foster. "Agricultural Chemicals and the Beekeeping Industry". *New Zealand Journal of Agriculture* October (1958).

Runlöf, M., G.K.S. Anbdersson, R. Bonmarco, I. Fries, V. Hederström, L. Herbertsson, O. Jonsson, B.K. Klatt, T.R. Pedersen, J. Yourstone, and H.G. Smith. "Seed Coating With a Neonicotinoid Insecticide Negatively Affects Wild Bees". *Nature* 521 (2015): 77–80.
Smith, F.G. *Beekeeping in the Tropics*. London: Longmans, 1960.
Stathers, R. *The Bee and the Stockmarket: An Overview of Pollinator Decline and its Economic and Corporate Significance*. Research Paper, Schroders, January 2014.
Turney, J. *Lovelock & Gaia: Signs of Life*. Canada: Allen & Unwin Pty. Ltd, 2003.
Wilson, Bee *The Hive: The Story of the Honeybee and Us*. London: John Murray, 2004.
Winston, M. L. *Nature Wars: People vs. Pests*. Cambridge, MA: Harvard University Press, 1997.
Winston, M.L. *Bee Time: Lessons from the Hive*. Cambridge, MA: Harvard University Press, 2014.

2

The historic, cultural and philosophical context of bee decline

Jill Atkins
University of Sheffield, UK

Barry Atkins
University of South Wales, UK

> Will a time come when we must learn from the honey-bee or perish?[1]

The story of bees and their evolving relationship with the human race dates back to the earliest human societies, as identified through anthropology and represented in cave paintings drawn by the earliest human civilizations. Our own reading of a wide array of literature, poetry, philosophy and other varied writings has led us to see a dynamic narrative which has evolved over time and which shows the changing nature of bees' relationship with humans and of the liquid attitudes of people towards bees. In this chapter, we discuss the historical, cultural and philosophical context of bee decline, spelling out the evolution of the relationship between bees and human beings and use this as a frame for seeking to explain and understand the current situation whereby bee populations are in serious decline globally. This provides us with a basis for appreciating the current situation for bees and their decline. We aim to establish a framework which may then be used to set the context for the discussions throughout the book in relation to bee decline in the context

1 Tickner Edwardes, *The Lore of the Honey-Bee* (London: Methuen & Co Ltd, 1908).

of business, nature, the ecosystem, ecology and the planet. Evolution of bee "narratives" has also gone hand in hand with the development and growth of a bee discourse. The discourse relating to bees, through words but also music, image and other media, has evolved over time, reflecting the attitude of people towards bees. This chapter serves to give a flavour of the deep and ancient influence of bees on the human race.

The evolution of bees and their relationship with the human race

Bees first evolved from wasps about 125 million years ago and were gathering nectar at the same time as pollinating flowers well before humans arrived on the planet.[2] Over 20,000 species of bees inhabit every continent except Antarctica.[3] Bee society predates human society by millions of years. Humans began to keep bees around 10,000 years ago, whereas social honey bee colonies have existed for around 20 million years.[4] Accounts date back to early civilization as well. It seems that when we go as far back in time as possible, from existing records, the relationship that has developed between bees and the human race has, from the outset, been one where humans have "benefited" from bees. We could call this an "anthropocentric" (human-centred) approach to bees, in other words, one where the relationship between the two species is characterized by "what bees do for people", or, "what's in it for us?". This may seem somewhat cynical but it demonstrates the fact that from the earliest contact between bees and humans, people have found ways of harnessing bees' activities and bees' "produce" to their own advantage. A relatively recent book, *Robbing the Bees*, has been described as a biography, history, celebration and love letter to bees and their magical produce. The author shadows a beekeeper to study their behaviour.[5] The title of the book itself betrays concerns about the way in which bees have been effectively exploited by the human race.

There is a continual tension throughout this book, with the emphasis alternating from chapter to chapter between an "anthropocentric" approach to bees and bee decline and a "deep ecology" approach. As we saw in the previous chapter, an anthropocentric approach appears to dominate any scientific or business

2 See Mark Winston, *Bee Time: Lesson from the Hive* (Cambridge, MA: Harvard University Press, 2014) for a full discussion of the early history of the bee.

3 *Ibid.*

4 See Bee Wilson, *The Hive: The Story of the Honeybee and Us* (London: John Murray Publishers, 2004).

5 H. Bishop, *Robbing the Bees: A Biography of Honey—The Sweet Liquid Gold that Seduced the World* (New York: Free Press, 2005).

discussion of bees whereas a deep ecology, or "bee-centric" approach dominates discussions of bee decline from a more systems or "planet" view of species and their intrinsic value. The burning issue surrounding bee decline from an anthropocentric perspective is the worry of pollinators disappearing and affecting our food chain and consequently our diet and our pockets. University research into issues of ecosystems services and "food security" again tends to focus on an anthropocentric view of bees and their place in the human-driven ecosystem. In this chapter, we explore the human-centred approach to bees from a historic perspective, looking at the ways in which people have "used" bees throughout the ages and how this exploitation has ultimately led to their decline. We then go on to explore the more esoteric aspects of bees and humans, discussing bees in culture throughout the ages, religion, philosophy and indeed all aspects of human history and activity. When we started this project neither of us had any idea how much had been written over the centuries about bees and researching the subject from a wide array of perspectives has been an education indeed.

Since the earliest times of human life on Earth, people have realized that bees are useful to them in a whole manner of ways. The many varied "uses" of bees to human society include: honey as a sweetener; honey as a medicine; royal jelly; bees wax; mead (a favourite drink of monks, who first created it); propolis; and, of course, pollination. Beeswax was used from earliest civilization in candle-making. In the Bronze Age, it was used in casting ornaments and weapons.[6] Wax was also used as an embrocation for joint pain and was thought to cure ulcers. Distilled wax, known as "oil of wax" was used worldwide to cure almost everything. A commonly used medieval medicine, Oxymel, made from honey, water and vinegar, was believed to cure gout, sciatica and may other illnesses. More modern "uses" of bees and their by-products include the use of venom from bee stings for the treatment of arthritis and the training of bees to detect bombs by the American military.[7]

It appears that bees and people had a comfortable and relatively symbiotic relationship, which benefited both species throughout history until more recent times, with beekeeping being a natural activity for humans and bees. In fact, early beekeepers expounded on the love they had for their bees and the need to ensure that beekeepers were caring and respectful towards their bees. We suggest, from our reading and research, that things changed only when bees' use as pollinators began to be harnessed as a corporate and commercial activity. When the natural business of bees became a commercial anthropocentric business, bee decline began. The turning point in the relationship between bees and humans appears to be the "industrialization" of pollination, when bees became a serious commercial activity. We were amazed to learn of the global trade in different bee species and the mass trade in hives for pollination purposes between large-scale farms and industrial agricultural producers. For example, we discovered from reading that in the

6 Edwardes, *The Lore of the Honey-Bee.*

7 See T. Horn, *Bees in America: How the Honey Bee Shaped a Nation* (Lexington, KY: University Press of Kentucky, 2005).

late 1990s, there were about 3.5 million managed honey bee colonies in the US and Canada and hundreds of thousands of hives are transported around the US to provide pollinator services for agriculture—bees for rent, basically.[8] The numbers around the turn of the century were estimated as 30% of US bee colonies being moved around the country each year.[9] In areas of agriculture and especially where there is "mono-agriculture" (just one crop) there are almost no wild bees at all:[10] commercialized and intensive farming has eradicated natural pollinators (lack of biodiversity and insecticides to name but a few problems) and therefore "rent-a-bee" for pollination has had to fill the space created by wild bee decline. Chapter 4, for example, discusses the bumblebee trade which is global. We, initially as lay people entering into the bees' world, had no idea that such a trade even existed. Indeed, trade in bumblebees has been established for some time. Dave Goulson, in his autobiographical book *A Sting in the Tale*, described how New Zealand farmers in the latter part of the 19th century discovered that red clover, which they had imported from the UK to feed livestock, was not growing because of a lack of pollination by bumblebees, as they were the primary pollinators of this flower. After a few failed attempts at importing bumblebees, 282 bumblebee queens were transported in a refrigerated container by steamship from Kent in southern England to Christchurch, New Zealand. A second consignment was sent soon afterwards. Within eight years, bumblebees had become widespread and are still present in the wild in New Zealand. A recent natural history book, rewritten from earlier editions, focuses on UK bumblebees and studies their ecology and details for identification. The book considers the threats to bumblebees and highlights problems associated with their commercial use.[11] Bumblebees are the subject of Chapters 4 and 5 of this book.

In 1853, Quinby[12] provided a natural history of bees which focused on obtaining the maximum quantity of pure surplus honey at the least possible expense, based on his life-long experiences as a beekeeper. This is typical of an anthropocentric, human-focused, view of beekeeping. Linking honey production to moneymaking in such an overt way, we believe, was the beginning of the end for bee populations and for the ancient, balanced, symbiotic connection between bees and people. Things seemed to go even more wrong for bees and their relationship with humans when people began to recognize their potentially massive commercial use as pollinators rather than honey producers. As people began to prioritize bees' pollinating activity and then to begin to commercialize pollination, the age-old relationship changed, possibly irrevocably. However, we have to remember that increasing

8 This statistic was provided by M.L. Winston, *Nature Wars: People vs Pests* (Cambridge, MA: Harvard University Press, 1997).

9 *Ibid.*

10 *Ibid.*

11 See O.E. Prys-Jones *et al.*, *Bumblebees: 6, Naturalist's Handbook* (Cambridge, UK: Cambridge University Press, 2011).

12 M. Quinby, *Mysteries of Bee-keeping Explained: Being a Complete Analysis of the Whole Subject* (New York: C.M. Saxton, Agricultural Book Publisher, 1853).

human population around the world as well as the challenges of trying to feed the developing world, make a human-focused approach on food security and food production crucial. If commercial pollination is the only way of mass-producing crops to feed the richer nations and the developing world, then it is hard to see any means of arguing against it. The issue is ultimately whether mass commercialized pollination is tipping the natural balance such that bees disappear to an extent which makes mass food production impossible. This seems an impossible conundrum for the human race, as we find ourselves "between the devil and the deep blue sea". No bees: no food.

To summarize, the mass commercialization of hive distribution for commercial reasons has altered dramatically humans' relationship with bees and the way that bees are viewed. Rather than being seen as natural providers of honey and goodness to whom "we" should be grateful, it seems that, sometime in the 20th century, humankind began to view bees as a natural asset or as a resource to be used and harnessed in a commercial manner at an industrial level. Instead of bees being natural providers, part of our ecosystem whose natural activities provide useful and fortunate by-products, humans decided at some point to see bees as a business opportunity, a species to be used and, dare we say, abused. We now trace back in time to the earliest forms of beekeeping in order to provide a historical context for discussions of bee commercialization throughout this book.

The business of bees through the ages

The business of beekeeping and allegorical interpretations of bee society have been a feature of human societies throughout history. Beekeeping has been described as the oldest craft under the sun.[13] Beekeepers are also called honey farmers, apiarists, or less commonly, apiculturists (both from the Latin *apis*, meaning bee). The term beekeeper refers to a person who keeps honey bees in hives, boxes or other receptacles. Honey bees are not domesticated and the beekeeper does not control the creatures. Although a full review of how bee society has been used metaphorically to interpret and inform human behaviour and society is beyond the scope of this book, we will mention a few applications to early and more modern forms of capitalism, as well as to other economic and political systems and patterns. Most of the later chapters are concerned primarily with the role of bees in stock markets, accounting and finance (i.e. in business) but we feel that it is important to place these commercial, anthropocentric discussions within a more interdisciplinary and holistic perspective.

Perceived similarities between bee society and human society have extended into the domains of politics, business and economics since earliest human civilizations.

13 Edwardes, *The Lore of the Honey-Bee.*

Pliny wrote of bees as inhabiting a nation of industrious creatures ruled over by a king, who wore a white spot on his forehead similar to a diadem.[14] Clearly, scientific understanding was at a very early stage but Pliny and other writers created a whole mystical interpretation of how bees laboured together as a society to produce honey. In later centuries, bees have featured metaphorically in writings across the economics and sociopolitical disciplines. Bees work tirelessly for their hive. As the writer, Bee Wilson, pointed out, one worker bee can visit up to 10,000 flowers a day.[15] Their capacity for work and "industry" has acted as an inspiration to economists and philosophers since ancient times. An exemplary summary of the perennial interpretation of human behaviour through the lens of bee behaviour is provided by Judith Halberstam, in her recent work, *The Queer Art of Failure*, where she says that:

> ...bees have long been used to signify political community; they have been represented as examples of the benevolence of state power (Virgil); the power of the monarchy (Shakespeare), the effectiveness of a Protestant work ethic, the orderliness of government...but bees have also represented the menacing power of the mob, the buzzing beast of anarchism, the mindless conformity of fascism, the organized and soulless labour structures proposed by communism, and the potential ruthlessness of matriarchal power.[16]

We seek to explore some of these interpretations of bee life below.

The famous Roman writer and poet, Virgil, who lived in the 1st century BC, devoted Book IV of his seminal work, *The Georgics*, to beekeeping. This is probably one of the best-known beekeeping books in history. Virgil begins by giving advice on where a beehive should be positioned:

> principio sedes apibus statioque petenda,
> quo neque sit ventis aditus—nam pabula venti
> ferre domum prohibent—neque oves headique petulci
> floribus insultent, aut errans bucula campo
> decutiat rorem et surgentes atterat herbas.

These lines have been translated, with artistic licence, by H.R. Fairclough as follows,

> First seek a settled home for your bees, whither the winds may find no access—for the winds let them not carry home their food—where no ewes or sportive kids may trample the flowers, nor straying heifer brush off the dew from the mead and bruise the spring blade.[17]

14 *Ibid.*
15 See Wilson, *The Hive.*
16 Judith Halberstam, *The Queer Art of Failure* (Duke University Press, 2011), 51.
17 "Virgil, Georgics", Loeb Classical Library, accessed 14 April 2016, http://www.loebclassics.com/view/virgil-georgics/1916/pb_LCL063.219.xml?readMode=recto

A principle underlying Virgil's *Georgics* is that:

> The bee-keeper must be first of all a bee-lover, or he will never succeed... Through the rich incrustation of poetic fancy, and the fragrant mythological garniture, we cannot fail to see the true bee-lover writing directly out of his own knowledge, fathered at first hand among his own bees.[18]

Reflecting on this attitude provides food for thought when we look at the situation bees currently find themselves in and the way in which the relationship between bees and humans has evolved and changed over the last two millennia.

In the UK, for example, there is a strong beekeeping heritage. In Anglo-Saxon times, beehives supplied the whole of ancient Britain with honey, mead and candles, from the king to the poorest peasants. Mead was the main alcoholic beverage served in ancient inns around Britain. Also, the growth of monasteries and monastic life in the 11th and 12th centuries has been linked strongly to the rise in beekeeping in the British Isles, as historical evidence shows monks to have been keen beekeepers. As described in a lovely book on the early roots of beekeeping in the south-west of England, Ogden states,

> [T]he keeping of bees played an integral part in the way of life cherished by the inhabitants in many of these houses [monasteries] as the bees provided honey to sustain life through food, drink and healing, and beeswax for other practical purposes, but principally to create a source of pure light, a matter of great doctrinal significance in their religious ceremonies.[19]

There are many early English books about beekeeping and in the Middle Ages one such publication was *Further Discovery of Bees,* written by King Charles II's "Bee-Master". The Bee-Master, Moses Rusden, used his book to support the "Divine Right of Kings", showing how bee society was always ruled by a king, demonstrating his wholehearted support for the restoration of the monarchy.[20] There was clearly some scientific uncertainty in the Middle Ages as to whether the Queen Bee was a Queen or a King! Another medieval writer, Butler, produced a book on honey bees entitled, *The Feminine Monarchie,* which was probably because he was writing during Queen Anne's reign.

Bees and bee society have clearly inspired human politics since the earliest times of human civilization, although it is interesting to see how interpretations of bee society have been moulded and manipulated to mirror the spirit of politics at different times. We can see how bee society moved away from supporting Divine Kingship or Queenship and towards supporting different forms of politics and society. In the mid-17th century, bee society was used to applaud the new society created in the aftermath of the English Civil War, when Oliver Cromwell became the Lord Protector, replacing the traditional monarchy. This new state of affairs resulted in

18 Edwardes, *The Lore of the Honey-Bee*, 2.
19 See R.B. Ogden, *In Pursuit of Liquid Gold* (Charlestown, UK: BBNO, 2001), 41.
20 *Ibid.*

Parliament and the notion of commonwealth. Bee society had clearly also been reformed from a monarchy to a republican-type organization. Indeed, John Day's *The Parliament of Bees*, published in 1641, is one of the Elizabethan dramatist's greatest and most well-known works.[21] The work is a poem which cameos 12 characters including Prorex (the Master Bee) and "The Noble Soldier". The writing takes the form of a dialogue and a parliament is held. Other similar attempts to draw synergies between bee society and the reformed state include Samuel Purchas' *A Theatre of Politicall Flying-Insects*[22] and *The Reformed Common-Wealth of Bees* by Samuel Hartlib.[23] In Shakespeare's Henry V Act I Scene II, the Archbishop of Canterbury makes a long analogy between a colony of honey bees and human society.

In 18th century Europe, various views of bee society began to emerge, given the age of Revolution, and especially the French Revolution which took place towards the end of the century.[24] *Le Gouvernement Admirable, ou La Republique des Abeilles,* translated as *The Admirable Government, or the Republic of Bees* was written in the 18th century by Jean-Baptise Simon. The book praised the republican tendencies of bee society. This is another example of how the positively viewed aspects of bee society have been used to support different political perspectives throughout history. Another French naturalist writer alluded to the republican nature of bees in his extensive works of natural history, saying that, "A beehive…is a republic in which the labour of each individual is devoted to the public good in which everything is ordered, distributed and shared with a foresight, an equity and a prudence which is really astonishing".[25]

21 John Daye, *The Parliament of Bees with their Proper Characters, or A Bee-hive Furnisht with Twelve Honycombes, as Pleasant as Profitable. Being an Allegorical Description of the Actions of Good and Bad Men in these our Daies* (1641 edition, republished in 1881 by A. H. Bullen).

22 The original publication is referenced as follows: *A theatre of politicall flying-insects wherein especially the nature, the worth, the work, the wonder, and the manner of right-ordering of the bee, is discovered and described: together with discourses, historical, and observations physical concerning them: and in a second part are annexed meditations, and observations theological and moral, in three centuries upon that subject / by* Purchas, Samuel, 1577?–1626 (London: Printed by R. I. for Thomas Parkhurst, 1657).

23 The original publication is referenced as follows: *The reformed Common-wealth of bees. Presented in severall letters and observations to Sammuel Hartlib Esq. With The reformed Virginian silk-worm. Containing many excellent and choice secrets, experiments, and discoveries for attaining of national and private profits and riches.* Hartlib, Samuel, d. 1662. Reformed Virginian silk-worm (London: Printed for Giles Calvert at the Black-Spread-Eagle at the West-end of Pauls, 1655).

24 See Wilson, *The Hive*, for a fuller discussion of the historical context.

25 This quote may be found on page 76 in Georges-Louis Leclerc, Comte de Buffon's Fifth Volume within *History of the Brute Creation*, from Buffon's Natural History containing a *Theory of the Earth, a General history of Man, of the Brute Creation, and of Vegetables and Minerals, &c. &c. &c.* in ten volumes. Translated from the French (London: H. D. Symonds, Pater-Noster Row, 1797).

Perhaps the most infamous, and controversial, literary work in the beehive-as-metaphor-for-human-society tradition was Bernard De Mandeville's *The Fable of the Bees: or, Private Vices, Publick Benefits*[26] and, in particular, the poem within it, *The Grumbling Hive: or, Knaves Turn'd Honest*. The poem initially attracted little attention upon its original publication as an anonymous pamphlet in 1705, but grew in notoriety once it had become embedded in *The Fable...*, so much so that, in its final edition (1724), a hounded and demonized De Mandeville had felt it necessary to add an additional chapter entitled "A Vindication of the Book". The work excited the wrath, scorn and anger of clergymen, politicians and the judiciary alike; unlike the rather more formal, and occasionally lazy, uses of the beehive as a metaphor for, say, communism or monarchical systems, it instead utilized the hive analogy in service of a meditation upon the hypocrisy, bankrupt morality and avarice underpinning a successful, "civilized" society.

> ...Luxury Employ'd a Million of the Poor,
> And odious Pride a Million more,
> Envy it self, and Vanity
> Were Ministers of Industry.

The poem depicted bees as little people, each occupying a human role in miniature, and posited De Mandeville's conviction that a successful society could only be built upon the foundations of vice, acquisitiveness and crime, with greed driving production, and criminality providing employment for lawyers, bailiffs and locksmiths, and so on.

> Thus every Part was full of Vice,
> Yet the whole Mass a Paradise.

It is a persuasive argument, and one only has to look, for example, at a person tossing a fast-food wrapper on to the pavement, while muttering, "Well, if there wasn't any litter, we wouldn't need any street-cleaners", to appreciate its continuing relevance in microcosm. As the "subtitle" of *The Grumbling Hive* implies, the poem concludes with the wish of the hypocritical inhabitants for ethics and morality to be restored granted unequivocally, leading, inexorably, to the withering of the bee/human society. The irony here, of course, is that, while De Mandeville's allegory still has immense pertinence today, its central metaphorical device—the beehive—*is* withering, but only because we are still driven by short-sighted, profit-driven motives.

A critical study of Mandeville's *Fable of the Bees*, considers the way in which this work helped to develop our understanding of the functioning of markets,[27] although Moss does not consider that Mandeville was one of the founders of "laissez-faire"

26 Bernard de Mandeville, *Fable of the Bees: Or Private Vices and Public Benefits* (Oxford: Clarendon Press, 1732).

27 Laurence S. Moss, "The Subjectivist Mercantilism of Bernard Mandeville", *International Journal of Social Economics* 14, Nos.3/4/5 (1987), 167–84.

economics. He does consider that Mandeville's work on the "grumbling hive" did help in advancing economics in general as he showed links between politicians, egos and society. This shows the power the bee metaphor has had in shaping the development of economic and political thought through the ages. The philosopher, Thomas Hobbes, however, begged to differ with the tendency to interpret bees as political animals. Instead, he emphasized that bees lack language and reason and therefore cannot need nor develop politics. He also argued that their mechanistic ways of working together have resulted in a lack of conflict and therefore political arguments and revolutions could not take place in the beehive.[28]

Probably the most comprehensive book on beekeeping ever produced, *Collins Beekeeper's Bible*, covers practical essentials for beekeepers as well as providing information on the multitude of uses of honey, beeswax and pollen, such as in cooking, medicines and domestic uses.[29] An American beekeeper, Langstroth, produced *The Hive and the Honey bee: The Classic Beekeepers' Manual*, which was one of the most significant publications in more modern beekeeping, published in 1853. This book established the development of modern beekeeping as he described how to construct a beehive, as he had invented it in 1851, with hanging movable frames. This was the beginning of frame hive management and the birth of beekeeping as a commercial enterprise. Langstroth's ideas were adopted quickly by beekeepers in Australia, Canada, France, Italy, New Zealand, Poland, Russia, Spain, Poland, the USA and across Europe, apart from in the UK and Germany. These last two countries resisted the new American methods for several decades, preferring more old-fashioned methods.[30] Maybe it was at this point that things changed for bees?

An 1867 etching by George Cruikshank, "The British Bee Hive", housed in the Victoria and Albert Museum in London, depicted Victorian society in Britain as a hive. The various layers of the hive represented the different strata of society including royalty, aristocracy, church and professionals, under which were placed industry and the working class at the base of the hive. From the perspective of the Protestant work ethic, it was only possible (but it was possible) to climb from a lower layer of the hive to a higher layer through good, honest hard work. Again, we see industry and capitalism as founded upon a work ethic and bees a suitable metaphoric means of interpreting them.

The 20th century was also no stranger to interpreting human behaviour through a bee-inspired lens. In 1908, *The Mother Hive*, a story written by Rudyard Kipling, was published in *Collier's Weekly* in the USA and a month later in the UK's *Windsor* magazine. It is a rather frightening story, or in fact a fable, as it shows how a retraction from hard work can lead to ruination. By allowing a wax moth into the hive, the bees' hive is destroyed. The wax moth seduces the bees into indolence and

28 See Wilson, *The Hive*, for a fuller discussion of Hobbes' views of bees and politics.
29 See Philip McCabe, *Collins Beekeeper's Bible: Bees, Honey, Recipes and Other Home Uses* (Glasgow: Collins, 2010).
30 This information is provided by F.G. Smith, *Beekeeping in the Tropics* (London: Longmans, Green & Co Ltd, 1960).

laziness. The story again highlights the virtue of a work ethic and promotes capitalism according to such Protestant-derived ideals. However, it also provides a rather horrible account of what happens when a wax moth does penetrate a hive.

At the beginning of the 20th century, Edwardes wrote about the similarities between people and bees, commenting that,

> There is something curiously human-like in their movements over the crowded combs, and the old comparison of a bee-hive to a city of men is never out of mind. There are the incessant hurryings to and fro; chance meetings of friends at odd street-corners; altercations where we can almost hear the surly complaint and tart reply; busy masons and tillers and warehouse-hands at work everywhere: a hundred different enterprises going forward in every thronging thoroughfare or narrow side-lane, from the great entrance to the remotest drone-haunted corner of the hive.[31]

This is a particularly colourful comparison of bees and people, which, we feel, shows how much and why writers and thinkers have been inspired by bees over the centuries.

Not all comparisons of bee society and human society present a friendly face. Maurice Maeterlinck in his work, *La Vie des Abeilles*, was critical of the ways in which bees work but also employed a bee's perspective to identify failings within human society and labour, most notably as follows,

> How should we marvel, for instance, were we bees observing men, as we noted the unjust, illogical distribution of work among a race of creatures that in other directions appear to manifest eminent reason! We should find the earth's surface, unique source of all common life, insufficiently, painfully cultivated by two or three tenths of the whole population; we should find another tenth absolutely idle, usurping the larger share of the products of this first labor; and the remaining seven-tenths condemned to a life of perpetual half-hunger, ceaselessly exhausting themselves in strange and sterile efforts whereby they never shall profit, but only shall render more complex and more inexplicable still the life of the idle.[32]

This view resonates today with issues relating to collapsing biodiversity, inequalities between developed and developing countries, income inequalities, concerns about climate change, desertification, migration of human populations, refugees and in fact all the human-caused and created unpleasantness which affects our planet.

Bees and bee society have also been associated with ideas inherent in Communism. Edwardes, in 1908, suggested that bee society represented a socialistic political economy in which all is sacrificed for the good of the State and that the

31 *Ibid*, 49.

32 This quotation is taken from M. Maeterlinck, *The Life of the Bee* (*La Vie des Abeilles*), trans. A. Sutro (London: Dodd, Mead and Company, 1928, published originally 1901), paragraph 111.

individual is "nothing". He explained that "bee-polity" demonstrates a merciless adherence to the demands of a system which is reminiscent of "Absolute Communism", which, in his terms, also encapsulates cruelty, "In the republic of bees, nothing is allowed to persist that is harmful or useless to the general good".[33] One of the aspects of bee society to which he refers is the ruthless slaughtering of the queen bee when she grows old or can no longer lay eggs. Edwardes wrote extensively about "the commonwealth of the hive".

In his seminal 1930 essay, "The Protestant Ethic and the Spirit of Capitalism", Max Weber explores the foundations of modern capitalism through a religious lens. First, he defines capitalism by stating explicitly that it is not purely about the greedy and rapacious pursuance of profit but has instead a more ethical foundation grounded in Protestantism and the "Protestant work ethic". Weber explains that,

> [T]he impulse to acquisition, pursuit of gain, of money, of the greatest possible amount of money, has in itself nothing to do with capitalism ... It should be taught in the kindergarten of cultural history that this naïve idea of capitalism must be given up once and for all. Unlimited greed for gain is not in the least identical with capitalism, and is still less its spirit.[34]

He shows how the form of capitalism which has evolved in the West is different from any other profit-seeking system in the world because of, "... the rational capitalistic organization of (formally) free labour".[35] By free labour, Weber draws a distinction between the voluntary offering of labour by society as opposed to forced labour or slavery. In terms of Puritanism and Protestant ethics (asceticism), people are believed to have a "calling" from God. If this calling is to work in industry, where labour enhances profit (private or public) then it is morally right to pursue this goal. The accumulation of wealth through hard work only becomes an unethical pursuit when the labourer is tempted into "idleness and sinful enjoyment of life, and its acquisition is bad only when it is with the purpose of later living merrily and without care".[36] A puritanical lifestyle lived by members of the Protestant church is traditionally viewed as one of hard work, joylessness and sobriety. Weber does, however, conclude that capitalism has become an "iron cage" which individuals are born into and from which they cannot escape, "...until the last ton of fossilized coal is burnt".[37] Although early capitalists may have chosen to pursue wealth accumulation on the basis of a Protestant work ethic, he suggests that later members of Western society are now trapped into the capitalist system. The notion of the capitalist "iron cage" is as relevant today as it was then and it seems bees are now trapped within the same iron cage as we are. Although Weber does not explicitly use

33 Edwardes, *The Lore of the Honey-Bee*, xvi.

34 This quotation is taken from M. Weber, *The Protestant Ethic and the Spirit of Capitalism* (London: Routledge Classics, 2001, first published 1930), xxxv.

35 *Ibid.*, xxxiv.

36 *Ibid.*, 108.

37 *Ibid.*, 123.

a bee metaphor, many later writers have associated a work ethic and capital accumulation with the way in which bee society functions—all the bees work together tirelessly to attain their goals of honey collection (and pollination), wax production and the nurturing of offspring and their society. For example, in a recent book, *Bees in America: How the Honey Bee Shaped a Nation,* the author discusses the ways in which "busy bees" have influenced American culture.[38] She discusses the way in which the Puritan work ethic could have been modelled on the beehive and how she believes this approach to industry continues to influence the American concept of success and prosperity.

If we contrast the Weberian view of capitalism with the image of business and trade proposed in *The Fable of the Bees,* we can see they are diametrically opposite. One depicts a (very early) form of capitalism (really mercantilism at that time) driven by greed and avarice whereas the later work interprets capitalism as good, honest accumulation of capital driven by a Protestant notion of hard work and ethical principles.

If we continue the tradition of drawing on bee society as a means of supporting and representing current trends in politics and society, how would we interpret bee society now? From a capitalist perspective perhaps a beehive has a nominal queen with little genuine power and a busy government, with power disseminated through local councils and manufacturing outside the hive. This could still be synonymous with the grumbling hive. From a communist perspective, bees all work together for the common good. There are still many different forms of society and political environment around the world to merit many very different interpretations of bee society to support these competing forms of social construction. Indeed, we can see that throughout history bee society has been likened and used as a supporter for the Divine Right of Kings (and Queens), republican movements, the Reformation, the Commonwealth, Parliament, communism, capitalism, the Protestant work ethic and socialism. How strange that movements which have such different aims, objectives, traditions and characters can all draw analogies from the same society which has in essence remained unchanged in nature throughout all of these human political movements. Taking such contrary systems as communism and capitalism yet finding their roots within bee society seems odd to say the least. It is as though the admirable qualities of bee society and their work ethic can be transplanted onto any human attempt to bring an improved order into the world and bees' activities can be interpreted and moulded to support any political or socio-economic agenda.

38 See Horn, *Bees in America.*

Bees in religion, mythology, folklore, superstition and philosophy

Bees have been used as symbols in religion, mythology and in philosophical writings since the dawn of human society. Research shows that bees are scattered throughout religious writings and holy books, providing symbolic significance and representations. Within the religious writings of the Judeo Christian tradition there are frequent references to bees and honey throughout the Bible. Jews used bees as a symbol in the celebration of Rosh Hashanah, by eating apples dipped in honey to symbolize hopes for a happy and healthy new year. Indeed, Christianity has often used bees to represent the character and gentleness of Jesus Christ. Honey has been interpreted as representing his gentle character whereas the bee's sting has been taken to represent the crucifixion. Bees are mentioned four times in the Bible, in Deuteronomy, Judas, the Psalms and Isaiah. Probably the best known reference is where the body of the lion which Samson killed becomes inhabited after his death by a beehive, "... he turned aside to see the carcass of the lion, and behold, there was a swarm of bees in the body of the lion, and honey" (Judges 14:8). The image on Lyle's Golden Syrup tin shows this scene implying that from "strength cometh sweetness". Bees also feature in Islam in a similar way to that seen in Christian and Jewish writings. Chapter 16 of the Quran, the revelation of God, is called "The Bee". It speaks about how a drink of different colours comes from bees which can be used for healing. Indeed, the healing qualities of honey are mentioned in several places throughout the Quran. Also, paradise is described as having rivers of honey, clear and pure (Muhammad 47:15).

Writers and theologians have linked religion and nature throughout history. There is a train of thought that effectively "blames" the Judeo-Christian tradition for our ecological crisis and for climate change. Some consider that because in the book of Genesis in the Bible, God granted the human race "dominion" over all species, this gave humanity the right to use and abuse other species.[39] Writers from over five decades ago have started to attack humanity on their maltreatment of nature.[40] However, this very negative view does not allow for a loving relationship between humans and other inhabitants of the planet, which also emanates from our Western religious traditions. Surely, the intention was not to give humans stewardship over nature so that she may be abused and ruined? Perhaps it makes more sense to blame the ecological crisis on human greed and vanity, vices which are

39 See, for example, a seminal paper by Lynn White, "The Historical Roots of our Ecologic Crisis [with discussion of St Francis; reprint, 1967]", in *Ecology and Religion in History*, ed. David and Eileen Spring (New York: Harper and Row, 1974).

40 See, for example, F. A. Schaeffer, *Pollution and the Death of Man* (Wheaton, IL: Tyndale House, 1970), for a discussion of the destruction of nature by the human race, with a focus on a Christian perspective. He states that the human calling is for a healing process to take place between people and nature to establish the balance which should be the essence of ecology.

not in keeping with any religious faith. Indeed, the recent encyclical issued by Pope Francis certainly does not associate Christianity, or more specifically the Roman Catholic faith, with humans being granted permission to decimate nature. The Pope calls for people to express their love of God through their treatment of fellow species, treating the whole planet and its inhabitants as a common "home". Having taken "Francis" as his name in reverence of St Francis, the patron saint of Nature, the Pope has devoted himself to calling on all peoples from all religions to alter their relationship with fellow species.[41]

If we turn to Eastern religions, the focus on a more holistic, systems-oriented approach towards nature is evident. Zen Buddhism possibly presents a more egalitarian view of the relationship between humans and other species. In specific relation to bees, Buddhists bring honey as gifts to monasteries during the festival of Madhu Purnima. In Hinduism, early Hindu Vedic scriptures, as old as 1500 BC, have references to pollen and honey, referring to honey as the "nectar of the sun". The Hindu scripture, Srimad Bhagavatam, likens a honey bee gathering honey to wise men searching for truth and seeing only good in all religions. Furthermore, the "five Nectars" of Hinduism include honey. The Taoist context and its close relationship to a deep ecology approach to bees and nature is explored in more depth in Chapter 5 in which the author presents a personal deep ecology "account" of the bumblebee. Perhaps bees and their society could be seen as a means of drawing religions together in their admiration of the bee throughout history? Bees certainly represent a commonality between the world's great religions, with similar interpretations and symbolic usages of this well-loved insect.

Bees have also had a presence in mythology, folklore and superstition since the most ancient beginnings of human society. From the earliest human civilizations, honey and nectar have been associated with eloquence. Throughout the ages bees have often been perceived as sacred, representing a bridge between this life and the next.[42] Mycenaean tombs were shaped as beehives. In ancient times, Aristaeus, son of Apollo and the nymph Cyrene, was the Greek God of beekeeping. One myth relates that when Plato was an infant, bees settled on his lips while he slept, a portent of the honeyed prose that he would bring to his philosophical discourse:[43]

> And as when Plato did i' the cradle thrive,
> Bees to his lips brought honey from their hive.

41 Interestingly, White (1974) ends his essay with a possible way forward from a religious perspective, "The profoundly religious, but heretical, sense of the primitive Franciscans for the spiritual autonomy of all parts of nature may point a direction. I propose Francis as a patron saint for ecologists".

42 McCabe, *Collins Beekeeper's Bible* provides a detailed summary of the honey bee in myth and symbol.

43 This comes from William Browne "Second Song", *Dictionary of Phrase and Fable* (E Cobham Brewer, 1898).

However, the use of this imagery is not unique to Plato; it represents a standard reference to literary distinction, possibly first employed in accounting for the poetic gifts of Pindar (518–438 BC). The Roman author Claudius Aelianus said in (c. 175–235 CE) *On the Nature of Animals*:

> Bees dislike foul odors and sweet perfume alike, just as modest young girls do.
>
> Titmice, swallows, snakes, spiders and other creatures are at war with the bees. Bees are afraid of them.[44]

Pliny the Elder, the Roman author and philosopher (AD 23–79) believed that honey was engendered from the air, and put the variation in quality of honey down to the influence of the stars.

Throughout history, bees have appeared in mythology and ancient beliefs to symbolize the soul flying away from the body in many parts of the world. Bees are scattered throughout mythology and historic symbolism. For example, the Egyptians had bees carved into their royal seals. Coins from the Ephesians carried bee images. In keeping with Pliny's views about honey and the stars, the Ancients believed that honey collected after the rising of Sirius was always the best. The star, Sirius, was seen as the honey star by all the ancient writers. Even the French Emperor Napoleon Bonaparte had a bee embroidered into his coat of arms to represent power, immortality and resurrection.[45] Medieval received wisdom held that bees were born from the bodies of oxen, or from the decaying flesh of slaughtered calves; worms form in the flesh, then they turn into bees.[46]

In early England, bees featured significantly in folklore and superstition, and still do. These superstitions appear to date back to Virgil's and others' beliefs that beekeepers should treat their bees with love and respect. A fascinating book, *In Pursuit of Liquid Gold*, investigates the beginnings of beekeeping in England, bringing out the rich cultural history of folklore embedded in the beekeeping community. From as early as the 9th century, the "Dark Ages" in the British Isles, there is evidence of a method of raising an alarm when a hive started to "swarm" (basically, the swarm decides to move on, usually as there is not enough space for all the bees in the existing hive). This method was called "tanging", which involves an ancient practice of making a noise with early musical instruments. Tanging was used in the Dark Ages to stop swarms leaving their "owners". However, in ancient England, instead of using musical instruments, people simply made a cacophony by beating pans and kettles. This was believed to encourage the bees to return to their owner and settle close to the original hive. Indeed, this seems to be one of the earliest indications, in Britain at least, that bees and hives were starting to be viewed as someone's property, i.e. a capitalist asset. Early laws started to emerge around this time regarding

44 Gregory McNamee, *Aelian's On the Nature of Animals* (San Antonio, TX: Trinity University Press, 2011).

45 We are grateful to Bishop, *Robbing the Bees*, for some of these historical snippets.

46 See the website "Medieval Bestiary", bestiary.ca

bees and beehive ownership. These early laws may have represented the turning point for human exploitation of bees.

Folklore around bees also extended into deep superstitions among beekeepers, which have persisted into modern times. It was, for example, believed to be extremely ill-advised to sell bees, and a superstition dating back to the 15th century suggested that, "to sell your bees is to sell your luck".[47] This made it difficult for a prospective beekeeper to acquire their first beehive. The only way this could be done was through bartering. Medieval traditions also persist into modern times, which appear to derive from the mythological and ancient belief that bees are sacred creatures. The tradition of "telling the bees" has been revered by beekeepers in England and likely elsewhere, since at least as far back as the 16th century. In short, if a beekeeper or any member of (usually, at that time) his family, grew sick or died, another family member must tap on the hive gently and tell the bees the news in a whisper. There are many other events which bees should have been informed of by their beekeeper: family marriages, births and the outbreak of war, to name but a few. Interestingly, there are a whole array of penalties and punishments which were believed to be exacted by bees on their "owners" if the beekeeper did not treat them with respect, or failed to observe the type of traditions mentioned here. A list is provided in *In Search of Liquid Gold*, which includes: the bees dying or leaving; the bees failing to produce honey; further death in the family or among domestic animals; and crops dying.[48] Is it possible, from the perspective of a superstitious person, that mass commercialization at an industrial level of beehives for pollination purposes and mass trading in bees and their hives have broken these age-old links between bees and their beekeepers and resulted in a rebellion? Surely not...

Cultural bees: how bees have inspired art, literature, poetry, music and film

Our research has shown what an immense and wide-ranging influence bees have had over the ages on all aspects of human culture and society. We cannot even begin to do justice to this influence and just hope that we can at least provide a flavour of the cultural and historical context. Indeed, we intend to encourage interest in bees and culture rather than try to provide a comprehensive coverage and apologize for the multitude of examples and illustrations which we are sure to have missed. From the *Man of Bicorp*, a cave painting near Valencia, Spain, depicting a human picking up honeycombs while bees fly around him, to Dali's *Dream Caused by the Flight of a Bee Around a Pomegranate a Second Before Awakening* (1944), and *From the Life of Bees* (1954) by post-modern artist and guru, Joseph Beuys and

47 Ogden, *In Pursuit of Liquid Gold*, 69.
48 *Ibid.*, 73.

beyond, art has been perhaps the most reliable cultural commentator on the relationship between humankind and bees. It offers up a constantly evolving archive that evokes through eras, cultures and artistic movements the symbiosis, myths, legends, beliefs and superstitions that have flourished within the unique connection between our species.

Art also describes an awe-inspiring timeline. The *Man of Bicorp*, discovered in the early 1900s, is dated at between 8,000 and 15,000 years old, placing it roughly at the boundary of the Paleolithic and Mesolithic eras. Later, in the Mesolithic era, or Middle Stone Age, countless examples of rock art depicting bees and honey-gathering, were created throughout Africa, Europe, India and Australia. Similarly, the various symbolisms and metaphors acquired by the bee, throughout the lifespans of the great early civilizations, particularly Ancient Egypt and Greece, found extensive representation through artistic expression. The bee was the official insignia of Lower Egypt, its form was routinely portrayed on tomb walls, and bee hieroglyphs on obelisks and sarcophagi are testament to their cultural significance. In Ancient Greece, the Melissae, which translates as "The Bees", were a triad of winged priestesses, who taught the God Apollo the art of divination. Gold plaques embossed with their images, dating back to the 7th century, were found in Rhodes.

From these ancient beginnings, the bee, together with its emblematic subtexts of, among many others, fertility, order, kingship, immortality and diligence, has continued to influence, entrance and inspire artists through the centuries. In the Middle Ages, the beehive was often employed in coats of arms and heraldic badges, as a symbol of industry. In the Renaissance era, the bee was frequently deployed as a metaphor for creativity and wealth. Through Baroque art to Romanticism, from Impressionism to Surrealism and on through Post-Modern to Pop Art, to the present day, to significant works such as *Queen Bee* (2013) by pop surrealist Mark Ryden, and the 21st century sculptural hives created by American artist Hilary Berseth, in collaboration with actual bee colonies, the bee has continued to pollinate the creative imagination. Finally, the recent street art project, *Save the Bees*, in which artists Louis Masai Michel and Jim Vision drew attention to colony collapse disorder by covering building walls in London with images of bees, can be seen as a definitive conflation of two of the major themes defining the relationship between our two species: our appreciation of the aesthetic beauty and symbolic richness of bees; and humanity's growing awareness of what we may be on the verge of losing. Photography as well as more traditional forms of art is also packed with bee projects. For example, an artistic book of bee photographs was published in 2010. The photographs depict all aspects of bee anatomy: eyes, wings, hair, sting, antennae and are aimed to stimulate more interest in bees and the need to attend to their needs in this period of decline in bee populations.[49]

49 The photographs are found in R-L. Fisher, *BEE* (New York: Princeton Architectural Press, 2010).

The influence of bees is also pervasive in music. Throughout the ages, bees have influenced composers, lyricists and musicians. Bees, honey and hives have inspired music from Thomas Tallis's early arrangement of, "Where the Bee Sucks so Suck I", to Rimsky-Korsakov's "Flight of the Bumble Bee", through to a recent release by the group, The Bohicas, entitled "Swarm", which uses electric guitars to evoke a menacing sound of angry bees swarming. A personal favourite of one of the authors is Alison Moyet's "Honey for the Bees". Classics such as Rimsky Korsakov's Flight of the Bumblebee and Francois Schubert's, The Bee, are generally well-known around the world. Less well-known examples include John Downland's, "It was a time when silly bees could talk" and John Duke's "Bee I'm Expecting You".

In architecture, honeycombs appear to have been a significant inspiration to architects throughout the ages. The development of mathematics has also featured bees and their allegedly mathematical abilities. Pappus of Alexandria, a Greek mathematician, published an eight-volume series entitled *The Collection* (c. 340 BC) in which he summarized all of Greek mathematics. In Book V, he refers to the *honeycomb conjecture*, where he suggested that bees understood geometry as they selected naturally the best structure for their combs: a hexagon. Indeed, the amazing and intricate design of a natural beehive has attracted the interest of human builders for centuries. Further, the ability of bee society to keep the hive clean and sanitized, despite being so highly populated, has also fascinated bee-ologists over the years. In *The Lore of the Honey-Bee*, Edwardes comments that,

> The comparison of a modern beehive with a building similar in construction, and as densely crowded with human beings, brings the whole problem to a sharp definition. In such a building, unless a through current of air could be established, the preservation of life must soon become impossible. Yet the bees have triumphantly overcome all difficulties. Whether in winter or summer, the air within the hive is almost as pure as that in the open, while the temperature can be regulated at will.[50]

The bees fan air through the hive to regulate temperature and provide air conditioning.

The Spanish architect Anton Gaudi's (1852–1926) Parabolic Arches "possess a shape exactly the same as that made by bees when they build a natural honeycomb…for him, the beehive provided a way of fusing his Catholic faith with an almost Moorish love of organic natural forms".[51] The Swiss-French architect, Le Corbusier (1887–1965), one of the pioneers of modern architecture, was similarly inspired by bees. Overseeing the post-Second World War reconstruction of Marseille, his housing estates on stilts were modelled on beehives.

50 Edwardes, *The Lore of the Honey-Bee*, 63.
51 See Wilson, *The Hive.*

Bees abound in literature and poetry and again, we can only skim the surface of this immense body of work here. We looked at many of the earlier writings inspired by bees in relation to economics and politics but there is also a wealth of literature in the form of novels and poems. For example, a recent novel, *The Secret Life of Bees*, tells a warming story of how a young girl who has lost her mother manages to rebuild her life, living with three eccentric beekeeping sisters.[52]

In poetry, Virgil, discussed above, provided an informative, educational and social overview of bees and beekeeping in his tireless and beautiful Latin poetic style. The British writer George Bernard Shaw wrote,

> Go to the bee,
> thou poet:
> consider her ways
> and be wise.[53]

The British poet laureate, Carol Ann Duffy wrote a whole book inspired by bees and bee society, with this lovely excerpt:

> Been deep, my poet bees,
> in the parts of flowers,
> in daffodil, thistle, rose, even
> the golden lotus; so glide,
> gilded, glad, golden, thus—[54]

Another contemporary poet adopted an anthropomorphic view of bees as follows,

> Bees in the roof, bees on the walls
> Stitching the house in a net of flightways,
> Just like surveillance, just like snoopers
> In the open air[55]

Bees have also inspired many films and television series over the years. "The Swarm", a classic horror film in which the characters deal with deadly swarms of killer bees, represents one of the more negative ways in which bees have influenced culture and the film industry. In the vastly popular (especially with one of the authors) American science fiction series of the 1990s, "The X-Files", two episodes, "Zero Sum" and "Herrenvolk" involved FBI special agents Fox Mulder (David Duchovny) and Dana Scully (Gillian Anderson) investigating the concealed case

52 See S. Monk Kidd, *The Secret Life of Bees* (New York: Viking Penguin, 2002).
53 From George Bernard Shaw, *Man and Superman* (New York: Brentano's, 1903).
54 This is taken from poet laureate Carol Ann Duffy, *The Bees* (Oxford, UK: Picador, 2011).
55 This extract is from Sean Borodale, 27 May: Geography, *Bee Journal* (London: Jonathan Cape, 2012).

of virus-carrying bees. Further, the new series, released after a 13 year "holiday" includes an episode devoted to investigating the causes of colony collapse disorder. There is a vast array of natural history documentary television relating to bees. One striking example is the BBC's documentary series, "The Wonder of Bees" and another is "Beetalker: The Secret World of Bees" from 2006.

Most if not all areas of human life are affected and inspired by bees. Even fashion has been influenced by bees. Just take, for example, the iconic look created by the "beehive" hairstyle, reintroduced in recent years by the late Amy Winehouse. Recently, a fashion designer, Ada Zanditon, designed a London Fashion Week collection inspired by colony collapse disorder. In 2015, Reebok announced its new "Bees and Honey" theme. Linked to this, its brand Sneakersnstuff has started a new initiative called Bee Urban in Stockholm, which aims to inspire people and businesses to sponsor beehives on city rooftops. The shoe has a black leather upper and a honey yellow lining.

Even gaming and technology have been inspired by bees with Capcom's Darkstalkers fighting game series, featuring Q-Bee, short for Queen Bee, who is Soul Bee born in the demon world, Makai. In technology, XBee radios were first introduced in 2005 under the MaxStream brand. XBee is the brand name from Digi International for a series of radio modules that can be used with minimum connections. Specifically, XBees are tiny blue chips that communicate with each other without wires. Harvard University has invented a robotic bee which it is testing as an alternative to the natural pollinator! An electric vehicle, called the Zbee, developed by Clean Motion is marketed as a super-efficient clean vehicle for urban transport in the 21st century. The Zbee received awards from WWF Climate Solver and the Zennsström Green Mentorship Award in 2013.

In a recent cameo of bees and human society from the science field, the writer compares the strategies and communication bees use to locate and retrieve nectar to the world of science and the scientist, arguing that if successful bees made assumptions, they would be similar to those of the scientist; flowers can be regarded as facts, nectar as knowledge, honey as technology and their "waggle dance" as communication of ideas (from abstract).[56] The analogies and comparisons continue in our contemporary society.

This chapter has provided some historical context to the age-old relationship between human society and bee society. Hopefully, the reader will feel that this broad-ranging discussion helps us to understand and appreciate how important bees are to humans and how bees have had an immense impact on culture, economics, politics, architecture and essentially on the way people think. The rest of

56 See Ben Trubody "The Bee-haviour of Scientists: An Analogy of Science from the World of Bees". *Between the Species* 14 (2011). digitalcommons.calpoly.edu.

this book, for the most part, explores bee decline from a business-world perspective. Some chapters investigate the influence of bee decline on companies, production, stock markets and investors. Other chapters research the extent to which companies are accounting for bee decline, analysing the corporate response to the bee crisis at an international level. However, before focusing on the "business of bees", we turn to a discussion of the scientific context of bee decline and a deep ecology perspective of bees and their intrinsic value.

Bibliography

Birchall, E. *In Praise of Bees: A Cabinet of Curiosities*. Shrewsbury: Quiller Publishing Ltd, 2014.
Bishop, H. *Robbing the Bees: A Biography of Honey—The Sweet Liquid Gold that Seduced the World*, New York: Free Press, 2005.
Borodale, S. *Bee Journal*. London: Jonathan Cape, 2012.
Browne, William. "Second Song". *Dictionary of Phrase and Fable*. E Cobham Brewer, 1898.
Daye, John. *The Parliament of Bees with their Proper Characters, or A Bee-hive Furnisht with Twelve Honycombes, as Pleasant as Profitable. Being an Allegorical Description of the Actions of Good and Bad Men in these our Daies*. 1641 edition, republished in 1881 by A. H. Bullen.
Duffy, C. A. *The Bees*. Oxford, UK: Picador, 2011.
Edwardes, T. *The Lore of the Honey-Bee*. London: Methuen & Co Ltd, 1908.
Fisher, R-L. *BEE*. New York: Princeton Architectural Press, 2010.
Goulson, D. *A Sting in the Tale*. Croydon, UK: Jonathan Cape, 2013.
Goulson, D. *A Buzz in the Meadow*. London: Vintage Books, 2014.

Halberstam, J. *The Queer Art of Failure.* Duke University Press, 2011.

Hartlib, Samuel. *The reformed Common-wealth of bees. Presented in severall letters and observations to Sammuel Hartlib Esq. With The reformed Virginian silk-worm. Containing many excellent and choice secrets, experiments, and discoveries for attaining of national and private profits and riches.* London: Printed for Giles Calvert at the Black-Spread-Eagle at the West-end of Pauls, 1655.

Horn, T. *Bees in America: How the Honey Bee Shaped a Nation.* Lexington, KY: University Press of Kentucky, 2005.

Huber, F. *New Observations on the Natural History of Bees.* Translated from the original French. Edinburgh: John Anderson and London: Longman, Hurst, Rees, and Orms, 1806.

Leclerc, Georges-Louis. Comte de Buffon's Fifth Volume. In *History of the Brute Creation,* from Buffon's Natural History containing a *Theory of the Earth, a General history of Man, of the Brute Creation, and of Vegetables and Minerals, &c. &c. &c.* in ten volumes. Translated from the French. London: H. D. Symonds, Pater-Noster Row, 1797.

Lee, J-D, H-J. Park, Y. Chae, and S. Lim. "An Overview of Bee Venom Acupuncture in the Treatment of Arthritis". *Evidence Based Complementary Alternative Medicine* 2(1) (2005): 79-84.

Loeb Classical Library. "Virgil, Georgics". Accessed 14 April 2016. http://www.loebclassics.com/view/virgil-georgics/1916/pb_LCL063.219.xml?readMode=recto

de Mandeville, Bernard. *Fable of the Bees: Or Private Vices and Public Benefits.* Oxford: Clarendon Press, 1732.

McCabe, Philip. *Collins Beekeeper's Bible: Bees, Honey, Recipes and Other Home Uses.* Glasgow: Collins, 2010.

McNamee, Gregory. *Aelian's On the Nature of Animals.* San Antonio, TX: Trinity University Press, 2011.

Maeterlinck, M. *The Life of the Bee* (*La Vie des Abeilles*). Translated by A. Sutro. London: Dodd, Mead and Company, 1928 (published originally 1901).

Monk Kidd, S. *The Secret Life of Bees.* New York: Viking Penguin, 2002.

Moss, Laurence S. "The Subjectivist Mercantilism of Bernard Mandeville". *International Journal of Social Economics* 14, Nos.3/4/5 (1987), 167–84.

Ogden, R. B. *In Pursuit of Liquid Gold.* Charlestown, UK: BBNO, 2001.

Pell, J. and I. van Staveren. *Handbook of Economics and Ethics.* Cheltenham, UK: Edward Elgar Publishing Limited, 2009.

Prys-Jones, O. E., S.A. Corbet, and T. Hopkins. *Bumblebees: 6, Naturalist's Handbook.* Cambridge, UK: Cambridge University Press, 2011.

Purchas, Samuel. *A theatre of politicall flying-insects wherein especially the nature, the worth, the work, the wonder, and the manner of right-ordering of the bee, is discovered and described: together with discourses, historical, and observations physical concerning them: and in a second part are annexed meditations, and observations theological and moral, in three centuries upon that subject.* London: Printed by R. I. for Thomas Parkhurst, 1657).

Quinby, M. *Mysteries of Bee-keeping Explained: Being a Complete Analysis of the Whole Subject.* New York: C.M. Saxton, Agricultural Book Publisher, 1853.

Schaeffer, F. A. *Pollution and the Death of Man,* Wheaton, IL: Tyndale House, 1970.

Seeley, T. D. *Honeybee Democracy.* Princeton, NJ: Princeton University Press, 2010.

Shaw, George Bernard. *Man and Superman.* New York: Brentano's, 1903.

Smith, F. G. *Beekeeping in the Tropics.* London: Longmans, Green & Co Ltd, 1960.

Trubody, Ben. "The Bee-haviour of Scientists: An Analogy of Science from the World of Bees". *Between the Species* 14 (2011). digitalcommons.calpoly.edu.

Vanbergen, A. J., Baude, M., Biesmeijer, J. C., Britton, N. F., Brown, M. J. F., Bryden, J., Budge, G. E., Bull, J. C., Carvell, C., Challinor, A. J., Connolly, C. N., Evans, D. J., Feil, E. J., Garratt, M. P., Greco, M. K., Heard, M. S., Jansen, V. A. A., Keeling, M. J., Kunin, W. E., Marris, G. C., Memmott, J., Murray, J. T., Nicolson, S. W., Osborne, J. L., Paxton, R. J., Pirk, C. W. W., Polce, C., Potts, S. G., Priest, N. K., Raine, N. E., Roberts, S.,Ryabov, E. V., Shafir, S., Shirley, M. D. F., Simpson, S. J., Stevenson, P. C., Stone, G. N., Termansen, M. and Wright, G. A. Threats to an Ecosystem Service: Pressures on Pollinators. *Frontiers in Ecology and the Environment* 11 (5) (2013): 251-259.

Weber, M. *The Protestant Ethic and the Spirit of Capitalism*. London: Routledge Classics, 2001 (first published 1930).

White, L. "The Historical Roots of our Ecologic Crisis [with discussion of St Francis; reprint, 1967]". In *Ecology and Religion in History*, edited by David and Eileen Spring. New York: Harper and Row, 1974.

Wilson, Bee. *The Hive: The Story of the Honeybee and Us*. London: John Murray Publishers, 2004.

Wilson-Rich, N. *The Bee: A Natural History*. Lewes, UK: Ivy Press, 2014.

Winston, M.L. *Nature Wars: People vs Pests*. Cambridge, MA: Harvard University Press, 1997.

Winston, M. L. *Bee Time: Lessons from the Hive*. Cambridge, MA: Harvard University Press, 2014.

3

Bee bio-basics and conservation benefits

Essential pieces in the pollinator puzzle

Scott Longing and Samuel Discua
Texas Tech University, USA

Chapters 1 and 2 have provided an overview of the contributions of honey bees to pollination as well as an introduction to the issues of pollinator decline and colony collapse disorder, including some of the factors affecting pollinators worldwide. In this chapter, we extend these topics by further highlighting specific contributions of managed honey bees and *wild* bees to pollination services in agricultural systems, a key benefit to humans as a result of responsible land-management practices. As it is important to understand *what* must be conserved, we provide information on the cast of characters—common types of bees and their general biology, including brief abstracts of the most common *families* of bees. Examples are given from the scientific literature regarding the positive relationships of least-affected landscapes and wild bee abundances and diversity. A biological and ecological basis is emphasized to foster an understanding of the essential interacting components—the bees and their altered and diminishing habitat resources. Information provided should support a basic foundation to develop strategies for biodiversity conservation and sustainable ecosystem services for the production of human goods and services, which is especially important for enterprises that directly or indirectly depend on pollinators.

Like all insects, bees have persisted for millions of years, yet extensive changes to terrestrial landscapes by modern humans and irresponsible environmental practices are leading to some alarming downward trends in pollinator communities,[1] with the capacity for pollinator shortages to result in economic risk to direct and indirect crop sectors.[2] We are yet to understand fully how future changes to pollinator biodiversity worldwide will manifest broadly in natural ecosystems, for example long-term or chronic effects on pollinator–plant networks, broader food webs, water, air and soil quality, and other benefits humans receive from ecosystems. We do know, however, that bees are an essential contributor to the successful reproduction of most of the world's flowering plants, a component of both the natural and human-modified world that is vital to humans.

Pollination services and bee contributions

Pollination is a process involving the transfer of pollen from the male anther to the female stigma of the same plant or between two plants. The movements of pollen via wind, water and animals is one of the most important processes of the natural world. The majority of flowering plants depend on one or more of approximately 200,000 vertebrate and invertebrate species of animals for pollination, with bees being one of the most effective pollinators. In addition to bees, several other insect groups (e.g. butterflies and moths, beetles and flies) as well as some vertebrates (e.g. hummingbirds and bats) contribute to the pollination requirements of at least three-quarters of over 250,000 species of flowering plants worldwide.[3] Insects are key contributors to pollination via regulating the successful reproduction of plants with high nutritional quality (when compared with major staple crops),[4] forage for animals consumed as part of the human diet, and food and habitat for wildlife. Failure to understand the importance of pollinator biodiversity would place humans in the precarious position of maintaining agricultural production solely through artificial and technological means, likely a non-prevailing strategy.

Pollination is both a natural process and '*ecosystem service*', which the Millennium Ecosystem Assessment (MA) defines simply as "benefits people obtain from ecosystems."[5] The MA categorizes ecosystem services as: 1) *provisioning* (e.g. tim-

1 National Research Council. *Status of Pollinators in North America*. Washington, DC: National Academies Press, 2007.

2 D.M. Bauer and I.S. Wing. "Economic Consequences of Pollinator Declines: A Synthesis". *Agricultural and Resource Economics Review* 39 (2010): 368–83.

3 A.M. Klein *et al.*, "Importance of Pollinators in Changing Landscapes for World Crops". *Proceedings of the Royal Society B* 274 (2007): 303–13.

4 E.J. Eilers *et al.*, "Contribution of Pollinator-Mediated Crops to Nutrients in the Human Food Supply". *PLoS ONE* 6(6) (2011): e21363, doi:10.1371/journal.pone.0021363.

5 Millennium Ecosystem Assessment (MA). *Ecosystems and Human Well-Being: Synthesis*. Washington, DC: Island Press, 2005.

ber and water); 2) *regulating* (e.g. those that moderate natural phenomena such as pollination and water quality); 3) *cultural* (e.g. those that provide recreational or spiritual benefits); and 4) *supporting* services (e.g. photosynthesis and nutrient cycling) (Figure 3.1). Pollination and insect pollinators are positioned within the *regulating* services category because they regulate crop production. Yet, insect pollinators also contribute to materials that can be extracted from nature (i.e. agricultural crops), thus contributing indirectly to *provisioning* services. Furthermore, bees (and pollination) depend on supporting services such as photosynthesis, soil formation and nutrient cycling in order to carry out ecosystem services; without supporting services, other ecosystem services would simply not exist. Although we receive direct and indirect benefits from natural and human-influenced ecosystem services, our mismanagement of ecosystems can result in ecosystem dis-services, such as habitat loss, nutrient runoff, and poisoning of beneficial insects from pesticide misuse.[6]

Figure 3.1 **Categories of ecosystem services as defined by the Millennium Ecosystem Assessment. Natural resources extracted for human use, such as food, fibre and fuel, are provisioning services typically produced from agriculture and that depend on many regulating and supporting services. Pollination is a regulating service that supports the provision of agricultural products.**

Source: Redrawn from Zhang *et al.*, "Ecosystems Services and Dis-services to Agriculture"

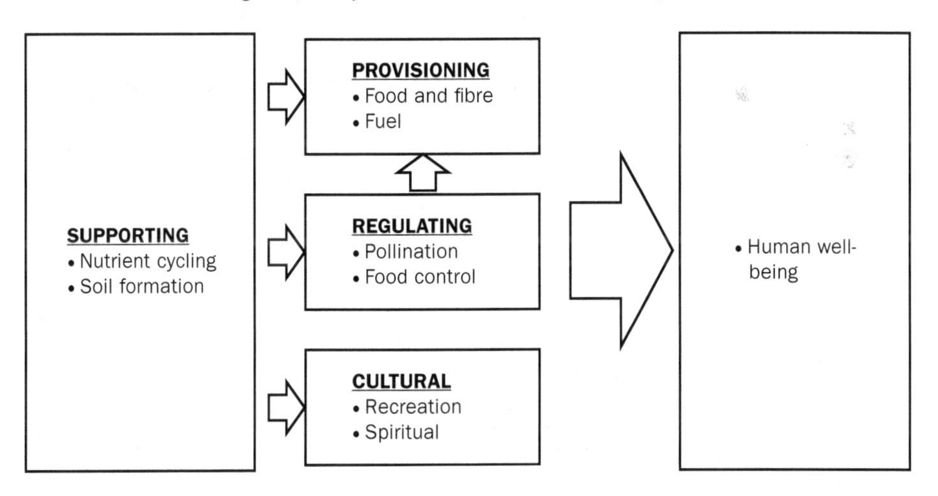

6 W. Zhang *et al.*, "Ecosystems Services and Dis-services to Agriculture". *Ecological Economics* 64(2) (2007): 261–8, doi: 10.1016/j.ecolecon.2007..02.024.

Approximately one in three bites of food eaten by humans is directly attributed to pollinators.[7] The global value of pollination services has been estimated at $212 billion, with the value of insect pollination in Europe and the US nearly equal at approximately $15 billion.[8] In addition to the immense value derived from one of the most common pollinators, managed honey bees, the value of wild pollinators in agricultural production in the US has been estimated at over $3 billion annually.[9] Pollination services are essential to yields and production of 75% of crop species globally, while up to 94% of wild flowering plants are dependent on pollinators.[10,11] Furthermore, the majority of crop plants globally benefit from pollinators for improving or stabilizing yields.[12]

Although some of the most widely grown staple crops such as corn, wheat and rice do not depend on pollinators, many fruits, vegetables, edible oil crops, seeds and nuts require their input. Meat and dairy products produced from animals fed insect-pollinated forage are an integral part of agriculture and the human diet. Across 200 countries, 87 of the leading global food crops, in terms of total production, were determined to be dependent on animal pollination, whereas only 28 crops did not rely on animal pollinators.[13] Because of large-scale monocultures and production of global staple crops, which account for approximately 60% of total production, it has been estimated that major pollinator losses would result in only a 4–8% reduction in total agricultural production.[14] Yet, while a vast amount of human food is made from corn, wheat and rice, it is unreasonable to believe that humans could survive with elimination of vital micronutrients and nutritional quality provided by pollinator-dependent crops, in addition to the products indirectly regulated by insect pollination. Furthermore, many of the pollinator-dependent crops have up to five times greater value when compared with those not dependent on pollinators.[15] From both a human nutrition perspective and an economic-value standpoint, crops requiring the inputs of pollinators are an important component of global economies and human welfare.

7 M. Ingram *et al.*, "Our Forgotten Pollinators: Protecting the Birds and the Bees". *Global Pesticide Campaigner* 6 (1996): 1–8.

8 N. Gallai *et al.*, "Economic Valuation of the Vulnerability of World Agriculture Confronted with Pollinator Decline". *Ecological Economics* 68 (2009): 810–21.

9 J.E. Losey and M. Vaughn. "The Economic Value of Ecological Services Provided by Insects". *Bioscience* 56 (2006): 311–23.

10 J. Ollerton *et al.*, "How Many Flowering Plants Are Pollination by Animals?" *Oikos* 120 (2011): 321–6.

11 A.J. Vanbergen and Insect Pollinators Initiative. "Threats to an Ecosystem Service: Pressures on Pollinators". *Frontiers in Ecology and the Environment* 2013, doi: 10.1890/120126.

12 Klein, "Importance of Pollinators in Changing Landscapes for World Crops", 303–13.

13 Klein, "Importance of Pollinators in Changing Landscapes for World Crops", 303–13.

14 M.A. Aizen *et al.*, "How Much Does Agriculture Depend on Pollinators? Lessons from Long Terms Trends in Crop Production". *Annals of Botany* 103 (2009): 1579–88.

15 Gallai *et al.*, "Economic Valuation of the Vulnerability of World Agriculture Confronted with Pollinator Decline", 810–21.

Significant threats exist to thc production of fruits, nuts and vegetable crops because of reductions in the number of colonies of managed honey bees in the US, Europe and elsewhere. In the US, the total number of managed honey bee colonies has dropped from 5–6 million in the mid-20th century to around half of that today. Most of the 2.5 million or so colonies in the US are used in migratory beekeeping to pollinate crops such as almonds, blueberries, pumpkins, apples and cherries in different regions of the country. Management pressures associated with these migratory activities, pests and diseases, pesticide toxicity and nutritional deficiencies because of reduced wildflower forage are some of the factors affecting honey bees. The consequences of these interacting factors have in part been related to extreme winter losses of colonies in the US, which threatens a vast number of speciality crops that benefit from pollination services by managed honey bees. In recent years, numerous governmental and non-governmental programmes in the US have been focused on both basic research to understand honey bee health in relation to environmental pressures and conservation practices to increase the areal coverage of nectar and habitat resources for pollinators, including those for managed honey bees used in migratory beekeeping.

Managed honey bees and wild bees are considered to be some of the most effective pollinators, with the latter receiving more attention in recent years because of an improved understanding of their roles in crop yields.[16,17] As one of the most effective pollinators, it is easy to observe honey bee individuals foraging among flowering plants. However, it is often common and abundant wild bees that contribute to crop pollination services. Using data from 90 different studies, one study found that wild bee communities contributed a mean of over $3,000 per hectare to the production of insect pollinated crops, a value similar to that attributed to honey bee pollination (mean = $2,913).[18] Wild bees were shown to have greater pollination efficiency that resulted in greater fruit set in sweet cherry when compared with honey bees.[19] Pollination measured across 23 watermelon farms showed that wild bees alone provided sufficient pollination services at >90% of the farms; total pollen deposition at flowers was strongly correlated with native bee visitations, but

16 R. Winfree *et al.*, "Native Bees Provide Insurance Against Ongoing Honey Bee Losses". *Ecology Letters* 10 (2007): 1105–13.

17 L.A. Garibaldi *et al.*, "From Research to Action: Enhancing Crop Yield through Wild Pollinators". *Frontiers in Ecology and the Environment* 12 (2014): 439–47.

18 D. Kleijn *et al.*, "Delivery of Crop Pollination Services is an Insufficient Argument for Wild Pollinator Conservation". *Nature Communications* 6 (2015): 7414, doi: 1038/ncomms8414.

19 A. Holzschuh *et al.*, "Landscapes with Wild Bee Habitat Enhance Pollination, Fruit Set and Yield of Sweet Cherry". *Biological Conservation* 153 (2012): 101–7.

not honey bee visitations.[20] Across 41 cropping systems globally, positive associations were determined between wild bee visits and fruit set, while fruit set attributed to honey bees increased significantly in only 14% of the systems surveyed.[21] Results from this study further suggested that because wild bees and honey bees promoted fruit set independently, practices to conserve both managed and wild bees would be most beneficial.[22] Moreover, the interactions of wild bees and honey bees can synergistically influence pollination and crop production. For example, behavioural interactions between wild bees and managed honey bees were found to increase the pollination of hybrid sunflower fivefold.[23] While honey bees have long been known to be the dominant pollinator for crop production, scientific evidence to date suggests that wild bees are essential contributors to pollination services, and integration of wild bees (through sufficient habitat resource conservation) with managed bees (e.g. honey bees, bumblebees and mason bees) would be an effective strategy to broadly enhance pollination services in agriculture.[24]

The cast of characters: common bees and their biology

General knowledge of bee diversity, biology and their habitats facilitates a broader understanding of the contributions of this important group to pollination services and their role in regulating the majority of Earth's flowering plants. From a business and production standpoint, integrating pollinator conservation strategies with management and ultimate production of goods and services should benefit from an understanding of *what* is being conserved for direct and indirect benefits.

What are bees?

The world's insects are grouped into approximately 29 **orders** of insects. Think of the taxonomic level of **order** as a practical, coarse grouping of the approximately 1 million insect species currently known to science. For example, all the species of

20 Winfree *et al.*, "Native Bees Provide Insurance Against Ongoing Honey Bee Losses", 1105–13.

21 L.A. Garibaldi *et al.*, "Wild Pollinators Enhance Fruit Set of Crops Regardless of Honey Bee Abundance". *Sciencexpress* 2013, doi: 10.1126/science.1230200.

22 *Ibid.*

23 SS. Greenleaf and C. Kremen. "Wild Bees Enhance Honey Bees' Pollination of Hybrid Sunflower". *Proceedings of the National Academy of Science* 103 (2006): 13890–5.

24 Garibaldi *et al.*, "From Research to Action: Enhancing Crop Yield through Wild Pollinators", 439–47.

butterflies and moths are in the order Lepidoptera, crickets and grasshoppers are in the order Orthoptera, and house flies and mosquitoes are in the order Diptera. Bees belong to the order **Hymenoptera**, which in addition to bees includes ants, horntails, sawflies and wasps. Wasps are the closest relatives to bees, but they differ in that bees are usually more robust and hairy than wasps, most wasps are carnivorous, and bees are adapted for feeding on nectar and pollen as their primary energy and protein sources.

Bees undergo **complete metamorphosis**, meaning that in their life-cycle individuals go through **egg**, **larva**, **pupa** and **adult** stages. A honey bee worker's development from egg to adult takes around 21 days. The process begins with the queen honey bee laying an individual egg into one hexagonal comb cell. The larva hatches from the egg after three days and is fed by other female worker bees. Bee larvae undergo five moults (i.e. shedding of their exoskeleton) before entering the pupal stage, where they will not eat and will exhibit little movement. During the pupal stage adult structures will start developing and after approximately 13 days a new worker bee emerges to begin its societal duties. Other bees, including solitary bees, also go through complete metamorphosis. However, solitary female bees will lay only a few eggs in their lifetime, with each egg developing into a larva, pupa and finally an adult bee that lives for a few weeks on average. In contrast, honey bee queens can live up to several years while laying thousands of eggs.

Like all adult insects, bees have a three-part body divided into the **head**, **thorax** and **abdomen**, with most of the body covered by a hardened exoskeleton made from a nitrogen-containing polysaccharide called chitin. The bee's head contains a pair of compound eyes, three simple eyes and a pair of antennae. The antennae of bees are covered with olfactory and tactile receptors. The bee's tongue, which contains touch and taste receptors, is used for lapping nectar from flowers. Bees are commonly classified at a coarse, higher level of taxonomic classification according to the average length of their tongue (i.e. long-tongued and short-tongued bees), and some tongues even match the general architecture of the flower in order to most efficiently lap nectar. The bee's thorax, housing two pairs of membranous wings and six legs, is the power-house for movement including rapid flight. The bee's abdomen contains much of the digestive systems and respiratory organs for gaseous transport and exhaust. In some bees the abdomen also contains wax glands, venom glands and a sting apparatus.

Evidence from the fossil record suggests that early bees appeared around 100 million years ago, shortly after the evolution and appearance of flowering plants (i.e. angiosperms).[25] As these early bees evolved from carnivorous wasps, they developed longer tongues, hairier bodies and specialized structures such as pollen baskets and brushes, allowing more efficient collection and transport of pollen and nectar. To attract pollinators, plants developed attractive, scented and brightly coloured flowers. The interaction of flowering plants and bees, over time, facilitated

25 G.O. Poinar and B.N. Danforth. "A Fossil Bee from Early Cretaceous Burmese Amber". *Science* 314 (5799) (2006): 614, doi: 10.1126/science.1134103.

the appearance of new species of plants and bees though a process known as co-evolution.

Bee diversity and classification

To understand better how bees are classified into different groups and the terminology used in this chapter, a comparative classification of humans and honey bees using a **taxonomic hierarchy** is presented (Table 3.1). A **taxonomic rank** is the relative level of an organism or group of organisms in the taxonomic hierarchy, following the formal system of binomial nomenclature introduced by the Swedish naturalist Carl Linnaeus in the late 18th century. Examples of taxonomic ranks are species, genus, family and kingdom. In a taxonomic hierarchy, ranks go from more general categories such as **kingdom** (Animalia or Plantae) to lower levels of classification such as **genus** or **species**, which are more basic taxonomic ranks. For the purposes of our explanation, **species** represents the lowest level of classification in a taxonomic rank, and is generally defined as a population of interbreeding individuals which produce viable offspring. In the following section we present some of the common bees at the taxonomic rank of **family**.

Table 3.1 **Taxonomic classification of the western honey bee, *Apis mellifera* and humans, *Homo sapiens***

Taxonomic rank	Western honey bee	Humans
Kingdom	Animalia	Animalia
Phylum	Arthropoda (jointed limbed invertebrates such as insects, arachnids, myriapods and crustaceans)	Chordata (backboned animals)
Subphylum	Hexapoda (six legged arthropods: insects and allies)	Vertebrata (vertebrates)
Class	Insecta (insects)	Mammalia (mammals)
Order	Hymenoptera (sawflies, ants, wasps and bees)	Primates (monkeys and apes)
Family	Apidae (solitary and social long- tongued bees)	Hominidae (great apes, humans)
Subfamily	Apinae (orchid bees, stingless honey bees, bumblebees and true honey bees)	Homininae (gorillas, chimpanzees, bonobos, humans)
Tribe	Apini (true honey bees)	Homini (chimpanzees, bonobos, humans)
Genus	*Apis*	*Homo*
Species	*mellifera*	*sapiens*

More than 20,000 species of bees worldwide have been formally described, more than all birds and mammals combined. In the United States, there are approximately 4,000 bee species and in the United Kingdom there are around 250 bee species. Entomologists and taxonomists classify bees (and some related wasp groups) in the superfamily **Apoidea**, which in part contains the seven families of bees: Apidae, Megachilidae, Halictidae, Andrenidae, Colletidae, Melittidae and Stenotritidae.[26] The following sections provide brief abstracts of these families of bees including descriptions of some representative lower taxonomic ranks within each bee family.[27]

Family Apidae: cuckoo bees, carpenter bees, digger bees, bumblebees and honey bees

Apidae is the largest family of bees, with more than 5,750 species in more than 200 genera worldwide. Apidae is one of only two families containing representatives of the long-tongued bee category, the other family being Megachilidae. The subfamily Apinae includes honeybees (genus *Apis*) and bumble bees (*Bombus*), the social, honey-producing stingless bees of South America (tribe Melipolini), orchid bees (tribe Euglossini), and the longhorned bees (tribe Eucerini). In the euglossine or orchid bees, males have specialized pouches on the hind legs where fragrances from flowers are stored and subsequently exposed for mating. Male orchid bees are often the primary or only pollinators to visit orchid plants where they acquire these aromatic compounds, therefore the conservation of this group is vital to plant reproductive success. Cuckoo bees of the subfamily Nomadinae are wasp-like bees that lack pollen transporting structures and parasitize other bees by stealing their habitat resources. The subfamily Xylocopinae, which includes carpenter bees (genus *Xylocopa*), are the largest bees in North America, often becoming structural pests that destroy wood as they chew holes for nesting.

Family Megachilidae: leafcutter bees, mason bees and allies

The Megachilidae is the other long-tongued family of bees in addition to the Apidae, with about 4,100 species in 80 genera worldwide. These are solitary bees, with large mandibles that they use for chewing and handling nest-building material that includes leaf fragments. Female bees in this family carry pollen under their abdomen (scopa) instead of on their hind legs or internally. *Megachile pluto* is the largest bee in the world, measuring 38 mm, and lives inside active termite mounds in Indonesia. Major groups of Megachilidae include the leafcutter bees (genus *Megachile*), which neatly cut pieces of leaves or flower petals in order to build their nests in cavities. In addition, another common megachilid are the mason bees (genus *Osmia*), which are often metallic blue in colour and are named for their habit of making nest compartments from dried mud. Species of Megachilidae like the blue orchard bee *Osmia lignaria* are one of the bees commercially managed for orchard crop pollination in North America.

26 C.D. Michener, *The Bees of the World* (2nd edn, Johns Hopkins University Press, 2007).
27 J.S. Ascher and J. Pickering. "Discover Life: Bee Species Guide and World Checklist (Hymenoptera: Apoidea: Anthophila)", accessed 3 March 2016, http://www.discoverlife.org/mp/20q?guide=Apoidea_species&flags=HAS.

Family Halictidae: sweat bees

The halictids are short- to medium-tongued bees that include 4,300 species in 80 genera worldwide. They are typically ground-nesting bees and exhibit a wide range of social behaviours ranging from solitary to eusocial. Some species are attracted to sweat. Many halictids are tiny, with coloration ranging from metallic green to dark brown or black, while others species resemble honey bees with a fuzzy yellow thorax and a striped abdomen. The alkali bee, *Nomia melanderi,* has beautiful iridescent stripes made of enamelled scales and is the only solitary ground-nesting bee commercially managed for alfalfa pollination in the western US.[28]

Family Andrenidae: mining bees

This family of short-tongued bees includes 3,000 species in 45 genera worldwide (except Australia). Mining bees are solitary ground-nesting bees, mostly belonging to the genera Andrena and Perdita. Mining bees are mostly dark, black or reddish, but can also be metallic blue and yellow, or red and yellow. They are medium- to large-sized bees and can be distinguished from other bees by visible facial structures between the eyes and the base of the antennae, called foveae.

Family Colletidae: plasterer and masked bees

Colletidae contains 2,500 species in 90 genera worldwide. Some Colletidae closely resemble mining bees (Andrenidae) and all are solitary with some species being communal. While the species of Colletidae occur worldwide, most are found in Australia and South America. They use a unique two-part tongue to create burrows lined with smooth cellophane-like secretions. This family includes crepuscular (i.e. twilight-active) bees, which have large simple eyes (ocelli) to see through dim light and to fly in the dark just before sunrise. The yellow-faced bees in the genus *Hylaeus* are unique among bees in that they transport pollen internally by swallowing and then regurgitating it.[29]

Family Melittidae: melittid bees

This is a small family of solitary ground-nesting bees that includes 180 species in 15 genera. They are found mainly in Africa and are associated with dry climates. They are small to medium-sized bees with or without abdominal bands of hair. Several species specialize on floral oils as larval food rather than pollen.[30]

28 J.H. Cane, "A Native Ground Nesting Bee, *Nomia melanderi,* Sustainably Managed to Pollinate Alfalfa across an Intensively Agricultural Landscape". *Apidologie* 39 (2008): 315–23.

29 N. Wilson-Rich, *The BEE: A Natural History.* Princeton, NJ: Princeton University Press, 2014.

30 *Ibid.*

Family Stenotritidae: stenotritid bees

This is the smallest family of bees, with only 21 species in two genera that are restricted to Australia. These are solitary ground-nesting bees similar to the Colletidae (and were generally considered to belong to the Colletidae), but differ in that they only have a one-part tongue.[31] They are large, densely-haired and with individual body coloration varying between black, black and yellow-striped, or metallic green.

Sociality of bees

Bees have varying degrees of sociality that ranges from solitary to completely social, termed **eusocial** (Table 3.2), and they also vary in nesting structure and habitat based on this social biology (Table 3.3). Eusociality in bees and other organisms is characterized by three main features: 1) reproductive division of labour;

Table 3.2 **Classification of sociality in insects**

Sociality	Common nest site	Cooperative brood care	Reproductive division of labour	Overlapping generations
Solitary	No	No	No	No
Communal	Yes	No	No	No
Quasisocial	Yes	Yes	No	No
Semisocial	Yes	Yes	Yes	No
Eusocial	Yes	Yes	Yes	Yes

2) overlapping generations; and 3) cooperative care of young.[32] Most species of bees are **solitary**, meaning that they do not make colonies like eusocial bumblebees or honey bees. Instead, a single female works alone in preparing the nest and in provisioning food for her offspring. In solitary bees, generations do not overlap so the female dies before her offspring become adults. Examples of solitary bees include species of mining bees (genera *Anthophora* and *Adrena*), leaf cutter bees (genus *Megachile*), mason bees (genus *Osmia*), carpenter bees (genus *Xylocopa*) and carder bees (genus *Anthidium*).

Sub-social bees are a step further from solitary living. These bees have more or less tolerance between groups of other nesting females. For example, **communal** groups are those with two or more females sharing a common nest entrance or

31 *Ibid.*
32 Bernard J. Crespi and Douglas Yanega. "The Definition of Eusociality". *Behav Ecol* 6 (1995): 109–15, doi:10.1093/beheco/6.1.109.

tunnels but having separate side branches and cells. Among the communal bee species is the Eurasian mining bee, *Adrena carantonica*.

Table 3.3 **Characteristics and nesting behaviour of the seven bee families**

Bee family	Common names	Tongue length	Sociality	Nesting behaviour
Andrenidae	mining bees	Short	Solitary	Ground nesting
Apidae	cuckoo, carpenter, digger bees, bumblebees and honey bees	Long	Solitary to eusocial	Ground nesting, cavity nesters, hives
Colletidae	plasterer and masked bees	Short	Solitary to communal	Ground and cavity nesting
Halictidae	sweat bees	Medium	Solitary to eusocial	Ground nesting
Megachilidae	leaf-cutter bees, mason bees and allies	Long	Solitary to communal	Ground and cavity nesting
Melittidae	melittid bees	Medium	Solitary	Ground nesting
Stenotritidae	stenotritid bees	Short	Solitary	Ground nesting

There are also bee species that are **facultatively communal**, meaning that these bees can be found living either a solitary or communal lifestyle. One such group includes the metallic green sweat bees of the genus *Augochloropsis*. **Quasisocial** and **semisocial** colonies are groups of bees that may provide, in addition to communal living, cooperative breeding systems where adult bees will help take care of the young even if they are not related. Examples include mining bees of the family Apidae (genus *Exomalopsis*) and some sweat bees (Family Halictidae).

Eusocial animals live in complex societies that have the three main characteristics previously mentioned. Eusocial bees generally have a reproductive division of labour including the queen and a caste of sterile females, which are the workers involved in cooperative brood care and other activities. Examples of eusocial bees include bumblebees (genus *Bombus*), stingless bees (Tribe *Melipolini*) and honey bees (genus *Apis*). In insects, eusociality is also found in all termites and ants and in some wasps, gall aphids, thrips and ambrosia beetles.

Causes, consequences and conservation

Here, we discuss findings of researchers who have explored bee decline and the impacts of conservation actions on bees. A new focus in the broader issue of pollinator decline has involved managed honey bees (discussed in Chapter 1), where colony collapse disorder around the year 2006 in the US began to shed light on major problems related to honey bee health. Numerous studies have been conducted to address the presumptive causes of colony collapse disorder, with habitat loss, pests and disease, pesticides and nutritional deficiencies being some of the key drivers.[33,34,35,36] While the influence of factors impacting honey bee decline have been studied less for wild bees, evidence suggests declines in wild bees are occurring as well,[37,38,39] with agricultural intensification and reductions of available natural habitat found to affect wild bees and other pollinators.[40,41,42] Furthermore, pathogen spillover from managed bees to wild populations has emerged as an important issue regarding the conservation of bumblebees,[43] which could further affect some bumblebees in the US that have undergone reduced population sizes and ranges over a relatively short time period.[44]

In addition to the factors discussed above, a consequence of pollinator decline is the loss of pollinator–plant networks. In a study spanning 120 years that focused on specific pollinator–plant interactions from one location, 50% of the bee species had been extirpated (i.e. had gone extinct) and 407 pollinator–plant interactions were lost including 183 that were lost because of bee extirpations.[45] Moreover, across the two time periods (late 1800s and 2009–2010), only 54 of the previously collected 109 species were found.[46] Using field experiments, investigators found that removing even one species from a pollinator community could have profound consequences on plant reproduction through reductions in floral fidelity—the

33 G. Allen-Wardell *et al.*, "The Potential Consequences of Pollinator Declines in the Conservation of Biodiversity and Stability of Food Crop Yields". *Conservation Biology* 12 (1998): 8–17.
34 D. vanEngelsdorp *et al.*, "Colony Collapse Disorder: A Descriptive Study". *PLoS ONE* 4(8) (2009): e6481, doi: 10.1371/journal.pone.0006481.
35 G. Lebuhn *et al.*, "Detecting Insect Pollinator Declines on Regional and Global Scales". *Conservation Biology* 27 (2012): 113–20.
36 Vanbergen and Insect Pollinators Initiative, "Threats to an Ecosystem Service: Pressures on Pollinators".
37 D. Goulson *et al.*, "Decline and Conservation of Bumble Bees". *Annual Review of Entomology* 3 (2008): 191–208.
38 S.G. Potts *et al.*, "Global Pollinator Declines: Trends, Impacts and Drivers". *Trends in Ecology and Evolution* 25 (2010): 345–53.
39 S.A. Cameron *et al.*, "Patterns of Widespread Decline in North American Bumble Bees". *Proceedings of the National Academy of Science* 108 (2011): 662–67.
40 C. Kremen *et al.*, "Crop Pollination from Native Bees at Risk from Agricultural Intensification". *Proceedings of the National Academy of Sciences* 99 (2002): 16812–6.
41 C.M. Kennedy *et al.*, "A Global Quantitative Synthesis of Local and Landscape Effects on Wild Bee Pollinators in Agroecosystems". *Ecology Letters* 2013, doi: 10.1111/ele.12082.
42 Garibaldi *et al.*, "From Research to Action: Enhancing Crop Yield through Wild Pollinators", 439–47.
43 S.R. Colla *et al.*, "Plight of the Bumble Bee: Pathogen Spillover from Commercial Wild Populations". *Biological Conservation* 129 (2006): 461–7.
44 Cameron *et al.*, "Patterns of Widespread Decline in North American Bumble Bees", 662–67.
45 L.A. Burkle *et al.*, "Plant-Pollinator Interactions over 120 years: Loss of Species, Co-occurrence and Function". *Science* 339 (2013):1611–5.
46 *Ibid.*

affinity of particular pollinators to individual plant species—among the remaining pollinators, and this can affect plant reproduction even when effective pollinators remain in the system.[47] While some recent studies have begun to show that abundant and common wild bees are responsible for the majority of pollination services in *crop production* systems,[48] the pollinator *diversity*, especially when considering those high-fidelity linkages between some plants and their specific pollinators, plays a major role in the regulation of flowering plant reproduction and the overall biodiversity of natural plant communities.[49,50]

It is important to understand the difference between conservation efforts focused on species-level biodiversity conservation and the conservation of ecosystem services for human benefit. Overall species diversity (in other words, the number of different species) is often the focus of conservation efforts for pollinators, which is appropriate considering that diversity has been shown to moderate interactions between insects and between insect and plants,[51,52] thereby potentially affecting the process of pollination and plant reproductive success. Furthermore, higher bee *diversity* has been shown to improve crop fruit set in coffee[53] and higher biodiversity has been shown to benefit the maintenance of ecosystem services in agroecosystems.[54] However, using data from 90 independent studies, it was determined that a relatively small number of abundant and common wild species dominate plant pollination, with only 2% of wild bees accounting for 80% of crop visits.[55] This finding has important implications in that strategies to promote the conservation of pollination services would differ from those aimed at the conservation of threatened or rare bees—those that potentially occur at lower densities and abundances.

Plant breeding programmes and trait development—selection and breeding of plant traits resistant to pests and disease—could have unforeseen consequences for pollinators. In large crop monocultures, mass-flowering events are typical which are often synchronous and in many cases can attract diverse nectar- and pollen-seeking insects. Plant nectar concentration, including that arising from

47 B.J. Brosi and H.M. Briggs, "Single Pollinator Species Losses Reduces Floral Fidelity and Plant Reproductive Function", *PNAS* 110 (2013): 13044–8.

48 Kleijn *et al.*, "Delivery of Crop Pollination Services is an Insufficient Argument for Wild Pollinator Conservation", 7414.

49 L. Burkle and R. Irwin. "Nectar Sugar Limits Larval Growth of Solitary Bees (Hymenoptera: Megachilidae)". *Environmental Entomology* 38 (2009): 1293–1300.

50 Brosi and Briggs, "Single Pollinator Species Losses Reduces Floral Fidelity and Plant Reproductive Function", 13044–8.

51 *Ibid.*

52 J. Frund *et al.*, "Bee Diversity Effects on Pollination Depend on Functional Complementarity and Niche Shifts", *Ecology* 94 (2013): 2042–54.

53 A.M. Klein *et al.*, "Fruit Set of Highland Coffee Increases with the Diversity of Pollinating Bees", *Proceedings of the Royal Society of London B* 270 (2003): 955–61.

54 M. Altieri, "The Ecological Role of Biodiversity in Ecosystems", *Agriculture, Ecosystems and Environment* 74 (1999): 19–31.

55 Kleijn *et al.*, "Delivery of Crop Pollination Services is an Insufficient Argument for Wild Pollinator Conservation", 7414.

both flowers and leaves, has been shown to be variable among different cotton (*Gossypium*) species.[56] In modern crop production systems, numerous plant varieties are available, each with its own qualitative characteristics that are typically advantageous for growth and pest resistance, but also could include ranges of nectar concentration and quality. Accordingly, a potentially threatening scenario might involve more or less successful foraging of bees attributed to nectar quality and sugar content, with impacts on bee fitness and ultimately local populations. While the effects of different sugar contents resulting from plant breeding on bee health has not been determined, in a non-crop environment scientists added varying sugar levels to subalpine solitary bee provisions and found that larval growth and development differed based on sugar content, with highest larval mass reached in the high-nectar sugar addition.[57] More studies are needed to determine how different crop plant varieties could affect bee health and pollination services.

Applied conservation for pollinators, involving both *active* restoration—generally the enhancement of floral resources and habitat conditions —and *passive* restoration—land conservation strategies such as leaving intact areas of undisturbed, wild habitat—have been a focus of research aimed at understanding landscape factors influencing bees. A growing body of evidence has shown the positive influence of semi-natural or upland wild habitat area on wild bee diversity in agroecosystems, with consensus that the proximity and area of high quality, semi-natural habitats such as perennial native grasslands strongly promotes bee diversity.[58,59,60,61,62] In an intensive agricultural area, canola fields with higher proportions of adjacent pastureland (relative to cropland) had relatively high diversity of bumblebees and other wild bees.[63]

While wild bee diversity has been shown to be positively related to the proportion of natural, upland habitat in the proximity of cultivated crops, enhancing farmland pollinator habitat by improving the coverage of floral resources (i.e. active restoration) might fail to show improvements in bee diversity within the context

56 G.D. Butler *et al.*, "Amounts and Kinds of Sugars in the Nectars of Cotton (*Gossypium* spp.) and the Time of their Secretion". *Agronomy Journal* 64 (1972): 364–368.

57 Burkle *et al.*, "Plant-Pollinator Interactions over 120 years: Loss of Species, Co-occurrence and Function", 1611–5.

58 Greenleaf and Kremen, "Wild Bees Enhance Honey Bees' Pollination of Hybrid Sunflower", 13890–5.

59 K.M. Krewenka *et al.*, "Landscape Elements as Potential Barriers and Corridors for Bees, Wasps and Parasitoids". *Biological Conservation* 144 (2011): 1816–25.

60 Holzschuh *et al.*, "Landscapes with Wild Bee Habitat Enhance Pollination, Fruit Set and Yield of Sweet Cherry", 101–7.

61 Kennedy *et al.*, "A Global Quantitative Synthesis of Local and Landscape Effects on Wild Bee Pollinators in Agroecosystems", doi: 10.1111/ele.12082.

62 A.B. Bennett and R. Isaacs. "Landscape Composition Influences Pollinators and Pollination Services in Perennial Biofuel Plantings". *Agriculture, Ecosystems and Environment* 193 (2014): 1–8.

63 L.A. Morandin *et al.*, "Can Pastureland Increase Wild Bee Abundance in Agriculturally Intense Areas?" *Basic and Applied Ecology* 8 (2007): 117–24.

of the local and surrounding landscape.[64] In addition to the influence of local and regional landscape structure, it is important to understand both feeding (nectar and pollen) and nesting resources utilized by wild bees, of which undisturbed and uncultivated ground-nesting sites are required for approximately 70% of bee species. Furthermore, in areas dominated by expansive row crop agriculture, strategies for conserving remaining areas of wild land should be supplemented with conservation strategies involving the restoration of native vegetation including wildflowers. Local and landscape level floral resources that might result from both of these actions have been shown to affect native bee occurrences across cropping systems in Europe,[65] while targeted, small-scale habitat strategies could enhance biodiversity of pollinator specialists that are vulnerable to habitat degradation.[66]

Are bees contributing to improved yields as a result of better quality habitat? Fruit set, berry weight and mature seeds—all terms that indicate the successful transition from flower to quality fruit—were significantly greater in blueberry fields adjacent to wildflower plantings 3 to 4 years after (wildflower) seeding.[67] In a study in the Central Valley of California, it was shown that if around 40% of an area within a 2.4 km radius or 30% of an area within a 1.2 km radius of a watermelon field was left undisturbed (i.e. in natural habitat), growers could achieve full pollination of watermelons.[68] Similarly, in canola and in the absence of honey bees, yield and profit was shown to be maximized when 30% of land was left uncultivated within 750 m of field edges.[69] While the proximity and cover of high-quality habitats appear to directly favour pollinators in agroecosystems, flowering characteristics of wild plants and crops could have detrimental effects on each other based on species' flowering phenologies and push–pull mechanisms between wild habitats and crop production systems. For example, steady-state flowering (i.e. continuous flowering and fruiting phenology) has a potentially negative impact on wild and honey bee densities on coffee blooms because native bees could be drawn away

64 T.J. Wood *et al.*, "Pollinator-Friendly Management Does Not Increase the Diversity of Farmland Bees and Wasps". *Biological Conservation* 187 (2015): 120–6.

65 Jeroen Scheper *et al.*, "Local and Landscape Level Floral Resources Explain Effects of Wildflower Strips on Wild Bees Across Four European Countries", *Journal of Applied Ecology* 52 (2015): 1165–75.

66 C. Kremen and L.K. M'Goningle, "Small-Scale Restoration in Intensive Agricultural Landscapes Supports More Specialized and Less Mobile Pollinator Species", *Journal of Applied Ecology* 52 (2015): 602–10.

67 B. Blaauw and R. Isaacs, "Flower Plantings Increase Wild Bee Abundance and the Pollination Services Provided to a Pollination-Dependent Crop", *Journal of Applied Ecology* 2014, doi: 10.1111/1365-2664.12257.

68 C. Kremen *et al.*, "The Area Requirements of an Ecosystem Service: Crop Pollination by Native Bee Communities in California", *Ecology Letters* 7 (2004): 1109–19.

69 L. Morandin and M.L. Winston, "Pollinators Provide Economic Incentive to Preserve Natural Land in Agroecosystems", *Agriculture, Ecosystems and the Environment* 116 (2006): 289-292.

from thc focal crop.[70] Whole-farm and landscape approaches that integrate information on the habitat requirements of bees, including the seasonal phenologies of pollinators, wild and cultivated flowering plants, are central to the development of effective conservation actions for bees while sustaining agricultural production.

Programme management strategies such as the Targeted Agri-environment Schemes (i.e. Higher Level Stewardship farms involving the management of uncropped lands) as part of the EU Common Agricultural Policy have been effective in some aspects of pollinator conservation, such as positive effects on bumblebee nest density and bumblebee population sizes following their implementation.[71] Similar strategies have been initiated in the US, primarily through various conservation programmes of the US Department of Agriculture Natural Resource Conservation Service, with additional guidance and support provided by non-governmental programmes such as the Xerces Society and the Pollinator Partnership. Farm-level research and producer outreach involving the conservation of wild bees and their contributions to pollination services has been led in the US by the Integrated Crop Pollination (ICP) project, funded by the USDA Specialty Crop Research Initiative. Overall, the underlying charge for improving pollinators has been a broad strategy for active restoration—improving the habitat resources for pollinators—while improving our understanding of wild bees' contributions to agricultural production. In addition to the primary and direct benefits of wild bee conservation through habitat improvements on farmland, secondary benefits are far reaching, including improved water quality, protection against soil erosion, pest population reduction and enhanced rural aesthetics.[72]

In addition to agricultural systems, gardens in urban areas have been a focus of bee diversity studies and pollinator conservation efforts, and even urban beekeeping has become a common activity in some high-density urban areas. In north-western Ohio residential gardens, a study found that local (native plants, floral abundance, vegetation height, grass cover and vegetables) and landscape level (forest, open-space and wetlands) characteristics influenced bee diversity in backyard gardens.[73] While habitat modifications using native plants have been the best strategy for promoting pollinator health in agricultural systems,[74] one

70 V. Peters *et al.*, "The Contribution of Plant Species with a Steady-State Flowering Phenology to Native Bee Conservation and Bee Pollination Services", *Insect Conservation and Diversity* 6 (2013): 45–56.

71 T.J. Wood *et al.*, "Targeted Agri-Environment Schemes Significantly Improve the Population Size of Common Farmland Bumblebee Species". *Molecular Ecology* 24 (2015):1668–80.

72 S.D. Wratten *et al.*, "Pollinator Habitat Enhancement: Benefits to Other Ecosystem Services". *Agriculture, Ecosystems and Environment* 159 (2012): 112–22.

73 G.L. Pardee and S.M. Philpott. "Native Plants are the Bee's Knees: Local and Landscape Predictors of Bee Richness and Abundance in Backyard Gardens". *Urban Ecosystems* 17 (2014): 641–59.

74 R. Isaacs *et al.*, "Maximizing Arthropod-Mediated Ecosystem Services in Agricultural Landscapes: The Role of Native Plants". *Frontiers in Ecology and the Environment* 2009, doi: 10.1890/080035.

urban study conducted in New York City found that native plant additions did not improve beneficial insects, with exotic plants heavily utilized by beneficial insects including bees.[75] Furthermore, in a similar study, the amount of landscape green space was not a significant predictor of bee species richness; however, local within-garden floral area and sunlight availability were the only factors included in highly supported predictive models.[76] Therefore, in intensive urban areas, isolated local gardens and rooftops with adequate sunlight could influence visitation by bees and other pollinators, regardless of whether the plants are native or exotic. Bee hotels, structures that provide nesting sites for cavity-nesting bees (e.g. leafcutter bees) are also a common means for habitat improvements in urban settings, yet their effectiveness for enhancing pollinator diversity is unfounded despite their wide use.[77] Furthermore, isolated and relatively large contiguous green spaces provided by golf courses, when environmentally sensitive management is used, can provide effective systems to apply pollinator habitat restoration and other conservation actions.[78]

Species occurrence information, including the total number and geographic extent of localities, supports the prioritization of conservation actions and assessment of threats within species' distributional ranges. The International Union for Conservation of Nature (IUCN) is a worldwide authority in the standard practices and criteria development for assignment of conservation status for insects based in part on this information. Lists produced from the IUCN methodologies for species' listing are known as IUCN Red Lists, which help guide and prioritize conservation actions. A recent red list showed that out of a total of 1,101 species, 9.2% of bees in the EU were threatened with extinction and 5.2% of bees were near-threatened.[79] The butterflies of the EU showed similar numbers; out of a total of 482 species of butterflies, approximately 9% were threatened and 10% were near threatened.[80] This study further noted that nearly 31% of the butterflies in the EU have declining populations. The IUCN's Status and Trends of European Pollinators (STEP) programme reported that nearly one in ten wild bee species face extinction in Europe, with 56.7% of the 1,965 species classified as data deficient, and therefore

75 K. Matteson and G.A. Langellotto. "Small Scale Additions of Native Plants Fail to Increase Beneficial Insect Richness in Urban Gardens". *Insect Conservation and Diversity* 4 (2011): 89–98.

76 K. Matteson and G.A. Langellotto. "Determinates of Inner City Butterfly and Bee Species Richness". *Urban Ecosystems* 13 (2010): 333–47.

77 J.S. MacIvor and L. Packer. "'Bee Hotels' as Tools for Native Pollinator Conservation: A Premature Verdict?" *PLoS ONE* 10 (2015): e0122126, doi: 10.1371/journal.pone.0122126.

78 E.K. Dobbs and D.A. Potter. "Forging Natural Links with Golf Courses for Pollinator-Related Conservation, Outreach, Teaching and Research". *American Entomologist* 61 (2015): 116–23.

79 A. Nieto *et al.*, *European Red List of Bees* (Luxembourg: Publication Office of the European Union, 2014).

80 C. Van Swaay *et al.*, *European Red List of Butterflies*. Luxembourg: Publications Office of the European Union, 2010.

with potential unknown threats for these species lacking adequate information.[81] The Xerces Society, a non-profit organization focused on the conservation of invertebrates and their habitat in North America, Canada and Mexico, currently lists 47 species of wild bees as vulnerable, imperilled, critically imperilled or possibly extinct. Currently, there are 75 insects on the list of threatened or endangered species of the US Fish and Wildlife Service (USFWS), including 34 butterflies, moths and skippers, but no bees. For some bee species, adequate information exists to classify them according to conservation criteria and to petition them for listing. However, as with many insect species of conservation concern, we simply don't know enough about their distributions and habitat resources to ascertain conservation status or to develop conservation actions and assessments of threats to populations. As more information is produced from research funded by state-level, national and international pollinator initiatives and programmes, it is likely that bees will begin to be included in conservation listings. For example, in the US, states have developed wildlife action plans that consist of preliminary lists of species of concern, with subsequent funding in place to develop further information to modify conservation priority rankings for these species.

How does conservation proceed? Across the numerous initiatives of both governmental and non-governmental sectors, a common strategy to promote pollinator health is to increase the coverage of habitat resources used by pollinators, essentially conserving and restoring native habitat when and where possible. In addition to bees, this is also critical for insects such as the monarch butterfly that depends on milkweed (*Asclepias* spp.), which is a primary host plant that has undergone a 58% decline in the major eastern US monarch migration zone (this zone also lies in the agriculturally intense "corn belt" region of the Midwestern US).[82] Remnant wild patches of land are important for sustaining the diversity of bees and other pollinators; further, aiming to restore larger portions of landscapes to native, high quality vegetation should be a common action to combat the issue of pollinator decline. A global quantitative synthesis of local and landscape effects on wild bee populations across 39 crop systems revealed that bee richness was higher in diversified and organic fields and in landscapes comprising more *high-quality* habitat.[83] Business sectors that directly and indirectly depend on insect pollinators have an important role in understanding and promoting the need for integrating agricultural production with high quality landscapes for conserving pollination and other ecosystem services.

81 Nieto *et al.*, *European Red List of Bees.*

82 J.M. Pleasants and K.S. Oberhauser. "Milkweed Loss in Agricultural Fields because of Herbicide Use: Effect on the Monarch Butterfly Population". *Insect Conservation and Diversity* 2012, doi: 10.1111/j.1752-4598.2012.00196.x.

83 Kennedy *et al.*, "A Global Quantitative Synthesis of Local and Landscape Effects on Wild Bee Pollinators in Agroecosystems", doi: 10.1111/ele.12082.

The International Convention on Biological Diversity (CBD) provides 12 conservation principles within its Ecosystem Approach.[84] Principle 4, which addresses management of ecosystems within an economic context, with relevance to businesses and economic markets, states the following:

> Principle 4: Recognizing potential gains from management, there is usually a need to understand and manage the ecosystem in an economic context. Any such ecosystem management programme should:
> a. Reduce those market distortions that adversely affect biological diversity
> b. Align incentives to promote biodiversity conservation and sustainable use
> c. Internalize costs and benefits in the given ecosystem to the extent feasible

A further description of Conservation Principle 4 of the CBD includes:

> The greatest threat to biological diversity lies in its replacement by alternative systems of land-use. This often arises through market distortions, which undervalue natural systems and populations and provide perverse incentives and subsidies to favor the conversion of land to less diverse systems. Often those who benefit from conservation do not pay the costs associated with conservation and, similarly, those who generate environmental costs (e.g. pollution) escape responsibility. Alignment of incentives allows those who control the resource to benefit and ensures that those who generate environmental costs will pay.

Ultimately, an effective level of social and environmental responsibility should influence positive actions towards biodiversity conservation and the conservation of ecosystem services provided by the natural world. Principle 1 of the CBD states the following: "The objectives of management of land, water and living resources are a matter of societal choices." Among overly consumptive modern human societies, we should certainly consider that biodiversity conservation and the protection of ecosystems services is a responsible choice, and vital to all of life on Earth.

Despite "exceptional coverage in pollination ecology ranging from basic ecological relationships to applied aspects of ecosystem services and ecosystem management",[85] broad strategies for pollinator conservation have not been developed to address critical priorities, such as "*Investing* in restoration and management of diversity of pollinators and their habitats adjacent to croplands in order to stabilize or improve crop yields".[86] As the human population soars towards 8 billion, intensifying stressors in the form of habitat loss and fragmentation, invasive species and environmental pollution will continue to threaten biodiversity. Failure to

84 https://www.cbd.int/ecosystem/principles.shtml

85 Stefan-Dewenter *et al.*, "Scale-Dependent Effects of Landscape Context on Three Pollinator Guilds", 1421–32.

86 Allen-Wardell *et al.*, "The Potential Consequences of Pollinator Declines in the Conservation of Biodiversity and Stability of Food Crop Yields", 8–17.

address the potentially disastrous scenarios of further pollinator decline would be a serious lack of vision towards sustainable management of our natural world and the natural capital it provides.[87] Further, we should ask ourselves how the decline of bee pollinators—that have persisted for millions of years—translates into losses across all biodiversity. The current Anthropocene era and sixth mass extinction attributed to humans—the "dominant animal"[88]—continues to seriously threaten Earth's biological and ecological fabric that is so essential to life.

A central aim regarding the protection of ecosystem services is to sustain those things that operate in the natural world and that humans depend on for their survival: an entirely anthropocentric concept. In such a narrow view, we degrade a necessary holistic, ecological perspective towards one that, in current political economies, mostly prioritizes our natural world at a level below the production of goods for human use. Because of the dominance of insects—which total approximately 60% of all known species of animals and plants on Earth— and their role in sustaining the natural world and important ecosystem services such as pollination, it is reasonable and responsible to focus on their conservation. As the well-known entomologist and sociobiologist E.O. Wilson stated,

> In two or three centuries, with humans gone, the ecosystems of the world would regenerate back to the rich state of equilibrium that existed ten thousand or so years ago, minus of course the many species that we have pushed into extinction. But if insects were to vanish, the terrestrial environment would soon collapse into chaos.[89]

The conservation of our natural world and its biodiversity, and the processes it supports, should certainly begin with the world's most efficient pollinators, the bees.

Further reading and sources of information

General books on the natural history of bees:

O'Toole, C. *Bees: A Natural History*. Buffalo, New York: Firefly Books Inc., 2013.
Wilson-Rich, N. *The BEE: A Natural History*. Princeton, NJ: Princeton University Press, 2014.

87 R. Costanza *et al.*, "The Value of the World's Ecosystem Services and Natural Capital". *Ecological Economics* 25 (1998): 3–15.
88 P.R. Ehrlich and A.E. Ehrlich. *The Dominant Animal: Human Evolution and the Environment*. Washington, DC: Island Press, 2008.
89 E.O. Wilson, *The Creation: An Appeal to Save Life On Earth* (New York: W.W. Norton and Company, 2006), 33–34.

Useful websites:

Ascher, J.S., and J. Pickering. "Discover Life: Bee Species Guide and World Checklist (Hymenoptera: Apoidea: Anthophila)". Accessed 3 March 2016. http://www.discoverlife.org/mp/20q?guide=Apoidea_species&flags=HAS [useful checklist of all bees of the world with online taxonomic key]

http://bugguide.net/node/view/117315 [useful overview of the bee families and photos]

Technical guides on bee identification and biology:

Goulet, H., and J. T. Huber. *Hymenoptera of the World: An Identification Guide to Families.* Canada Communication Group – Publishing, 1993.

Michener, C.D. *The Bees of the World,* 2nd edn. The Johns Hopkins University Press, 2007.

Michener, C.D., R.J. McGinley, and Danforth. *The Bee Genera of North and Central America (Hymenoptera: Apoidea).* Smithsonian Institution Press, 1994.

Wilson, J.S. and O.M. Carril. *The Bees in Your Backyard.* Princeton University Press, 2016.

Information on native pollinators and habitat restoration:

Mader, E., M. Shepherd, M. Vaughan, S.H. Black, and G. LeBuhn. *The Xerces Society Guide: Attracting Native Pollinators: Protecting North America's Bees and Butterflies.* Storey Publishing, 2011.

Technical information on pollination and floral ecology:

Wilmer, P. *Pollination and Floral Ecology.* Princeton University Press. 2011.

Bibliography

Aizen, M.A., L.A. Garibaldi, S.A. Cunningham, and A.M. Klein. "How Much Does Agriculture Depend on Pollinators? Lessons from Long Terms Trends in Crop Production". *Annals of Botany* 103 (2009): 1579–88.

Allen-Wardell, G., P. Bernhardt, R. Bitner, A. Burquez, S. Buchmann, J. Cane, P.A. Cox, V. Dalton, P. Feinsinger, M. Ingram, D. Inouye, C.E. Jones, K. Kennedy, P. Kevan, H. Koopowitz, R. Medellin, S. Medellin-Morales and G.P. Nabhan. "The Potential Consequences of Pollinator Declines in the Conservation of Biodiversity and Stability of Food Crop Yields". *Conservation Biology* 12 (1998): 8–17.

Altieri, M. "The Ecological Role of Biodiversity in Ecosystems". *Agriculture, Ecosystems and Environment* 74 (1999): 19–31.

Ascher, J.S., and J. Pickering. "Discover Life: Bee Species Guide and World Checklist (Hymenoptera: Apoidea: Anthophila)". Accessed 3 March 2016. http://www.discoverlife.org/mp/20q?guide=Apoidea_species&flags=HAS

Bauer, D.M. and I.S. Wing. "Economic Consequences of Pollinator Declines: A Synthesis". *Agricultural and Resource Economics Review* 39 (2010): 368–83.

Bennett, A.B. and R. Isaacs. "Landscape Composition Influences Pollinators and Pollination Services in Perennial Biofuel Plantings". *Agriculture, Ecosystems and Environment* 193 (2014): 1–8.

Blaauw, B. and R. Isaacs. "Flower Plantings Increase Wild Bee Abundance and the Pollination Services Provided to a Pollination-Dependent Crop". *Journal of Applied Ecology* 2014, doi: 10.1111/1365-2664.12257.

Brosi, B.J. and H.M. Briggs. "Single Pollinator Species Losses Reduces Floral Fidelity and Plant Reproductive Function". PNAS 110 (2013): 13044–8.
Burkle, L. and R. Irwin. "Nectar Sugar Limits Larval Growth of Solitary Bees (Hymenoptera: Megachilidae)". *Environmental Entomology* 38 (2009): 1293–1300.
Burkle, L.A., J.C. Marlin and T.M. Knight. "Plant-Pollinator Interactions over 120 years: Loss of Species, Co-occurrence and Function". *Science* 339 (2013):1611–5.
Butler, G.D.., G.M. Looper, S.E. McGregor, J.L. Webster and H. Margolis. "Amounts and Kinds of Sugars in the Nectars of Cotton (*Gossypium* spp.) and the Time of their Secretion". *Agronomy Journal* 64 (1972): 364–368.
Cameron, S.A., J.D. Lozier, J.P. Strange, J.B. Koch, N. Cordes, L.F. Solter, and T, Griswold. "Patterns of Widespread Decline in North American Bumble Bees". *Proceedings of the National Academy of Science* 108 (2011): 662–67.
Cane, J.H. "A Native Ground Nesting Bee, *Nomia melanderi*, Sustainably Managed to Pollinate Alfalfa across an Intensively Agricultural Landscape". *Apidologie* 39 (2008): 315–23.
Cariveau, D.P. and R. Winfree. "Causes of Variation in Wild Bee Responses to Anthropogenic Drivers". *Current Opinion in Insect Science* 10 (2015): 104–9.
Colla, S.R., M.C. Otterstatter, R.G. Gegear, and J.D. Thompson. "Plight of the Bumble Bee: Pathogen Spillover from Commercial Wild Populations". *Biological Conservation* 129 (2006): 461–7.
Costanza, R., R. d'Arge, R. de Groot, S. Farber, M. Grasso, B. Hannon, K. Limburg, S. Naeem, R.V. O'Neill, J. Paruelo, R.G. Raskin, P. Sutton and M. van den Belt. "The Value of the World's Ecosystem Services and Natural Capital". *Ecological Economics* 25 (1998): 3–15.
Crespi, Bernard J., and Douglas Yanega. "The Definition of Eusociality". *Behav Ecol* 6 (1995): 109–15, doi:10.1093/beheco/6.1.109.
Dobbs, E.K. and D.A. Potter. "Forging Natural Links with Golf Courses for Pollinator-Related Conservation, Outreach, Teaching and Research". *American Entomologist* 61 (2015): 116–23.
Ehrlich, P.R. and A.E. Ehrlich. *The Dominant Animal: Human Evolution and the Environment.* Washington, DC: Island Press, 2008.
Eilers, E.J., C. Kremen, S.S. Greenleaf, A.K. Garber and A-Maria Klein. "Contribution of Pollinator-Mediated Crops to Nutrients in the Human Food Supply". *PLoS ONE* 6(6) (2011): e21363, doi:10.1371/journal.pone.0021363.
Frund, J., C.F. Dormann, A. Holzschuh and T. Tscharntke. "Bee Diversity Effects on Pollination Depend on Functional Complementarity and Niche Shifts". *Ecology* 94 (2013): 2042–54.
Gallai, N., J-M Salles, J. Settele and B. Vaissiere. "Economic Valuation of the Vulnerability of World Agriculture Confronted with Pollinator Decline". *Ecological Economics* 68 (2009): 810–21.
Garibaldi, L.A., and 49 co-authors. "Wild Pollinators Enhance Fruit Set of Crops Regardless of Honey Bee Abundance". *Sciencexpress* 2013, doi: 10.1126/science.1230200.
Garibaldi, L.A., L.G. Carvalheiro, S.D. Leonhardt, M.A. Aizen, B.R. Blaauw, R. Isaacs, M. Kuhlmann, D. Kleijn, A.M. Klein, C. Kremen, L. Morandin, J. Scheper, and R. Winfree. "From Research to Action: Enhancing Crop Yield through Wild Pollinators". *Frontiers in Ecology and the Environment* 12 (2014): 439–47.
Goulson, D., G.C. Lye and B. Darvill. "Decline and Conservation of Bumble Bees". *Annual Review of Entomology* 3 (2008): 191–208.
Greenleaf, S.S., and C. Kremen. "Wild Bees Enhance Honey Bees' Pollination of Hybrid Sunflower". *Proceedings of the National Academy of Science* 103 (2006): 13890–5.
Holzschuh, A., J-Henrick Dedenhoffer, T. Tscharntke. "Landscapes with Wild Bee Habitat Enhance Pollination, Fruit Set and Yield of Sweet Cherry". *Biological Conservation* 153 (2012): 101–7.
Ingram, M., G.C. Nabham, and S.L. Buchmann. "Our Forgotten Pollinators: Protecting the Birds and the Bees". *Global Pesticide Campaigner* 6 (1996): 1–8.

Isaacs, R, J. Tuell, A. Fiedler, M. Gardiner and D. Landis. "Maximizing Arthropod-Mediated Ecosystem Services in Agricultural Landscapes: The Role of Native Plants". *Frontiers in Ecology and the Environment* 2009, doi: 10.1890/080035.

Kennedy, C.M. and 41 coauthors. "A Global Quantitative Synthesis of Local and Landscape Effects on Wild Bee Pollinators in Agroecosystems". *Ecology Letters* 2013, doi: 10.1111/ele.12082.

Kleijn, D., R. Winfree, I. Bartomeus, L.G. Carvalheiro, M. Henry, R. Isaacs, A-M. Klein, C. Kremen, L.K. M'Gonigle, R. Rader, T.H. Ricketts, N.M. Williams, N.L. Adamson, J.S. Ascher, A. Báldi, P. Batáry, F. Benjamin, J.C. Biesmeijer, E.J. Blitzer, R. Bommarco, M.R. Brand, V. Bretagnolle, L. Button, D.P. Cariveau, R. Chifflet, J.F. Colville, B.N. Danforth, E. Elle, M.P.D. Garratt, F. Herzog, A. Holzschuh, B.G. Howlett, F. Jauker, S. Jha, E. Knop, K.M. Krewenka, V. Le Féon, Y. Mandelik, E.A. May, M.G. Park, G. Pisanty, M. Reemer, V. Riedinger, O. Rollin, M. Rundlöf, H.S. Sardinas, J. Scheper, A.R. Sciligo, H.G. Smith, I. Steffan-Dewenter, R. Thorp, T. Tscharntke, J. Verhulst, B.F. Viana, B.E. Vaissiere, R. Veldtman, C. Westphal and S.G. Potts. "Delivery of Crop Pollination Services is an Insufficient Argument for Wild Pollinator Conservation". *Nature Communications* 6 (2015): 7414, doi: 1038/ncomms8414.

Klein, A-M, B.E. Vaissiere, J.H. Crane, I. Stefan-Dewenter, S.A. Cunningham, C. Kremen and T. Tscharntke. "Importance of Pollinators in Changing Landscapes for World Crops". *Proceedings of the Royal Society B* 274 (2007): 303–13.

Klein, A-M, I. Steffan-Dewenter and T. Tscharntke. "Fruit Set of Highland Coffee Increases with the Diversity of Pollinating Bees". *Proceedings of the Royal Society of London B* 270 (2003): 955–61.

Kremen, C. and L.K. M'Goningle. "Small-Scale Restoration in Intensive Agricultural Landscapes Supports More Specialized and Less Mobile Pollinator Species". *Journal of Applied Ecology* 52 (2015): 602–10.

Kremen, C, N.M. Williams, R.L. Bugg, J.P.Fay and R.W. Thorp. "The Area Requirements of an Ecosystem Service: Crop Pollination by Native Bee Communities in California". *Ecology Letters* 7 (2004): 1109–19.

Kremen, C., N.M. Williams and R.W. Thorp. "Crop Pollination from Native Bees at Risk from Agricultural Intensification". *Proceedings of the National Academy of Sciences* 99 (2002): 16812–6.

Krewenka, K.M., A. Holzschuh, T. Tscharnke, and C.F. Dormann. "Landscape Elements as Potential Barriers and Corridors for Bees, Wasps and Parasitoids". *Biological Conservation* 144 (2011): 1816–25.

Lebuhn, G., S. Droege, E.F. Connor, B. Gemmill-Herren, S.G.Potts, R. Minckly, T. Griswold, R. Jean, E. Kula, D.W. Roubik, J. Cane, K.W. Wright, G. Frankie, and F. Parker. "Detecting Insect Pollinator Declines on Regional and Global Scales". *Conservation Biology* 27 (2012): 113–20.

Losey, J.E. and M. Vaughn. "The Economic Value of Ecological Services Provided by Insects". *Bioscience* 56 (2006): 311–23.

Maclvor, J.S., and L. Packer. "'Bee Hotels' as Tools for Native Pollinator Conservation: A Premature Verdict?" *PLoS ONE* 10 (2015): e0122126, doi: 10.1371/journal.pone.0122126.

Matteson, K. and G.A. Langellotto. "Determinates of Inner City Butterfly and Bee Species Richness". *Urban Ecosystems* 13 (2010): 333–47.

Matteson, K., and G.A. Langellotto. "Small Scale Additions of Native Plants Fail to Increase Beneficial Insect Richness in Urban Gardens". *Insect Conservation and Diversity* 4 (2011): 89–98.

Michener C.D. *The Bees of the World*, 2nd edn. The Johns Hopkins University Press, 2007.

Millennium Ecosystem Assessment (MA). *Ecosystems and Human Well-Being: Synthesis*. Washington, DC: Island Press, 2005.

Morandin, L.A. and M.L. Winston. "Pollinators Provide Economic Incentive to Preserve Natural Land in Agroecosystems". *Agriculture, Ecosystems and Environment* 116 (2006): 289–92.

Morandin, L.A., M.L. Winston, V.A. Abbot and M.T. Franklin. "Can Pastureland Increase Wild Bee Abundance in Agriculturally Intense Areas?" *Basic and Applied Ecology* 8 (2007): 117–24.

National Research Council. *Status of Pollinators in North America.* Washington, DC: National Academies Press, 2007.

Nieto, A., Roberts, S.P.M., Kemp, J., Rasmont, P., Kuhlmann, M., García Criado, M., Biesmeijer, J.C., Bogusch, P., Dathe, H.H., De la Rúa, P., De Meulemeester, T., Dehon, M., Dewulf, A., Ortiz-Sánchez, F.J., Lhomme, P., Pauly, A., Potts, S.G., Praz, C., Quaranta, M., Radchenko, V.G., Scheuchl, E., Smit, J., Straka, J., Terzo, M., Tomozii, B., Window, J. and Michez, D. *European Red List of Bees.* Luxembourg: Publication Office of the European Union, 2014.

Ollerton, J. R. Winfree and S. Tarrant. "How Many Flowering Plants Are Pollination by Animals?" *Oikos* 120 (2011): 321–6.

Pardee, G.L. and S.M. Philpott. "Native Plants are the Bee's Knees: Local and Landscape Predictors of Bee Richness and Abundance in Backyard Gardens". *Urban Ecosystems* 17 (2014): 641–59.

Peters, V., C.R. Carroll, R.J. Cooper, R. Greenberg, and M. Solis. "The Contribution of Plant Species with a Steady-State Flowering Phenology to Native Bee Conservation and Bee Pollination Services". *Insect Conservation and Diversity* 6 (2013): 45–56.

Pleasants, J.M. and K.S. Oberhauser. "Milkweed Loss in Agricultural Fields because of Herbicide Use: Effect on the Monarch Butterfly Population". *Insect Conservation and Diversity* 2012, doi: 10.1111/j.1752-4598.2012.00196.x.

Poinar, G.O., and B.N. Danforth. "A Fossil Bee from Early Cretaceous Burmese Amber". *Science* 314 (5799) (2006): 614, doi: 10.1126/science.1134103.

Potts, S.G., J.C. Biesmeijer, C. Kremen, P. Neumann, O. Schweiger, and W.E Kunin. "Global Pollinator Declines: Trends, Impacts and Drivers". *Trends in Ecology and Evolution* 25 (2010): 345–53.

Scheper, Jeroen, Riccardo Bommarco, Andrea Holzschuh, Simon G Potts, Verena Riedinger, Stuart PM Roberts, Maj Rundlöf, Henrik G Smith, Ingolf Steffan Dewenter, Jennifer B Wickens, Victoria J Wickens and David Kleijn. "Local and Landscape Level Floral Resources Explain Effects of Wildflower Strips on Wild Bees Across Four European Countries". *Journal of Applied Ecology* 52 (2015): 1165–75.

Stefan-Dewenter, I., U. Munzenberg, C. Burger, C. Thies, and T. Tscharntke. "Scale-Dependent Effects of Landscape Context on Three Pollinator Guilds". *Ecology* 83 (2002): 1421–32.

Van Swaay, C., Cuttelod, A., Collins, S., Maes, D., López Munguira, M., Šaši , M., Settele, J., Verovnik, R., Verstrael, T., Warren, M., Wiemers, M. and Wynhof, I. *European Red List of Butterflies.* Luxembourg: Publications Office of the European Union, 2010.

Vanbergen, A.J. and Insect Pollinators Initiative. "Threats to an Ecosystem Service: Pressures on Pollinators". *Frontiers in Ecology and the Environment* 2013, doi: 10.1890/120126.

vanEngelsdorp, D., J.D. Evans, C. Saegerman, C. Mullin, E. Haubruge, B.K. Nguyen, M. Frazier, J. Frazier, D. Cox-Foster, Y. Chen, R. Underwood, D.R. Tarpy, and J.S. Pettis. "Colony Collapse Disorder: A Descriptive Study". *PLoS ONE* 4(8) (2009): e6481, doi: 10.1371/journal.pone.0006481.

Wilson, E.O. *The Creation: An Appeal to Save Life On Earth.* New York: W.W. Norton and Company, 2006.

Wilson-Rich N. *The BEE: A Natural History.* Princeton, NJ: Princeton University Press, 2014.

Winfree, R., N.M. Williams, J. Dushoff, and C. Kremen. "Native Bees Provide Insurance Against Ongoing Honey Bee Losses". *Ecology Letters* 10 (2007): 1105–13.

Wood, T.J., J.M. Holland and D. Goulson. "Pollinator-Friendly Management Does Not Increase the Diversity of Farmland Bees and Wasps". *Biological Conservation* 187 (2015): 120–6.

Wood, T.J., J.M. Holland, W.O. Hughes and D. Goulson. "Targeted Agri-Environment Schemes Significantly Improve the Population Size of Common Farmland Bumblebee Species". *Molecular Ecology* 24 (2015):1668–80.

Wratten, S.D., M. Gillespie, A. Decourtye, E. Mader and N. Desneux. "Pollinator Habitat Enhancement: Benefits to Other Ecosystem Services". *Agriculture, Ecosystems and Environment* 159 (2012): 112–22.

Zhang, W., T.R. Ricketts, C. Kremen, K. Carney and S.M. Swinton. "Ecosystems Services and Dis-services to Agriculture". *Ecological Economics* 64(2) (2007): 261–8, doi: 10.1016/j.ecolecon.2007..02.024.

4

From corporate social responsibility to accountability in the bumblebee trade

A Japanese perspective*

Carol Reade
San José State University, USA

Koichi Goka
National Institute for Environmental Studies, Japan

Robbin Thorp
University of California, Davis, USA

Masahiro Mitsuhata
Arysta LifeScience Corporation, Japan

Marius Wasbauer
University of California, Davis, USA

There has been growing pressure in recent years for firms to behave in more socially responsible and accountable ways. A timely issue concerns the effects of global corporate activities on local ecosystems, and how firms are responding to stakeholder pressures to be more accountable towards protection of the natural environment. A case in point is the global bumblebee trade and biodiversity loss. The global bumblebee trade provides mass-reared hives for the pollination of greenhouse crops

* An earlier version of this paper was published as "CSR, Biodiversity and Japan's Stakeholder Approach to the Global Bumble Bee Trade" in *The Journal of Corporate Citizenship* Issue 56, December 2014 (Greenleaf Publishing).

worldwide. Through this trade, commercial bumblebees are introduced in ecosystems where they are not native and may be considered invasive species. From the standpoint of general systems theory (see Chapter 1), this trade provides an example of how corporate activities aided by global supply chains can have negative effects on local ecosystems worldwide such as biodiversity loss. These effects include a decline in native bumblebee species which can herald longer-term domino effects including declines in other native flora and fauna, ultimately affecting human food supply.[1]

This chapter focuses on the effects of this global trade on Japanese ecosystems as well as Japanese stakeholder responses, including movement towards greater corporate social responsibility and accountability. Japan provides an interesting perspective on this global issue because of the unique legislation that evolved from relationships between the government, corporations, scientists, environmentalists and the general public. There is recognition from the Imperial Household down to the grass-roots level of society that the impact of invasive species is particularly detrimental to an island ecology such as Japan. This is evident in Japan in terms of the impact on biodiversity associated with the global trade in bumblebees.

In Japan, firms have long been aware of the potential impact of their activity on the natural environment.[2,3,4] Japan's Global Environment Charter was established in 1991 by Keidanren, Japan's primary business group which is currently known as Nippon Keidanren, to guide member companies on environmental issues. The formulation of the Charter was prompted by pollution problems in the 1960s, a time of high industrial growth, and the two oil crises of the 1970s. Through Keidanren an appeal was made to member companies to put Japan's highly developed technology to work to safeguard the natural environment. Through a set of voluntary guidelines, member companies are encouraged to protect ecosystems and conserve resources, protect the global environment and improve the local living environment, ensure the environmental soundness of products, and protect the health and safety of employees and citizens.[5]

1 Carol Reade *et al.*, "Invisible Compromises: Global Business, Local Ecosystems, and the Commercial Bumble Bee Trade", *Organization & Environment* 28 (2015): 436–57, doi: 10.1177/1086026615595085.

2 Kyoko Fukukawa and Jeremy Moon. "A Japanese Model of Corporate Social Responsibility? A Study of Website Reporting*", *The Journal of Corporate Citizenship* 16 (2004): 45–59.

3 K. Horiuchi and M. Nakamura, "Environmental Issues and Japanese Firms". In *The Japanese Business and Economic System: History and Prospects for the 21st Century*, edited by Masao Nakamura, 364-384 (New York, NY: Palgrave, 2001).

4 Kanji Tanimoto, "Changes in the Market Society and Corporate Social Responsibility". *Asian Business & Management* 3 (2004): 151–72.

5 Keidanren, "Guidelines for Corporate Action", 1991. Accessed May 17, 2015. http://www.keidanren.or.jp/english/speech/spe001/s01001/s01b.html.

These are the early roots of corporate social responsibility in Japan, which has been growing in popularity over the past decade.[6] Corporate social responsibility (CSR) is defined as "the efforts corporations make above and beyond regulation to balance the needs of stakeholders with the need to make a profit".[7] CSR embodies the notion of the "triple bottom line" where companies are expected to embrace environmental and social sustainability in addition to sustainability of corporate profits.[8] An appreciation of CSR has developed largely in response to pressures from environmentalists, scientists, NGOs and other key stakeholders of the firm.[9,10,11]

While voluntary guidelines such as Japan's Global Environment Charter are useful to promote CSR, they are not always adequate to ensure accountability for protection of the natural environment.[12] Questions have been raised, for instance, as to whether the voluntary efforts of individual firms will necessarily result in sustainable development of the entire socioeconomic system.[13] Stated differently, voluntary action does not guarantee corporate accountability on the part of individual firms. Corporate accountability refers to enforceable strategies to influence firm behaviour. It emphasizes "giving CSR some teeth" through legal regulation and empowering stakeholder groups.[14] As we will elaborate below, the Global Environment Charter guidelines provide a baseline for environment-related CSR efforts but legislation was ultimately enacted to ensure corporate accountability and to further protect biodiversity in Japan. Development of the legislative framework entailed stakeholder involvement of the government, firms, scientists, environmentalists and the general public.

The relationship between biodiversity protection and CSR is an understudied yet important relationship. While firms have paid increasing attention to their visible and direct impact on the natural environment, biodiversity issues are difficult

6 K. Tanimoto, "Structural Change in Corporate Society and CSR in Japan", In *Corporate Social Responsibility in Asia,* edited by K. Fukukawa (New York, NY: Routledge, 2012), 45–66.

7 Deborah Doane and Naomi Abasta-Vilaplana. "The Myth of CSR. (Undetermined)". *Stanford Social Innovation Review* 3 (2005): 23.

8 K. Fukukawa and Y. Teramoto, "Understanding Japanese CSR: The Reflections of Managers in the Field of Global Operations". *Journal of Business Ethics* 85 (2009): 133-46.

9 Tanimoto, "Changes in the Market Society and Corporate Social Responsibility", 151–72.

10 Tanimoto, "Structural Change in Corporate Society and CSR in Japan", 45–66.

11 Sandra A. Waddock *et al.*, "Responsibility: The New Business Imperative". *Academy of Management Executive* 16 (2002): 132–48, doi:10.5465/AME.2002.7173581.

12 Horiuchi and Nakamura, "Environmental Issues and Japanese Firms", 364-384.

13 Tanimoto, "Structural Change in Corporate Society and CSR in Japan", 45–66.

14 P. Utting, "CSR and Policy Incoherence" In *Fair Trade, Corporate Accountability and Beyond: Experiments in Globalizing Justice,* edited by K. Macdonald and S. Marshall (Farnham, UK: Ashgate Publishing, 2013), 171.

to detect and may have indirect linkages to business activity (Reade et al., 2015).[15] In short, firms may not realize the harm they are causing. Japan is addressing this challenge through the development of a legislative framework regarding invasive species and the balancing of stakeholder interests, which has a positive influence on business behaviour. We conclude that Japanese firms are moving from CSR to accountability, where there is a blend of voluntary and legalistic approaches to protect the environment. This includes the development and use of native bumblebee species for commercial crop pollination which we believe is the ultimate form of corporate social responsibility and accountability in the commercial bumblebee industry. After all, native bumblebees are integral to the health and sustainability of local ecosystems. To foster their development for commercial pollination within their natural distribution range is more supportive of ecosystem health than introducing alien and potentially invasive bumblebees for pollination purposes.

Biodiversity protection and CSR

While CSR encompasses environmental sustainability,[16] little attention has been given to the effects of business activity on biodiversity despite its fundamental role in human well-being.[17] Thus, biodiversity is a relatively unfamiliar concept among corporate actors. Biodiversity is defined as the "variety of genes, species, and ecosystems that constitute life on Earth".[18] It involves interactions among organisms and between organisms and the physical environment; that is, soil, water and air. Biodiversity allows for the provision of numerous essential services to society. Essential services include material goods such as food and non-material benefits such as recreation. It also includes pollination and pest control services for agriculture and longer-term resilience to disturbance in the environment.[19] The ability of an ecosystem to adapt to changing conditions is aided through reservoirs of biodiversity.[20] The loss of biodiversity, in the form of shrinkage or disappearance of genes and species, means that ecosystem services are diminished or compromised.

15 Reade *et al.*, "Invisible Compromises: Global Business, Local Ecosystems, and the Commercial Bumble Bee Trade".
16 Thomas Dyllick and Kai Hockerts. "Beyond the Business Case for Corporate Sustainability". *Business Strategy and the Environment* 11(2) (2002): 130.
17 Monika I. Winn and Stefano Pogutz, "Business, Ecosystems, and Biodiversity: New Horizons for Management Research". *Organization & Environment* 26 (2013): 203–29, doi: 10.1177/1086026613490173.
18 Michael R.W. Rands *et al.*, "Biodiversity Conservation: Challenges Beyond 2010". *Science* 329 (2010): 1298, doi:10.1126/science.1189138.
19 *Ibid.*
20 R. Slootweg, "Biodiversity Assessment Framework: Making Biodiversity Part of Corporate Responsibility". *Impact Assessment and Project Appraisal* 23 (2005): 37–46.
Rands *et al.*, "Biodiversity Conservation: Challenges Beyond 2010", 1298.

There has been a rapid decline in biodiversity in recent years[21,22] as discussed in the opening chapters of this book.

The key drivers of biodiversity loss include human population growth, international trade, economic specialization and invasive species. Continued growth of the human population results in the increased demand for consumption and land development to create living space. It also results in the intensification of agriculture to accommodate higher demand for food products.[23] International trade contributes to loss of biodiversity worldwide through the spread of disease, parasites and invasive species.[24,25] Invasive species, or the introduction of non-native species into local ecosystems, pose one of the greatest threats to biodiversity worldwide.[26] The introduction of non-native species through international trade can result in competition with native species, predatory behaviour, and the disruption of ecosystem relationships such as the unique pollinating activity between particular species of insects and plants.[27,28] This can threaten the existence of native species and compromise the health of local ecosystems by reducing future evolutionary potential and ecosystem resilience.[29]

Biodiversity protection poses a challenge to CSR because damage to ecosystems is poorly understood, often lacks visibility, and generally has indirect linkages to business activity.[30] Further, a high degree of stakeholder engagement and local respon-

21 Paul R. Ehrlich and Robert M. Pringle, "Where Does Biodiversity Go from Here? A Grim Business-as-Usual Forecast and a Hopeful Portfolio of Partial Solutions". *Proceedings of the National Academy of Sciences* 105 (Supplement 1) (2008): 11579–86, doi:10.1073/pnas.0801911105.

22 Winn and Pogutz, "Business, Ecosystems, and Biodiversity: New Horizons for Management Research", 203–29.

23 Ehrlich and Pringle, "Where Does Biodiversity Go from Here? A Grim Business-as-Usual Forecast and a Hopeful Portfolio of Partial Solutions", 11579–86.

24 Jouni Korhonen, "On the Paradox of Corporate Social Responsibility: How Can We Use Social Science and Natural Science for a New Vision?" *Business Ethics: A European Review* 15 (2006): 200–14, doi:10.1111/j.1467-8608.2006.00442.x.

25 M. Lenzen *et al.*, "International Trade Drives Biodiversity Threats in Developing Nations". *Nature* 486 (2012): 109–12, doi: 10.1038/nature11145.

26 Peter Graystock *et al.*, "The Trojan Hives: Pollinator Pathogens, Imported and Distributed in Bumblebee Colonies". *Journal of Applied Ecology* 50 (2013): 1207–15, doi:10.1111/1365-2664.12134.

27 Ehrlich and Pringle, "Where Does Biodiversity Go from Here?", 11579–86.

28 K. Goka, "Influence of Invasive Species on Native Species: Will the European Bumblebee, *Bombus terrestris*, Bring Genetic Pollution into the Japanese Native Species?" *Bulletin of Biogeographical Society of Japan* 53 (1998): 91-101.

29 Ehrlich and Pringle, "Where Does Biodiversity Go from Here?", 11579–86.

30 C. Reade *et al.*, "CSR, Biodiversity, and Japan's Stakeholder Approach to the Global Bumble Bee Trade". *The Journal of Corporate Citizenship* 56 (2014): 53–66.

siveness is required to address biodiversity protection.[31,32] Biodiversity loss can, of course, be visible and result from direct effects of business on the environment such as the loss of marine life from an oil spill. Corporate responsibility is evident in such a case. Responsibility is more difficult to ascertain when the environmental impact of business activity has low visibility and is diffused on a global scale.[33] Biodiversity loss that occurs through the global spread of invasive species provides such an example. Companies are likely to be sceptical of their indirect impact on the environment and to demand sufficient scientific evidence before acknowledging responsibility. As a result, the burden of proof is generally placed on the scientific community so that environmental degradation such as biodiversity loss often goes unrecognized as a responsibility of firms. In the absence of legislation or other enforceable means to influence firm behaviour, there may be little accountability for environmental damage that accrues while sufficient evidence is gathered.

The commercial bumblebee industry is an apt illustration of the complex relationships between business, CSR and biodiversity protection. This industry, discussed in detail below, rears and supplies bumblebees for crop pollination worldwide. Bumblebees have an important role as pollinators. About 8% of the world's known 250,000 species of flowering plants rely exclusively on bumblebees for pollination.[34] Flowering plants around the world include food crops and some that provide fibre, drugs and fuel. Bee species have disappeared in some countries and are under threat of decline in various other countries including Japan. This has far-reaching implications for biodiversity due to the role of native bees as pollinators of indigenous flora. Scientists have identified a number of reasons for the decline in bee populations. These include exposure to pesticides, loss of suitable habitat, climate change and the transmission of diseases and parasites through the international transport of bumblebees for commercial purposes.[35,36,37] A full discussion of the scientific causes is presented in Chapter 3.

31 Reade *et al.*, "CSR, Biodiversity, and Japan's Stakeholder Approach to the Global Bumble Bee Trade", 53–66.

32 Reade *et al.*, "Invisible Compromises: Global Business, Local Ecosystems, and the Commercial Bumble Bee Trade".

33 Horiuchi and Nakamura, "Environmental Issues and Japanese Firms", 364–84.

34 Stephen L. Buchmann and Gary Paul Nabhan, *The Forgotten Pollinators* (Island Press, 1997).

35 Koichi Goka *et al.*, "Bumblebee Commercialization Will Cause Worldwide Migration of Parasitic Mites". *Molecular Ecology* 10 (2001): 2095–99, doi:10.1046/j.0962-1083.2001.01323.x.

36 R.W. Thorp, "Bumble bees (Hymenoptera: Apidae): Commercial Use and Environmental Concerns". In *For Non-native Crops, Whence Pollinators of the Future?* Proceedings of Thomas Say Publications in Entomology, 21–40 (Lanham, MD, 2003).

37 K. Winter *et al.*, "Importation of Non-Native Bumble Bees into North America: Potential Consequences of Using *Bombus terrestris* and other Non-Native Bumble Bees for Greenhouse Crop Pollination in Canada, Mexico, and the United States". A White Paper of the North American Pollinator Protection Campaign, 2006.

The commercial bumblebee trade

Economic and social rationale

The commercial bumblebee rearing business began in 1985 with the discovery in Belgium of the value of using the Eurasian bumblebee *Bombus terrestris* for the pollination of greenhouse tomatoes.[38] Bumblebees pollinate a range of crops and are particularly effective pollinators of tomatoes. Compared with the traditional hormone or mechanical pollination for greenhouse crops, bumblebee pollination is cheaper, more natural and leads to higher fruit quality, more abundant fruit and higher prices for producers.[39]

The growing demand for greenhouse crop pollination, particularly tomatoes, has led to the substantial development of the commercial bumblebee trade over the past several decades. Bumblebee colonies can be produced year-round. This has facilitated the global expansion of the tomato greenhouse industry by providing a more cost effective method of crop pollination. There are over 30 producers of commercial bumblebee hives worldwide, although most of the market share is covered by three companies: Biobest of Belgium, Koppert Biological Systems and Bunting Brinkman Bees, both of the Netherlands. The Eurasian *Bombus terrestris,* commonly called the large earth bumblebee, is the main species of bumblebee that is mass-reared and exported on a commercial basis to provide pollination services for agricultural crops. It is indigenous to Europe, coastal North Africa and West and Central Asia. This bee can now be found worldwide as a result of its commercial trade. The large earth bumblebee was introduced into Japan in 1991 for experimental purposes.[40] In 1992 it began to be used commercially for greenhouse crop pollination.[41,42] The companies that distribute bumblebee colonies in Japan are BioBest, Cats Agrisystems, Koppert (associated with Arysta LifeScience Corporation, formerly Tomen Corporation), Tokaibussan Co., Ltd and API Company, Ltd.

While the industry provides economic and social benefits as a natural and cost-effective pollination service for greenhouse crops, there are ecological concerns

38 H.H.W. Velthuis and A. van Doorn, "A Century of Advances in Bumblebee Domestication and the Environmental Aspects of its Commercialization for Pollination". *Apidologie* 37 (2006): 421–51.

39 *Ibid.*

40 Yuya Kanbe *et al.*, "Interspecific Mating of the Introduced Bumblebee *Bombus terrestris* and the Native Japanese Bumblebee *Bombus hypocrita sapporoensis* Results in Inviable Hybrids". *Naturwissenschaften* 95 (2008): 1003–8, doi: 10.1007/s00114-008-0415-7.

41 M. Ono, "Ecological Implications of Introduced *Bombus terrestris,* and Significance of Domestication of Japanese Native Bumblebees (*Bombus* spp.)". In *Proceedings of the International Workshop on Biological Invasions of Ecosystem by Pests and Beneficial Organisms: National Institute of Agro-Environmental Science, Ministry Agriculture, Forestry, and Fisheries, Japan, Tsukuba, Japan February 25–27, 1997,* 244–52.

42 Naoki Inari *et al.*, "Spatial and Temporal Pattern of Introduced *Bombus terrestris* Abundance in Hokkaido, Japan, and its Potential Impact on Native Bumblebees". *Population Ecology* 47 (2005): 77–82.

when the bees escape from greenhouse enclosures and interact with native bees and plants.[43]

Ecological concerns of invasive species

Concerns have been raised by entomologists in Japan and other countries that the large earth bumblebee (and other introduced bumblebees) can be invasive and harm the flora and fauna of local ecosystems. The main entomological research findings in Japan can be summarized as follows.[44]

Competition for foraging and nesting

The large earth bumblebee has been found to compete with native bumblebees for foraging and nesting sites in Japan.[45,46] Research findings indicate that the large earth bumblebee has a rapid rate of reproduction that could potentially displace native bees from foraging and nest sites. These findings substantiate research done in other parts of the world.

Disruption of pollinating behaviour

Japanese entomologists have found that the large earth bumblebee disrupts the pollinating behaviour of native bumblebee species, resulting in a decrease of nectar availability and consequent reduction in seed production.[47,48] The results suggest that bumblebee-pollinated native plants are relatively specialized to native bumblebee pollinators.

43 K. Goka, "Introduction to the Special Feature for Ecological Risk Assessment of Introduced Bumblebees: Status of the European Bumblebee, *Bombus terrestris*, in Japan as a Beneficial Pollinator and an Invasive Alien Species". *Applied Entomology and Zoology* 45 (2010):1-6.

44 Reade *et al.*, "CSR, Biodiversity, and Japan's Stakeholder Approach to the Global Bumble Bee Trade", 53–66.

45 Maki N. Inoue *et al.*, "Displacement of Japanese Native Bumblebees by the Recently Introduced *Bombus terrestris* (L.) (Hymenoptera: Apidae)". *Journal of Insect Conservation* 12 (2007): 135–46, doi:10.1007/s10841-007-9071-z.

46 Teruyoshi Nagamitsu *et al.*, "Foraging Interactions between Native and Exotic Bumblebees: Enclosure Experiments Using Native Flowering Plants". *Journal of Insect Conservation* 11 (2006): 123–30, doi:10.1007/s10841-006-9025-x.

47 Ikumi Dohzono *et al.*, "Alien Bumble Bee Affects Native Plant Reproduction through Interactions with Native Bumble Bees". *Ecology* 89 (2008): 3082–92.

48 Tanaka Kenta *et al.*, "Commercialized European Bumblebee Can Cause Pollination Disturbance: An Experiment on Seven Native Plant Species in Japan". *Biological Conservation* 134 (2007): 298–309, doi:10.1016/j.biocon.2006.07.023.

Reproductive disturbance and genetic contamination

Evidence has been found in Japan showing that interbreeding between non-native and native species may lead to reproductive disturbance and genetic contamination of native bumblebee species.[49] A laboratory examination of interbreeding revealed a low hatching rate of eggs; further, the eggs were found to be inviable due to genetic mechanisms which prevent normal egg development.[50] Similar results of introduced commercial bumblebees have been obtained in other parts of the world.

Transmission of parasites and diseases

The transport of bumblebees across international borders, or otherwise outside of their natural distribution range, is of great concern because of the transmission of parasites and disease to native bumblebees. In Japan, tracheal mites have been detected in imported bumblebee colonies.[51] Also, the presence of microsporidian (spore forming, unicellular fungal) pathogens in these bees have been determined to infect native species of bumblebees.[52] Similar findings have been observed in other parts of the world, for instance in the United States,[53] Mexico,[54] Argentina,[55] and the United Kingdom.[56]

Japan's stakeholder approach

The above findings from entomologists underscore concerns that non-native bumblebees may contribute to the decline or disappearance of local species of bumblebee, with implications for biodiversity loss. Japan's stakeholder approach to these

49 Koji Tsuchida *et al.*, "Reproductive Disturbance Risks to Indigenous Japanese Bumblebees from Introduced *Bombus terrestris*". *Applied Entomology and Zoology* 45 (2010): 49–58, doi:10.1303/aez.2010.49.

50 Kanbe *et al.*, "Interspecific Mating of the Introduced Bumblebee *Bombus terrestris* and the Native Japanese Bumblebee *Bombus hypocrita sapporoensis* Results in Inviable Hybrids", 1003–8.

51 Goka *et al.*, "Bumblebee Commercialization Will Cause Worldwide Migration of Parasitic Mites", 2095–99.

52 S. Niwa et al., "A Microsporidian Pathogen Isolated from a Colony of the European Bumblebee, *Bombus terrestris*, and Infectivity on Japanese Bumblebee", *Jpn. J. Appl. Entomol. Zool.* 48 (2004): 60–64.

53 Thorp, "Bumble bees (Hymenoptera: Apidae): Commercial Use and Environmental Concerns", 21–40.

54 Winter *et al.*, "Importation of Non-Native Bumble Bees into North America".

55 Arbetman, Marina P., Ivan Meeus, Carolina L. Morales, Marcelo A. Aizen, and Guy Smagghe. "Alien Parasite Hitchhikes to Patagonia on Invasive Bumblebee". *Biological Invasions* 15 (2013): 489–94, doi:10.1007/s10530-012-0311-0.

56 Graystock *et al.*, "The Trojan Hives", 1207–15.

concerns is embodied in legislation, the balancing of stakeholder interests, and the influence on business behaviour.

Legislative framework

The Global Environment Charter, as noted above, was established as a set of voluntary guidelines for Japanese firms to safeguard the environment while pursuing their business objectives. However, in some business domains, voluntary guidelines are thought to be inadequate to protect the environment.[57] Businesses may not necessarily be negligent, but may not be fully aware of the implications of their actions on the environment.[58,59] This was considered to be the case regarding the import of non-native wildlife into Japan.

In 2001, the Japanese Cabinet Office conducted a public opinion poll that showed a growing public concern about imported non-native species of plants and animals. In 2002, the New National Biodiversity Strategy was adopted, and alien species was identified as a critical factor affecting biodiversity which needed immediate attention. In 2003, a three-year plan for promoting regulation reform was initiated by the Cabinet, in which the issue of alien species was considered. It was acknowledged that there was no law that dealt with the conservation of biodiversity, and that "the public and business sectors do not fully understand the problem of alien species".[60]

In 2004, the Japanese Cabinet Office submitted a bill concerning Invasive Alien Species (IAS) to the Diet, Japan's national legislature. The bill was introduced to prevent adverse effects on ecosystems, human safety, agriculture, forestry and fisheries caused by IAS. The Diet passed the bill without amendments and the Invasive Alien Species Act came into effect in June 2004.[61] Enforcement of the law began in June 2005.[62] The law prohibits the raising, planting, storing or carrying, importing and other handling of IAS, with the exception of specified cases that require permission of the relevant authorities and securing IAS under special facilities. Substantial penalties are imposed on both individuals and corporations for violation of the law. The Invasive Alien Species Act stipulates that persons shall be imprisoned for up to three years and/or fined up to three million yen.[63]

57 Horiuchi and Nakamura, "Environmental Issues and Japanese Firms", 364–84.

58 Dror Etzion, "Research on Organizations and the Natural Environment, 1992-Present: A Review". *Journal of Management* 33 (2007): 637–64, doi:10.1177/0149206307302553.

59 Tanimoto, "Changes in the Market Society and Corporate Social Responsibility", 151–72.

60 MOE, "Measures to be Taken Against Invasive Alien Species in Japan" (Tokyo: Ministry of the Environment, Government of Japan, 2004). Accessed May 17. 2015. http://www.env.go.jp/en/nature/as/040326.pdf.

61 MOE, "Annual Report on the Environment and the Sound Material-Cycle Society in Japan 2007", 2004 (Tokyo: Ministry of the Environment Government of Japan).

62 *Ibid.*

63 MOE, "Invasive Alien Species Act (Law No. 78) (Tokyo: Ministry of the Environment, Government of Japan, 2004). Accessed May 17, 2015. http://www.env.go.jp/en/nature/as/040427.pdf.

In 2006, the large earth bumblebee (*Bombus terrestris*) was added to the list of Invasive Alien Species.[64] Importation is prohibited without permission from the relevant ministers. When permitted, a protective covering must be placed over the greenhouses so that the bees cannot escape. The two ministries most involved are the Ministry of the Environment (MOE) and the Ministry of Agriculture, Forestry and Fisheries (MAFF). MAFF has strong lobbies within the farming communities. It oversees Japanese agriculture the core of which is formed by many small farming households structured into producers' cooperatives. MOE has good working relations with scientists and conservationists. Entomologists made their research findings known to MOE about the harmful effects of the imported large earth bumblebee. The fate of the large earth bumblebee became caught between those who support agricultural productivity and those who support conservationism.[65]

Balancing stakeholder interests

The legislation that resulted from Japan's stakeholder process is relatively unique in that it balances commercial needs with ecological concerns through a permission system that was erected by MOE.[66] In the United States, for instance, the Eurasian large earth bumblebee is completely banned, and a native species has been developed for commercial use. The current regulatory challenge in the US is the intra-country transport of native bumblebees, where bees native to one part of the country are thought to be harmful to bees in other parts of the country.[67] In the case of Australia, the large earth bumblebee is not formally banned at the national level, yet the Australian Hydroponic and Greenhouse Association was denied an application to import the large earth bumblebee. This decision was apparently influenced by the bee being listed as a threatening species in two Australian states.[68] The large earth bumblebee is traded throughout Eurasia where it is native. However, there is growing concern among European entomologists about the spread of parasites[69] and the effects of the various Eurasian subspecies of this bumblebee on

64 MOE, "Annual Report on the Environment and the Sound Material-Cycle Society in Japan 2007".

65 For a detailed review of the politics surrounding the designation of this bumblebee as an invasive species, see Goka, "Introduction to the Special Feature for Ecological Risk Assessment of Introduced Bumblebees", 1-6.

66 *Ibid.*

67 Thorp, "Bumble bees (Hymenoptera: Apidae): Commercial Use and Environmental Concerns", 21–40.
Winter *et al.*, "Importation of Non-Native Bumble Bees into North America".

68 Cameron Alastair Moore and Caroline Gross, "Great Big Hairy Bees! Regulating the European Bumblebee, *Bombus terrestris* L. What Does it Say about the Precautionary Principle?" *International Journal of Rural Law and Policy* 2012. Accessed 4 March 2016. http://epress.lib.uts.edu.au/journals/index.php/ijrlp/article/view/2627.

69 Graystock *et al.*, "The Trojan Hives", 1207–15.

local European ecosystems.[70] By contrast to the above examples, Japan's permission system allows farmers to import the banned large earth bumblebee if they can demonstrate that their greenhouses are secure with netting.[71] In this way, the stakeholder process and resulting legislation in Japan has balanced the interests of conservationists and businesses.

At the same time, MAFF has advocated the commercial rearing of indigenous bumblebees as an alternative to the Eurasian large earth bumblebee.[72] In March 2015, MAFF and MOE announced plans to switch to native species for crop pollination, at least on a trial basis. And in April 2015, MOE formulated a new national strategy for controlling invasive alien species which includes a road map for the utilization of native bumblebee species as alternative biological agents. To this end, a project is under way to promote the development of a native species (*Bombus hypocrita sapporoensis*) in the northern island of Hokkaido where the Eurasian large earth bumblebee is widely used for greenhouse crop pollination. A related and critical issue to be examined is the suitability of commercializing native species for widespread use in Japan. This is because there is risk of disturbing genetic endemism in local bee populations when using bees from outside the natural distribution range, even if they are the same species. In a preliminary study of one bumblebee species in Japan (*Bombus ignitus*), it was found that there is genetic variation within this species depending on geographic location.[73] These studies entail the cooperation between government ministries and private companies. The results will inform policy and any subsequent legislation.

Influence on business behaviour

There is evidence of a change in behaviour in both the suppliers of bumblebees and the customers of bumblebees, the greenhouse crop growers. The bumblebee-rearing suppliers have begun to shift production to colonies of native Japanese bumblebees for use in Japan. In fact, some companies have been in the process of researching the use of native bees for the pollination of greenhouse crops well before the promulgation of the Invasive Species Act of 2004 and addition of the large earth bumblebee in 2006. Arysta LifeScience Corporation, for instance, succeeded in increasing colonies of a native bumblebee (*Bombus ignitus*) to commercial

70 F. Kraus *et al.*, "Greenhouse Bumblebees (*Bombus terrestris*) Spread Their Genes into the Wild". *Conservation Genetics* 12 (2011): 187–92, doi:10.1007/s10592-010-0131-7.

71 M. Yoneda *et al.*, "Preventing *Bombus terrestris* from Escaping with a Net Covering Over a Greenhouse in Hokkaido". *Japanese Journal of Applied Entomology and Zoology* 51 (2007): 39–44.

72 Goka, "Introduction to the Special Feature for Ecological Risk Assessment of Introduced Bumblebees", 1-6.

73 Satoshi Tokoro *et al.*, "Geographic Variation in Mitochondrial DNA of *Bombus ignitus* (Hymenoptera: Apidae)". *Applied Entomology and Zoology* 45 (2010): 77–87, doi:10.1303/aez.2010.77.

quantities in 1999.[74] Other companies such as Biobest and Tokaibussan Co., Ltd have also been in the process of replacing colonies of the large earth bumblebee with colonies of a native species for use in Japan. This can be attributed to the findings from early investigations of the large earth bumblebee by Japanese entomologists[75,76] as well as a lengthy stakeholder dialogue process that resulted in legislation.

Greenhouse crop growers have begun to use native species, although the preference for the large earth bumblebee remains strong. The large earth bumblebee is a hearty bumblebee that provides consistent high-yielding crops. Educational efforts have been under way through the producer cooperatives, where the ecological risks of using the large earth bumblebee are discussed.[77] Training is also provided on how to secure greenhouses with netting when using the large earth bumblebee, as required by law. The practice in small farming businesses has been to use one colony to pollinate several, clustered tomato greenhouses where the bees are allowed to freely move between greenhouse enclosures.[78] The use of native bees has been endorsed by large companies such as Kagome Company, Ltd that entered the fresh tomato business following agricultural reforms in 2001.[79] Kagome, one of the largest food companies in Japan and the largest tomato processor in the Pacific Rim, has tomato-growing contracts with multiple corporations and farms. It is encouraging its agricultural partners to use native bees as an environmental safety precaution.

The recent announcements by MOE and MAFF on the use of native species will further influence business behaviour. The bumblebee rearing companies will need to put more effort and resources into developing the capacity of native species to be effective pollinators of greenhouse crops. This is likely to be a relatively costly endeavour since the development of native species would be for a limited market, or geographical area. For instance, *Bombus ignitus* can be used in areas of the main island of Honshu where it is native, but would be invasive if used in the northern island of Hokkaido. Companies will need to be attuned to the natural distribution ranges of native species, and intensify their local responsiveness according to this criterion.[80] Likewise, it appears that customers will increasingly use native species

74 M. Mitsuhata, "Utilization of the Japanese Native Bumble Bee Species, *Bombus ignitus*", 2006. Accessed May 17, 2015. http://www.agrofrontier.com/guide/t_111c.htm.

75 Goka, "Influence of Invasive Species on Native Species", 91-101.

76 Ono, "Ecological Implications of Introduced *Bombus terrestris*, and Significance of Domestication of Japanese Native Bumblebees (*Bombus* spp.)", 244–52.

77 Goka, "Introduction to the Special Feature for Ecological Risk Assessment of Introduced Bumblebees", 1-6.

78 Hiroshi S. Ishii *et al.*, "Habitat and Flower Resource Partitioning by an Exotic and Three Native Bumble Bees in Central Hokkaido, Japan". *Biological Conservation* 141 (2008): 2597–607, doi:10.1016/j.biocon.2008.07.029.

79 JETRO, "More Enterprises Move into Agricultural Sector", 2005. Accessed May 17, 2015. http://www.jetro.go.jp/en/reports/market/pdf/2005_15_h.pdf.

80 Reade *et al.*, "Invisible Compromises: Global Business, Local Ecosystems, and the Commercial Bumble Bee Trade".

for their pollination needs particularly if further legislation is forthcoming. This may pose an initial financial risk to farmers unless the native species can yield crops as bountiful as the large earth bumblebee. Entomological research is progressing on the optimum greenhouse conditions, such as lighting and temperature, under which the pollination efficiency of native species can be maximized. Recent studies suggest that the pollination effectiveness of *Bombus ignitus* is approaching that of the large earth bumblebee as a greenhouse pollinator.[81]

From CSR to accountability

The commercial bumblebee trade in Japan exemplifies the role of stakeholders in bringing low-visibility biodiversity issues to the fore and forging clearer pathways of responsibility and accountability for corporations and small businesses.[82] The developments leading up to and following the designation of the large earth bumblebee as an invasive species are indicative of the challenges that biodiversity protection poses to businesses. The scientific community, government, civil society and businesses worked together towards biodiversity protection while supporting commercial interests. Japan's approach is consistent with CSR because of the breadth and integration of stakeholder engagement to arrive at stakeholder outcomes that balance the interests of conservationists and businesses. It is also consistent with greater accountability in that stakeholder outcomes include an enforceable means to influence business behaviour though legislation.

Japan's Global Environment Charter of 1991 set forth a broad guideline for companies to work with key stakeholders to preserve the environment. However, it was specific government regulation in the form of the Invasive Alien Species Act that subsequently limited the importation of the large earth bumblebee. The Invasive Alien Species Act came into being with input from various stakeholders including corporations, scientists and the general public. In sum, the legislation that restricts importation of the large earth bumblebee grew out of concerns by entomologists and others that biodiversity and ecosystem health in Japan was being compromised. Japan's legislation balances commercial interests and ecological concerns in a way that supports businesses to be accountable and responsible to the environment by changing business behaviours that had unintended consequences on local ecosystems.

Articles of the Invasive Species Act call for collaboration among stakeholders and the sharing of information. This includes allowing scientific research to be conducted on company activities, reporting results to government authorities and international agencies and NGOs. The Act also identifies public education as a

81 Mitsuhata, "Utilization of the Japanese Native Bumble Bee Species, *Bombus ignitus*".

82 Reade *et al.*, "CSR, Biodiversity, and Japan's Stakeholder Approach to the Global Bumble Bee Trade", 53–66.

significant priority. The Act is consistent with the operating guidelines offered by the Convention on Biological Diversity (CBD).[83] The CBD specifies that all relevant sectors of society and scientific disciplines should be involved in, and cooperate towards, addressing challenges related to maintaining biodiversity and the health of ecosystems. Indeed, the breadth of Japan's stakeholder engagement spans a full range of actors from the grass-roots community level to the Imperial Household.

Japan's regulation of the large earth bumblebee has been acknowledged by entomologists in other parts of the world as an important step towards protecting biodiversity not only in Japan, but globally. Japanese entomologists were among the early investigators of the impact of the large earth bumblebee on local ecosystems.[84,85,86] A substantial amount of research on the large earth bumblebee has been undertaken in Japan. These studies are increasingly cited globally as the adverse ecological effects of invasive bumblebees become more evident in other countries.[87] At the same time, agricultural businesses are able to continue using a commercially desirable bumblebee while exercising precaution, experimenting with the use of native species, and cooperating with scientists. Tomato-growing enterprises in Hokkaido and elsewhere in Japan have allowed entomologists to conduct research on their bumblebee pollination activities. Indicative of the range of stakeholder engagement, beyond that of scientists, businesses and government as described above, community volunteers have monitored the extent of alien bumblebees in rural areas and contributed to research efforts.[88] Further, a televised visit by the Emperor and Empress of Japan to the National Institute for Environmental Studies to discuss longer-term global environmental concerns, including invasive species and the deterioration of ecosystems, provided recognition of the importance of these issues.[89]

Returning to our definition of corporate social responsibility as "the efforts corporations make above and beyond regulation to balance the needs of stakeholders

83 CBD, Convention on Biological Diversity, Decision V/6 Ecosystem Approach. Decision VII/11 Ecosystem Approach, 2004. Accessed May 17, 2015. http://www.biodiv.org/decisions/default.aspx?m=COP-07&id=7748&lg=0.

84 S. Asada and M. Ono, "Tomato Pollination with Japanese Native Bumblebees (*Bombus spp.)*". ISHS *Acta Horticulturae* 437 (1996): VII International Symposium on Pollination.

85 Goka, "Influence of Invasive Species on Native Species", 91-101.

86 Ono, "Ecological Implications of Introduced *Bombus terrestris*, and Significance of Domestication of Japanese Native Bumblebees (*Bombus* spp.)", 244–52.

87 Graystock *et al.*, "The Trojan Hives", 1207–15.

88 N. Kojima, "Collaborating with Volunteer Citizens to the Exclusion of an Invasive Alien Bumblebee *Bombus terrestris*". *Japanese Journal of Conservation Ecology* 11 (2006): 61–9.

89 NIES, "National Institute for Environmental Studies, Japan", 2013. Accessed May 17, 2015. http://www.nies.go.jp/gaiyo/pamphlet/nies2013-e.pdf.

with the need to make a profit",[90] one may question whether the outcome would have been different without the Invasive Species Act. That is, would companies have acted in a precautionary manner on the concerns of scientists and changed their business behaviour? Firms tend to display more accountability when government regulation is in place.[91] Indeed, corporate accountability in the case of Japan is enforced through potential fines and prison sentences. There are indications that companies and small businesses in Japan are making efforts to go beyond what is required by legislation. The legislation generally prohibits the import of the large earth bumblebee, though allows the import and use under certain conditions, thus balancing environmental and commercial concerns. Yet, companies are going beyond what is legally mandated by experimenting with, and promoting, the use of native bumblebee species as an eventual replacement of the large earth bumblebee for commercial pollination in Japan. This is an acknowledgement not only of the seriousness of biodiversity loss but of the responsibility that all stakeholders have in its prevention.

In essence, there has been a gradual progression from CSR to accountability among businesses in the bumblebee trade in Japan. The voluntary guidelines of the 1990s, Japan's Global Environment Charter, were augmented a decade later by enforceable legislation in the Invasive Species Act. The recent announcements by MOE and MAFF to require the use of native species will move companies further towards accountability particularly if such a requirement becomes legislation. Firms can no longer deny their impact on the environment and their responsibility for its protection, even if the impact is not visible or fully established through research. The implication for companies in the bumblebee industry is the need to promote and use native species for business purposes within the geographical bounds of their natural distribution range. This, we believe, is the ultimate form of accountability in the commercial bumblebee trade because it removes threats to biodiversity and local ecosystems posed by invasive species.

There are several implications applicable to any business regardless of the specific biodiversity issue.[92] One is that the CSR challenges related to biodiversity are often invisible to corporate actors. Two, the challenges are voiced by scientists so it is important to keep abreast of the scientific debate. Three, the solution to these challenges is likely to be a synthesis of various stakeholder interests, so being prepared to collaborate with different stakeholder groups should be beneficial. These should be valid even if a particular biodiversity issue is not readily quantifiable in

90 Doane and Abasta-Vilaplana, "The Myth of CSR. (Undetermined)", 23.

91 A. Paulraj, "Environmental Motivations: A Classification Scheme and its Impact on Environmental Strategies and Practices", *Business Strategy and the Environment* 18 (2009): 453–68.

92 Reade *et al.*, "CSR, Biodiversity, and Japan's Stakeholder Approach to the Global Bumble Bee Trade", 53–66.

economic terms. After all, ecosystems and their biodiversity play a vital role in the provision of products and services that sustain and enhance human life, whether or not they are known, recognized or quantified.[93] The full value of biodiversity to humankind is yet to be discovered. The loss of biodiversity, including insect populations, has far-reaching consequences for present and future generations. Balancing biodiversity protection and commercial activities that together serve humankind, as Japan's stakeholder engagement has produced through blending voluntary and legalistic approaches, appears to be a new wave of corporate social responsibility infused with accountability.

Bibliography

Arbetman, Marina P., Ivan Meeus, Carolina L. Morales, Marcelo A. Aizen, and Guy Smagghe. "Alien Parasite Hitchhikes to Patagonia on Invasive Bumblebee". *Biological Invasions* 15 (2013): 489–94, doi:10.1007/s10530-012-0311-0.

Asada, S. and Ono, M. "Tomato Pollination with Japanese Native Bumblebees (*Bombus spp.)*". ISHS *Acta Horticulturae* 437 (1996): VII International Symposium on Pollination.

Buchmann, Stephen L., and Gary Paul Nabhan. *The Forgotten Pollinators*. Island Press, 1997.

CBD, Convention on Biological Diversity. Decision V/6 Ecosystem Approach. Decision VII/11 Ecosystem Approach, 2004. Accessed May 17, 2015. http://www.biodiv.org/decisions/default.aspx?m=COP-07&id=7748&lg=0.

Doane, Deborah, and Naomi Abasta-Vilaplana. "The Myth of CSR. (Undetermined)". *Stanford Social Innovation Review* 3 (2005): 22–29.

Dohzono, Ikumi, Yoko Kawate Kunitake, Jun Yokoyama, and Koichi Goka. "Alien Bumble Bee Affects Native Plant Reproduction through Interactions with Native Bumble Bees". *Ecology* 89 (2008): 3082–92.

Dyllick, Thomas, and Kai Hockerts. "Beyond the Business Case for Corporate Sustainability". *Business Strategy and the Environment* 11(2) (2002): 130.

Ehrlich, Paul R., and Robert M. Pringle. "Where Does Biodiversity Go from Here? A Grim Business-as-Usual Forecast and a Hopeful Portfolio of Partial Solutions". *Proceedings of the National Academy of Sciences* 105 (Supplement 1) (2008): 11579–86, doi:10.1073/pnas.0801911105.

Etzion, Dror. "Research on Organizations and the Natural Environment, 1992-Present: A Review". *Journal of Management* 33 (2007): 637–64, doi:10.1177/0149206307302553.

Fukukawa, K. and Teramoto, Y. "Understanding Japanese CSR: The Reflections of Managers in the Field of Global Operations". *Journal of Business Ethics* 85 (2009): 133-46.

Fukukawa, Kyoko, and Jeremy Moon. "A Japanese Model of Corporate Social Responsibility? A Study of Website Reporting*". *The Journal of Corporate Citizenship* 16 (2004): 45–59.

Goka, K. 'Influence of Invasive Species on Native Species: Will the European Bumblebee, *Bombus terrestris*, Bring Genetic Pollution into the Japanese Native Species?" *Bulletin of Biogeographical Society of Japan* 53 (1998): 91-101.

Goka, K. "Introduction to the Special Feature for Ecological Risk Assessment of Introduced Bumblebees: Status of the European Bumblebee, *Bombus terrestris*, in Japan as a Beneficial Pollinator and an Invasive Alien Species". *Applied Entomology and Zoology* 45 (2010):1-6.

93 *Ibid.*

Goka, Koichi, Kimiko Okabe, Masahiro Yoneda, and Satomi Niwa. "Bumblebee Commercialization Will Cause Worldwide Migration of Parasitic Mites". *Molecular Ecology* 10 (2001): 2095–99, doi:10.1046/j.0962-1083.2001.01323.x.

Graystock, Peter, Kathryn Yates, Sophie E. F. Evison, Ben Darvill, Dave Goulson, and William O. H. Hughes. "The Trojan Hives: Pollinator Pathogens, Imported and Distributed in Bumblebee Colonies". *Journal of Applied Ecology* 50 (2013): 1207–15, doi:10.1111/1365-2664.12134.

Horiuchi, K., and Nakamura, M. "Environmental Issues and Japanese Firms". In *The Japanese Business and Economic System: History and Prospects for the 21st Century*, edited by Masao Nakamura, 364-384. New York, NY: Palgrave, 2001.

Inari, Naoki, Teruyoshi Nagamitsu, Tanaka Kenta, Koichi Goka, and Tsutom Hiura. "Spatial and Temporal Pattern of Introduced *Bombus terrestris* Abundance in Hokkaido, Japan, and its Potential Impact on Native Bumblebees". *Population Ecology* 47 (2005): 77–82.

Inoue, Maki N., Jun Yokoyama, and Izumi Washitani. "Displacement of Japanese Native Bumblebees by the Recently Introduced *Bombus terrestris* (L.) (Hymenoptera: Apidae)". *Journal of Insect Conservation* 12 (2007): 135–46, doi:10.1007/s10841-007-9071-z.

Ishii, Hiroshi S., Taku Kadoya, Reina Kikuchi, Shin-Ichi Suda, and Izumi Washitani. "Habitat and Flower Resource Partitioning by an Exotic and Three Native Bumble Bees in Central Hokkaido, Japan". *Biological Conservation* 141 (2008): 2597–607, doi:10.1016/j.biocon.2008.07.029.

JETRO. "More Enterprises Move into Agricultural Sector", 2005. Accessed May 17, 2015. http://www.jetro.go.jp/en/reports/market/pdf/2005_15_h.pdf.

Kanbe, Yuya, Ikuko Okada, Masahiro Yoneda, Koichi Goka, and Koji Tsuchida. "Interspecific Mating of the Introduced Bumblebee *Bombus terrestris* and the Native Japanese Bumblebee *Bombus hypocrita sapporoensis* Results in Inviable Hybrids". *Naturwissenschaften* 95 (2008): 1003–8, doi:10.1007/s00114-008-0415-7.

Keidanren. "Guidelines for Corporate Action", 1991. Accessed May 17, 2015. http://www.keidanren.or.jp/english/speech/spe001/s01001/s01b.html.

Kenta, Tanaka, Naoki Inari, Teruyoshi Nagamitsu, Koichi Goka, and Tsutom Hiura. "Commercialized European Bumblebee Can Cause Pollination Disturbance: An Experiment on Seven Native Plant Species in Japan". *Biological Conservation* 134 (2007): 298–309, doi:10.1016/j.biocon.2006.07.023.

Kojima, N. "Collaborating with Volunteer Citizens to the Exclusion of an Invasive Alien Bumblebee *Bombus terrestris*". *Japanese Journal of Conservation Ecology* 11 (2006): 61–9.

Korhonen, Jouni. "On the Paradox of Corporate Social Responsibility: How Can We Use Social Science and Natural Science for a New Vision?" *Business Ethics: A European Review* 15 (2006): 200–14, doi:10.1111/j.1467-8608.2006.00442.x.

Kraus, F., Bernhard, H. Szentgyörgyi, E. Rożej, M. Rhode, D. Moroń, M. Woyciechowski, and R. F. A.. "Greenhouse Bumblebees (*Bombus terrestris*) Spread Their Genes into the Wild". *Conservation Genetics* 12 (2011): 187–92, doi:10.1007/s10592-010-0131-7.

Lenzen, M., D. Moran, K. Kanemoto, B. Foran, L. Lobefaro, and A. Geschke. "International Trade Drives Biodiversity Threats in Developing Nations". *Nature* 486 (2012): 109–12, doi:10.1038/nature11145.

Mitsuhata, M. "Utilization of the Japanese Native Bumble Bee Species, *Bombus ignitus*", 2006. Accessed May 17. 2015. http://www.agrofrontier.com/guide/t_111c.htm.

MOE. "Invasive Alien Species Act (Law No. 78)". Tokyo: Ministry of the Environment, Government of Japan, 2004. Accessed May 17. 2015. http://www.env.go.jp/en/nature/as/040427.pdf.

MOE. "Measures to be Taken Against Invasive Alien Species in Japan". Tokyo: Ministry of the Environment, Government of Japan, 2004. Accessed May 17. 2015. http://www.env.go.jp/en/nature/as/040326.pdf.

MOE. "Annual Report on the Environment and the Sound Material-Cycle Society in Japan 2007", 2004. Tokyo: Ministry of the Environment Government of Japan.

Moore, Cameron Alastair, and Caroline Gross. "Great Big Hairy Bees! Regulating the European Bumblebee, *Bombus terrestris* L. What Does it Say about the Precautionary Principle?" *International Journal of Rural Law and Policy* 2012. Accessed 4 March 2016. http://epress.lib.uts.edu.au/journals/index.php/ijrlp/article/view/2627.

Nagamitsu, Teruyoshi, Tanaka Kenta, Naoki Inari, Haruka Horita, Koichi Goka, and Tsutom Hiura. "Foraging Interactions between Native and Exotic Bumblebees: Enclosure Experiments Using Native Flowering Plants". *Journal of Insect Conservation* 11 (2006): 123–30, doi:10.1007/s10841-006-9025-x.

NIES. "National Institute for Environmental Studies, Japan", 2013. Accessed May 17, 2015. http://www.nies.go.jp/gaiyo/pamphlet/nies2013-e.pdf.

Niwa, S., H. Iwano, S.I. Asada, M. Matsuura, and K. Goka. "A Microsporidian Pathogen Isolated from a Colony of the European Bumblebee, *Bombus terrestris*, and Infectivity on Japanese Bumblebee". *Jpn. J. Appl. Entomol. Zool.* 48 (2004): 60–64.

Ono, M. "Ecological Implications of Introduced *Bombus terrestris*, and Significance of Domestication of Japanese Native Bumblebees (*Bombus* spp.)". In *Proceedings of the International Workshop on Biological Invasions of Ecosystem by Pests and Beneficial Organisms: National Institute of Agro-Environmental Science, Ministry Agriculture, Forestry, and Fisheries, Japan, Tsukuba, Japan February 25–27, 1997*, 244–52.

Paulraj, A. "Environmental Motivations: A Classification Scheme and its Impact on Environmental Strategies and Practices". *Business Strategy and the Environment* 18 (2009): 453–68.

Rands, Michael R. W., William M. Adams, Leon Bennun, Stuart H. M. Butchart, Andrew Clements, David Coomes, Abigail Entwistle, Ian Hodge, Valerie Kapos, Jörn P.W. Scharlemann, William J. Sutherland, and Bhaskar Vira. "Biodiversity Conservation: Challenges Beyond 2010". *Science* 329 (2010): 1298–303, doi:10.1126/science.1189138.

Reade, C., Goka, K., Thorp, R., Mitsuhata, M, and Wasbauer, M. "CSR, Biodiversity, and Japan's Stakeholder Approach to the Global Bumble Bee Trade". *The Journal of Corporate Citizenship* 56 (2014): 53–66.

Reade, Carol, Robbin Thorp, Koichi Goka, Marius Wasbauer, and Mark McKenna. "Invisible Compromises: Global Business, Local Ecosystems, and the Commercial Bumble Bee Trade". *Organization & Environment* 28 (2015): 436–57. doi: 10.1177/1086026615595085.

Slootweg, R. "Biodiversity Assessment Framework: Making Biodiversity Part of Corporate Responsibility". *Impact Assessment and Project Appraisal* 23 (2005): 37–46.

Tanimoto, K. "Structural Change in Corporate Society and CSR in Japan", In *Corporate Social Responsibility in Asia*, edited by K. Fukukawa, 45–66. New York, NY: Routledge, 2012.

Tanimoto, Kanji. "Changes in the Market Society and Corporate Social Responsibility". *Asian Business & Management* 3 (2004): 151–72.

Thorp, R. W. "Bumble bees (Hymenoptera: Apidae): Commercial Use and Environmental Concerns". In *For Non-native Crops, Whence Pollinators of the Future?* Proceedings of Thomas Say Publications in Entomology, 21–40. Lanham, MD, 2003.

Tokoro, Satoshi, Masahiro Yoneda, Yoko Kawate Kunitake, and Koichi Goka. "Geographic Variation in Mitochondrial DNA of *Bombus ignitus* (Hymenoptera: Apidae)". *Applied Entomology and Zoology* 45 (2010): 77–87, doi:10.1303/aez.2010.77.

Tsuchida, Koji, Natsuko Ito Kondo, Maki N. Inoue, and Koichi Goka. "Reproductive Disturbance Risks to Indigenous Japanese Bumblebees from Introduced *Bombus terrestris*". *Applied Entomology and Zoology* 45 (2010): 49–58, doi:10.1303/aez.2010.49.

Utting, P. "CSR and Policy Incoherence" In *Fair Trade, Corporate Accountability and Beyond: Experiments in Globalizing Justice*, edited by K. Macdonald and S. Marshall, 169–86. Farnham, UK: Ashgate Publishing, 2013.

Velthuis, H. H. W., and van Doorn, A. "A Century of Advances in Bumblebee Domestication and the Environmental Aspects of its Commercialization for Pollination". *Apidologie* 37 (2006): 421–51.

Waddock, Sandra A., Charles Bodwell, and Samuel B. Graves. "Responsibility: The New Business Imperative". *Academy of Management Executive* 16 (2002): 132–48, doi:10.5465/AME.2002.7173581.

Winn, Monika I., and Stefano Pogutz. "Business, Ecosystems, and Biodiversity: New Horizons for Management Research". *Organization & Environment* 26 (2013): 203–29, doi:10.1177/1086026613490173.

Winter, K., Adams, L., Thorp, R., Inouye, D., Day, L., Ascher, J., & Buchmann, S. "Importation of Non-Native Bumble Bees into North America: Potential Consequences of Using *Bombus terrestris* and other Non-Native Bumble Bees for Greenhouse Crop Pollination in Canada, Mexico, and the United States". A White Paper of the North American Pollinator Protection Campaign, 2006.

Yoneda, M., Yokoyama, J., Tsuchida, K., Osaki, T., Itoya, S., and Goka, K. "Preventing *Bombus terrestris* from Escaping with a Net Covering Over a Greenhouse in Hokkaido". *Japanese Journal of Applied Entomology and Zoology* 51 (2007): 39–44.

5

Bombus terrestris

A personal deep ecology account

Jack Christian
Manchester Metropolitan University Business School, UK

> The rambler made his way across the Pennine Moor. It was late February and a grey, uninspiring day yet the rambler's heart was light. The lapwings were back and their flopping flight and peewit calls told him that spring, and life anew, was just around the corner. As he watched them a small fuzzy, black and yellow ball flew towards him bobbing and weaving through the air seemingly in search of something; a new home perhaps? It was a *Bombus terrestris* queen and she was set to start a family.

This chapter takes a deep ecology perspective on accounting for bees. It focuses on the inter-connectedness of life and notes which stakeholders have an interest in *Bombus terrestris* and whose lives would be impacted should the queen and others like her fail to bring up a family. It starts from the premise that a bees' nest is not, and cannot be reduced to, a pollinating plant or honey factory; it is a living, adaptive community full of individuals who can themselves learn and adapt. Further it is part of an ecological supra-system often involving numerous ecological communities each of which can, in the longer term at least, adapt and change. In the face of such complexity any attempts at control simply ask for unintended consequences.

The following section outlines the life of a *Bombus terrestris* queen and her family. In particular their relationship with other flora and fauna is investigated. The section thereafter provides an overview of deep ecology, focusing on the concepts of inter-connectedness, ecological egalitarianism and diversity. These concepts, particularly inter-connectedness, are now seen as integral to general systems theory as noted in Chapter 1 of this book. The penultimate section discusses stakeholder engagement and argues the case for Nature and ultimately bees to be considered as stakeholders.[1] The final section argues that our current understanding of *Bombus terrestris* is at best shallow and most likely just plain wrong. I argue for an alternative view that demands we leave space for *Bombus terrestris* and Nature in general, one which also has lessons for humanity.

The life and times of *Bombus terrestris*

Benton[2] provides a comprehensive work on British bumblebees. He draws on his own observations and the work of numerous other bumblebee enthusiasts and scientists. It is no exaggeration to describe his work as seminal.

The book is broken down into 11 chapters as follows:

1. Introducing Bumblebees
2. The Bumblebee Life Cycle
3. Bumblebee Psychology
4. The Usurper Bumblebees
5. Predators, Parasites and Lodgers
6. Bumblebees and Flowers: Foraging Behaviour
7. Bumblebees and Flowers: Flower Arrangements
8. What Bumblebee is That?
9. The British Species
10. Agricultural Change and Bumblebee Decline: Explaining the Patterns
11. Back from the Brink? Bumblebee Conservation.

By and large the book is scientific; that is, it is empirical and analytical in its approach though there are occasions when Benton's passion for bumblebees shines through. For example the inclusion of a paragraph entitled "Bumblebee Psychology" might raise eyebrows in some circles.

As it is my intention to explore the life of *Bombus terrestris* I have scoured the index of Benton's book for references to this bee and, in the first part of this section, constructed a summary of the data Benton provides. In the second part of this section I look for assumptions implicit in the way the data is presented and where

1 Stakeholders in this context refer to organizational stakeholders. These are defined as any being—human or otherwise—that has an interest in the organization or whose interests are affected by the organization.

2 T. Benton, *Bumblebees* (Fulham: Collins, 2006).

possible I offer alternative assumptions that lead to a new interpretation of how or why *Bombus terrestris* exists and indeed how we might view Nature generally.

The empirics

Table 5.1 **Biology, ecology and conservation status of *Bombus terrestris***

Source: T. Benton, *Bumblebees* (Fulham: Collins, 2006).

Category	Type of data	*Bombus terrestris*
Nest	Pocket maker/pollen storer	Pollen storer
Nest	Simple/Complex	Complex
Nest	Nest position	Below ground
Nest	Colony size	Large
Nest	Nesting habitat	Woodland edge/urban
Nest	Colony cycle	Short-medium
Nest	Early or late nesting	Early
Nest	Colony dispersal	Medium
Nest	Length of season	Long
Nest	Nesting density	Medium?
Nest	*Predators, parasites and pests*	
Foraging	Tongue length	Short
Foraging	Pollen preference	General
Foraging	Foraging range	Large
Foraging	*Perception*	
Foraging	*Communication and memory*	
Range	Dispersal ability	Good
Range	Geographical range	Northern edge
Range	Status before 1960	Mainland, ubiquitous but local in the north
Range	Change 1960 to 1980	Unchanged
Range	Current trends as at 2006	Northerly expansion

On pages 448 to 451 of his book Benton provides a table listing biological, ecological and conservation data in respect of British bumblebees. Table 5.1 is drawn from Benton's table. I have categorized his data types in column 1 and re-ordered them in column 2 adding three other types (shown in italics), which I believe are pertinent. In column 3 I have reproduced Benton's analysis of how *Bombus terrestris* fits within the data types.

Nesting behaviour

Bombus terrestris queens are among the first to emerge from hibernation and can be seen flying as early as February searching for a nesting site. They generally seek a site that is underground often in the disused nest of a small mammal, which they may disguise or fortify in some way, such as by constructing a false nest nearby or narrowing the entrance hole. Such sites may be located in a wide variety of habitats including roadside verges and hedges, woodland rides, gardens, parks and other public spaces.

Having established a nest the queen then begins to build the stores of pollen that provide the protein for larvae to develop into bees. When she is ready she will lay between 6 and 16 eggs which she will rear herself. These will develop into worker bees which will help her rear subsequent broods until the colony may number in the hundreds. *Bombus terrestris* is a complex species which means it is believed the queen maintains control of the colony by means of chemical signals as well as physical dominance.

Worker bees are all female but they do not lay eggs themselves. It is thought that the queen may control this by means of chemical signals but this is not known for certain. At some point though the queen will start laying some eggs that are destined to become future queens. Again it is not known for certain why these eggs are chosen to become future queens; it seems the decision is actually made in the first week of the larval stage and thereafter the larvae receive greater attention and grow stronger and more substantial than their peers.

The queen will also start to produce males at some point. Male bumblebees are born of unfertilized or haploid eggs and are therefore genetically more closely related to their mother. Some colonies appear to produce more daughter queens while others produce more males. Again it is not understood why.

Shortly after the appearance of daughter queens and males, dissent becomes apparent within the colony. Some workers may try to lay their own (unfertilized and hence male) eggs and the other workers may assist or resist their efforts. In any event the queen appears to lose control of the colony which will begin to break down and disperse. Once again it is not known why the breakdown occurs and why it should appear at this particular point in the colony's history. Some researchers have tried to link it to colony size but the results are inconclusive.

Bombus terrestris colonies tend to break down after around three months, which is a short to medium colony cycle. However, their overall season is quite long: they can be seen from February to September. It may be that daughter queens from one nest take time to find males from another nest. Eventually all but daughter queens will die (though there is some evidence of continued colony activity in parts of southern England) and these will look for a place to hibernate. They have been known to excavate holes in rotten wood, construct chambers from wood chewings or simply burrow into piles of grass cuttings.

Worker bees of course do more than tend the larvae. They can be divided into house bees and foragers though there is some evidence that individual bees can retrain, as it were, should the nest need more of one than the other. House bees may be taken by mammalian predators attacking the nest; these include badgers and weasels as well as smaller creatures such as voles and mice. However, guard bees often repel attacks by such predators and observers have seen partially destroyed nests being rebuilt by the house bees.

Both house bees and foragers are susceptible to attack by mites, bacteria and other insects including the wax moth whose larvae infect and attack them. House bees may also be killed in defending the nest against the cuckoo bee *Bombus vestalis* whose queens may attempt to take over the nest. Finally, foragers face danger from birds, such as blue and great tits, which will take them in flight.

Foraging behaviour

Bombus terrestris foragers are generalists: they take nectar and pollen from a wide variety of plants. On page 326 Benton lists the plants the bees are known to visit, this list is summarized in Table 5.2.

Table 5.2 **Foraging patterns**

Visitor	Season	Total number of plants	Of which are garden plants	Of which pollen collected
Queen	Winter	6	4	
Queen	Spring	14	8	
Workers	Winter	6	6	
Workers	Spring	7	1	4
Workers	Summer	31*	9	11
Workers	Autumn	7	1	2
Daughter queens		15*	5	
Males		15	2	

The asterisks in Table 5.2 indicate that the number of actual plant species is greater than that shown because in some instances Benton indicates family groups rather than individual species within the family. For example, he refers to thistles, clovers and woundworts rather than individual plants within these families. Daughter queens and males may fly in summer and/or autumn but in their case Benton does not differentiate between seasons. These individuals are also uninterested in collecting pollen, as are queens for the most part, because they play no part in supplying food for larvae, this being the job of the worker bees. A further notable point is the importance of garden plants in winter and spring. Those few colonies that remain active in winter seem totally dependent on garden plants.

Foraging *Bombus terrestris* can cover a large area and are seemingly willing to fly up to 1.5 kilometres in search of nourishment though 200 to 600 metres is more normal. They appear to have a general landscape memory, which they learn in their first trips from the nest, flying in increasing circles to build this memory. Thereafter it seems they can fly directly from any point to any point within their landscape memory. Interestingly in experiments where bees have been released up to 10 kilometres from their nest some have been able to find their way back.

Unlike honey bees *Bombus terrestris* foragers tend to work as individuals finding their own patches of flowers, sources of nectar and pollen. These sources may however be used by other species of bumblebee or *Bombus terrestris* from other colonies or even other insects entirely. While some insects are territorial by and large the resources are generally shared. Indeed some insects, including *Bombus terrestris* leave chemical messages to inform future prospectors that the plant has just been visited and its nectar taken, hence they need not waste their time on it.

It appears that *Bombus terrestris* recognize plants by their shape rather than their colour. Bumblebees are sensitive to a different range of light waves than humans; their primary colours are ultraviolet, blue and green, which make flower shapes more obvious against leafy backgrounds. They learn different shapes—to this end they have a large number of hairs on their legs and antennae, which seem to relay messages to their brain allowing them to remember three-dimensional shapes—and how to harvest them. As a short-tongued bumblebee, they sometimes have to take nectar by drilling a hole through the side of a flower or bud, but generally they will take pollen and nectar by climbing on or even in the flower.

The last five rows of Table 5.1 refer to the range of *Bombus terrestris* in the UK. It is fairly clear that *Bombus terrestris* is expanding its range northwards. As strong flyers, its new queens can disperse widely, so it is not surprising that in a period of global warming it can find new breeding grounds further north.

Deep ecology

In this section I offer my interpretation of the deep ecology movement based on my ecosophy or personal ecocentric philosophy,[3] drawing specifically on the work of Arne Naess, founder of the deep ecology movement.[4] I work from seven key points outlined by Naess in a lecture in Bucharest in 1972 and subsequently published in *Inquiry* the following year.

3 For a more detailed exposition of ecosophy see A. Naess, "The Shallow and the Deep, Long-Range Ecology Movements: A Summary", *Inquiry* 16 (1973): 95–100 and A. Drengson and B. Devall, "The Deep Ecology Movement: Origins, Development & Future Prospects", *The Trumpeter* 26 (2) (2010): 48–69.

4 See D. Pepper, *Modern Environmentalism: An Introduction* (London: Routledge, 1996) and C. Belshaw, *Environmental Philosophy: Reason, Nature and Human Concern* (Chesham: Acumen Publishing, 2001) regarding Naess's role in founding deep ecology.

Naess[5] described deep ecology as built on a platform underpinned by a variety of religious and philosophical views. This broad foundation is exemplified in the work of Sivaraksa, Weiming, Suzuki, Jung and McIntosh all of whom,[6] from different global perspectives, talk of inter-connectedness and complexity in ways that support deep ecology principles. Naess himself draws on the work of philosophers such as Spinoza, Whitehead and Heidegger in creating his ecosophy. I do not intend to suggest that these examples are the only way or even the best way to underpin deep ecology, which is very much a pluralist movement, but rather to offer a personal, normative interpretation. I will then use this interpretation to reflect on the many connections between bees and humans as highlighted in Chapter 2.

Naess[7] characterized deep ecology by identifying seven key points:

1. Rejection of the man-in-environment image in favour of the relational total-field image.
2. Biospherical egalitarianism in principle.
3. Principles of diversity and of symbiosis.
4. Anti-class posture.
5. Fight against pollution and resource depletion.
6. Complexity not complication.
7. Local autonomy and decentralization.

He then noted, "the norms and tendencies of the deep ecology movement are not derived from ecology by logic or induction".[8] Rather ecological knowledge and experience "have *suggested, inspired and fortified* the perspective of the deep ecology movement". He also noted that the "significant tenets of the deep ecology movement are clearly and forcefully *normative.* They express a value priority system only based in

5 See A. Naess, "Intuition, Intrinsic Value, and Deep Ecology", *The Ecologist* 14 (1984): 201–3.

6 S. Sivaraksa, in *Conflict, Culture, Change Engaged Buddhism in a Globalizing World* (Somerville: Wisdom Publications, 2005, 71) talks of the Net of Indra and describes interdependence as the crux of Buddhism. T. Weiming, in "The Continuity of Being: Chinese Visions of Nature", in *Confucianism and Ecology: The Interrelation of Heaven, Earth and Humans,* edited by M. E. Tucker and J. Berthrong (Cambridge, MA: Harvard University Press, 1998), 108, talks of the Chinese concept of ta-hua and all modalities as a continuum. D.T. Suzuki, in *An Introduction to Zen Buddhism* (London: Rider, 1991) describes "satori" which C.G. Jung, in "Foreword", in *An Introduction to Zen Buddhism,* edited by D. T. Suzuki (London: Rider), 28, describes as becoming whole—regaining connection with the universe. Finally A. McIntosh, a Quaker, in *Soil and Soul: People versus Corporate Power* (London: Aurum Press, 2004), 45, talks of Celtic Christianity as community and nature in a triune confluence with God.

7 Naess, "The Shallow and the Deep, Long-Range Ecology Movements", 95–100.

8 All quotes in this paragraph cited from N. Witoszek and A. Brennan, *Philosophical Dialogues: Arne Naess and the Progress of Ecophilosophy* (Lanham: Rowan & Littlefield Publishers, 1999), 6.

part on results (or lack of results, see point 6) of scientific research". Further he said that the movement is ecophilosophical rather than ecological. The latter is described as a limited science whereas philosophy is seen as a more general forum of debate. He introduces ecosophy as a philosophy of ecological harmony. A philosophy he claims "is a kind of *sophia* wisdom, it contains both norms, rules, postulates, value priority announcements and hypotheses concerning the state of affairs of our universe".

The first of Naess's key points rejects anthropocentrism. Naess describes "organisms as knots in a biospherical net or field of intrinsic relations"[9] and places humankind in this field alongside all other organisms. He also notes that it is the intrinsic relation between two things that define them and that without the relation the two things would be different. This idea of a field of intrinsic relations is fundamental to the Chinese[10] view of reality, "All modalities of being, from a rock to Heaven, are integral parts of a continuum…". But this is just part of a more widespread belief, according to Sivaraksa "The concept of interdependent co-arising is the crux of Buddhist understanding. Nothing is formed in isolation and, like the jewelled net of Indra, each individual reflects every other infinitely".[11] As Zimmerman explains "Mahayana Buddhism holds that the phenomenal world is akin to such an interplay of reflected appearances, in which each thing is aware of its relation to all other things".[12]

Zimmerman also links Heidegger and Buddhist ideas and claims these are congruent with deep ecology. In particular, he suggests that seeing humans not as entities but clearings "in which entities (including thoughts, feelings, perceptions, objects, others) appear eventually helped Heidegger overcome not only dualism but also anthropocentrism". In claiming that human beings only exist in so far as they allow others to appear or manifest themselves as themselves Heidegger showed that we are nothing without others. The connection with deep ecology is further underlined by Sivaraksa:

> Attachment to an atomised sense of self and a self/other dualism are the antithesis of interdependence and is an obstacle to achieving the peace of enlightenment. A commitment to nature and a deep respect for all life can help foster a change from an individualized self to a self as interbeing.[13]

Biospheric egalitarianism exists "in principle" because as Naess explains, realistic praxis—that is the living of one's own life—will necessitate some killing or exploitation. What Naess is trying to capture is an intuition that everything has "an equal right to live and blossom" and that we must recognize that our quality of life

9 *Ibid.* 3.
10 See Weiming, "The Continuity of Being: Chinese Visions of Nature", 108.
11 See Sivaraksa, *Conflict, Culture, Change Engaged Buddhism in a Globalizing World*, 71.
12 See M.E. Zimmerman, "Heidegger, Buddhism, and Deep Ecology", in *The Cambridge Companion to Heidegger*, edited by C. B. Guignon (New York: Cambridge University Press, 2006). This quotation concerning Mahayana Buddhism can be found on page 306 and the quotation about Heidegger in the next paragraph can be found on page 295.
13 See Sivaraksa, *Conflict, Culture, Change Engaged Buddhism in a Globalizing World*, 71.

depends on "the deep pleasure and satisfaction we receive from close partnership with other forms of life".[14]

This point has been the subject of much debate[15] with many writers finding biospheric egalitarianism ambiguous and looking for clarification; perhaps by means of some sort of ethical system,[16] or by way of allocating different levels of intrinsic value to different beings.[17] Naess rejects a single correct system because life is too complex, arguing we must make judgements and decisions based on our intuition.

The principle of diversity is built around the claim that diversity enhances the chances of survival and increases the richness of experience. Symbiosis, mutually beneficial relationships among beings, is a natural corollary to the concept of intrinsic relations.[18] It is in direct contrast to the survival of the fittest and similar competitive strategies constructed by Darwin and others.[19] The anti-class posture follows from a combination of ecological egalitarianism and an encouragement of diversity in human ways of life. Under the banner of symbiosis it explicitly argues against the exploitation and suppression that underlie class systems.

Naess calls on the deep ecology movement to fight against pollution and resource depletion—but proponents are warned against alliances that may address this battlefield at the expense of the other key points. Complexity not complication reminds us that ecosystems are complex, often to an astonishingly high level, yet they are part of a unity. This compares to complication which has no unifying

14 Cited in Witoszek and Brennan, *Philosophical Dialogues: Arne Naess and the Progress of Ecophilosophy*, 4.

15 See R. Watson, "A Critique of Anti-Anthropocentric Biocentrism", *Environmental Ethics* 5 (1983): 245–56.
W. Fox, "Deep Ecology: A New Philosophy of Our Time". *The Ecologist* 14 (1984): 194–200.
W. Fox, "On Guiding Stars of Deep Ecology". *The Ecologist* 16 (1984): 203–4.
W.C. French, "Against Biospherical Egalitarianism". *Environmental Ethics* 17 (1995): 39–57. All of these papers are reproduced with minor revisions by Witoszek and Brennan, *Philosophical Dialogues: Arne Naess and the Progress of Ecophilosophy*, together with replies from Naess.

16 Such as those proposed by Desjardins: J.R. Desjardins, *Environmental Ethics: An Introduction to Environmental Philosophy* (Belmont: Wadsworth Publishing, 1993) and J.R. Desjardins, *Business, Ethics, and the Environment* (Upper Saddle River, New Jersey: Pearson Education, 2007) and by L.H. Newton, *Ethics and Sustainability: Sustainable Development and the Moral Life* (Upper Saddle River, NJ: Prentice-Hall, 2003).

17 As mooted by K.E. Goodpaster, "On Being Morally Considerable". *Journal of Philosophy* 75 (6) (1978): 308–25 and P. Singer, *Practical Ethics* (Cambridge: Cambridge University Press, 1993).

18 Together the principles of diversity and symbiosis support the argument for protecting biodiversity. Genetic diversity enhances the chances of a species surviving while symbiotic relationships between species enhance the survival chances of all the creatures within a biospheric community.

19 See S. Hinchliffe, *Geographies of Nature: Societies, Environments, Ecologies* (London: Sage, 2007), 28–33, for a critique of Darwin's theory *On The Origin of Species* which was first published in 1859 and has recently been republished as C. Darwin, *The Origin of Species* (Oxford, UK: Oxford University Press, 1998).

principles and can be described as chaotic. Disturbances to complex systems such as ecosystems can have unforeseeable effects across other parts of the system[20] and we should be sensitive to our ignorance. Finally deep ecology calls for more local autonomy and decentralization. I see this as another argument against exploitation and suppression and for (biospherical) egalitarianism.

Later interpretations of Deep Ecology[21] place less emphasis on the relationship total-field image and ecological egalitarianism, both of which are fundamental to my ecosophy. Further, many of these interpretations place undue emphasis on the preservation of wilderness and the controlling of the human population leading to accusations of imperialism and misanthropy.[22] To this extent they appear to have missed the point that humans are an integral part of Nature.

Others[23] have called for a more specific set of ethical principles. However Naess was at home in complexity and diversity. He did not seek to dominate but would welcome alternative views and seek to reconcile them as necessary. He did not define deep ecology preferring to describe it as an "intuition". As I suggested earlier, I believe intuition is a direct communication from our worldview and I would describe deep ecology as a way of understanding the world: as ontology. Fundamental to this ontology is the idea of inter-connectedness and a total-field view. From this it is possible to build the norms and values captured in the other key points elucidated in 1972 and in the deep ecology platform. In particular I refer to ecological egalitarianism and diversity and symbiosis (rather than competition.)

Stakeholder engagement

Traditionally organizational accounts are produced by one group of stakeholders reporting how they have used resources provided by other stakeholders. In the UK they are typically financial accounts, produced by the directors, trustees or managers of the organization primarily for those who have invested funds in the company. These quantitative, economic accounts are often accompanied by a narrative report, for example the UK Companies' Acts have always required a Directors' Report, but it is probably fair to say this typically remains focused on financial

20 Chaos theory, general systems theory and the "butterfly effect" are discussed in Chapter 1.

21 See for example B. Devall and G. Sessions, *Deep Ecology: Living As If Nature Mattered* (Layton: Gibbs Smith, 2007).

22 See R. Guha, "Radical American Environmentalism and Wilderness Preservation: A Third World Critique". *Environmental Ethics* 11(1) (1989): 71–83 and M. Bookchin, "Social Ecology versus Deep Ecology", *Green Perspectives, Newsletter of the Green Program Project*, 4/5 (1987), respectively.

23 See Fox, "Deep Ecology", 194–200; Fox, "On Guiding Stars of Deep Ecology", 203–4; and P. Curry, *Ecological Ethics: An Introduction* (Cambridge, Polity Press, 2006).

matters. That said, recent changes in UK company law, such as the Companies Act 2006, now require them to consider other stakeholders.

In the remainder of this section I shall first discuss some aspects of stakeholder theory. I will then move onto considering Nature as a stakeholder and, in the last sub-section, I imagine bees as stakeholders as a pre-conception to the concluding part of this chapter.

Stakeholder theory

In their seminal work Donaldson and Preston argued that stakeholder theory is managerial requiring managers to give "...simultaneous attention to the legitimate interests of all appropriate stakeholders".[24] While not defining who the stakeholders of any specific organization might be, they were quite clear that managerial responsibility goes beyond serving the shareowners/investors.

For a time, stakeholder theory, together with legitimacy theory, held the attention of social and environmental accounting researchers. Deegan and Unerman[25] gave considerable space to them in their book *Financial Accounting Theory*. Similarly Unerman *et al.* included two chapters on what had now become stakeholder engagement rather than stakeholder management.[26]

In the first of these chapters, Tilt describes a stakeholder as an individual or group having a legitimate claim on the firm in the sense that they can affect or be affected by the firm's activities. Both she and Unerman write about these individuals and groups. Tilt focuses on stakeholders' perception of sustainability reporting, noting that economic stakeholders tend to have less interest in sustainability issues than other stakeholders. She further notes that there is quite a high level of scepticism towards corporate reporting; indeed many see it as a form of greenwash. Unerman addresses how to engage stakeholders in meaningful dialogue, how their needs might be prioritized and consensus negotiated. He concludes by suggesting a broader range of stakeholders need to be empowered. Failing that stakeholder dialogue will be used to disguise a social and environmental reporting that has little to do with sustainability.

The movement towards stakeholder engagement compared with stakeholder management can be identified in the work of Thomson and Bebbington and Bebbington *et al.* who call for a dialogic approach with stakeholders.[27] Indeed

24 See T. Donaldson and L.E. Preston, "The Stakeholder Theory of the Corporation: Concepts, Evidence and Implications". *The Academy of Management Review* 20(1) (1995): 65.

25 See C. Deegan and J. Unerman, *Financial Accounting Theory* (Maidenhead, McGraw-Hill Education, 2006).

26 See J. Unerman *et al.*, *Sustainability Accounting and Accountability* (Abingdon: Routledge, 2007) and the chapters by Tilt and Unerman in particular.

27 See I. Thomson and J. Bebbington, "Social and Environmental Reporting in the UK: A Pedagogic Evaluation". *Critical Perspectives on Accounting* 16 (2005): 507–33 and J. Bebbington *et al.*, "Theorizing Engagement: The Potential of a Critical Dialogic Approach". *Accounting, Auditing and Accountability* 20 (3) (2007): 356–81.

Unerman's conclusion echoes that of Thomson and Bebbington who note that stakeholder engagement can be used to "explain" and justify the reporting organization's actions or alternatively it can be used to inform and educate the stakeholder. In the first instance it is often used as a defensive mechanism to maintain power relationships, in the second instance it can be emancipatory and transformative. They concluded that currently social and environmental reporting essentially acts to suppress criticism and maintain the status quo. Bebbington *et al.* return to this subject to look at the potential for a "critical dialogic" approach to social and environmental accounting that would require an open and honest dialogue between stakeholders.

Antonacopoulou and Meric[28] were even more specific about the relationship between an organization and its stakeholders. They called for a more feminist interpretation of stakeholder theory, which "places relationships at the heart of what organizations do, and, concretely rather than abstractedly, promotes a personal connection to relationships". Sadly, this call seems to have gone unheeded, and interest in stakeholder theory waned as social and environmental researchers moved on to other theories such as institutional and governmental theory. Meanwhile in praxis stakeholder management supported by a "business case" is much more in the vogue. AccountAbility's SES1000 for example suggests that the reasons for engaging with stakeholders include better risk management, improved resources and opportunities for "educating" them. In fairness AccountAbility also mentions the potential for a more equitable and sustainable society but this point is lost amid the business case arguments.

More recently however some writers[29] have suggested working with stakeholders to develop tools and ideas that may lead to more sustainable societies. They were discussing research rather than praxis but the concept of stakeholder engagement was revived and given a new lease of life. Despite criticism from some quarters[30] others such as Bebbington and Larrinaga[31] see stakeholder engagement as fundamental to sustainability accounting.

My own view is that Antonacopoulou and Meric had it right. Human beings live in and through personal relationships, not economic acquisitiveness. The latter simply turns them into functional parts equating more with good, an equation that suits businesses as they strive for growth at all costs, but one that robs humans

28 See E.P. Antonacopoulou and J. Meric, "A Critique of Stake-Holder Theory: Management Science or Sophisticated Ideology of Control?" *Corporate Governance* 5(2) (2005): 30.

29 See R.L. Burritt, "Environmental Performance Accountability: Planet, People, Profits", *Accounting, Auditing and Accountability Journal* 25(2) (2012): 370–405 and R. Gray, *Accountability, Social Responsibility and Sustainability: Accounting for Society and the Environment* (Harlow, Pearson Education, 2014).

30 See R. Gray and R. Laughlin, "It Was 20 Years Ago Today: Sgt Pepper, Green Accounting and the Blue Meanies", *Accounting, Auditing and Accountability Journal* 25(2) (2012): 228–55 regarding the limitations of stakeholder theory.

31 See J. Bebbington and C. Larrinaga, "Accounting and Sustainable Development: An Exploration", *Accounting, Organizations and Society* 39 (2014): 395–413.

of their humanity. Stakeholder engagement addressed at a personal level is germane to a sustainable society. In particular, as Thomson and Bebbington suggest, such engagement has an emancipatory potential which I believe is the key to sustainability.

Nature as a stakeholder

In proposing Nature as a stakeholder, Tilt's description of a stakeholder as an individual or group having a legitimate claim on an organization in the sense that they can affect or be affected by the organization's activities, is apposite.[32] Nature, I will argue, is a provider of services and therefore has a right to be considered as a stakeholder and is entitled to an account in respect of the services she provides. I also acknowledge that she would have to be represented by us: that is, by humans.

As a first step to considering Nature as a stakeholder I stop to reflect on the idea of bees as a resource. For all the reasons I argued in my earlier deep ecology section, in particular ecological egalitarianism, they should not be seen as such; however in the world of business as currently practised that is precisely what they are seen as. In fact they are seen and utilized as a primary resource in the agricultural industry. Their labour is a resource just like human labour but with the commercial advantage that they do not have to be paid. They pollinate crops for free so unlike human labour they do not even appear as a cost in the financial reports.[33]

Now it is a moot point whether labour should appear as a cost. I would argue the providers of labour are investors in a company as much as any financier and the company's financial statements should reflect this. However that is not the present way of things so we will stay with labour, human and bee, as a cost. My next thought is why shouldn't bees be paid if they are providing a service? Because, I guess it could be argued, they aren't providing a service, they are just doing what bees do. So why do bees do what bees do? I don't think anyone knows the answer to that although there are surely theories, most of which would fall into one of two categories. The first is, that is the way nature has evolved and the second is, that is the way our deity or deities planned it. It seems to me then we owe something to Nature or to our deities for the work bees do.

As this chapter represents a personal deep ecology account I will stay with the idea that we owe something to Nature and leave it to others to take up a deist approach. This is not to suggest an atheist argument; I am an agnostic and simply do not have the knowledge to pursue a religious argument. So, what do we owe to

32 See C.A. Tilt, "External Stakeholders' Perspectives on Sustainability Reporting", in *Sustainability Accounting and Accountability*, edited by J. Unerman, *et al.* (Abingdon: Routledge, 2007).

33 In some parts of the agricultural industry, for example the almond industry in California, it has now become necessary to transport bees to the almond plantations to pollinate the trees. The hire and transportation of the bees and their hives is of course a cost; however the bees still do their work for free, it is the farming practice that gives rise to the cost.

Nature? Frankly I do not know that either, not least because it is so hard to define what Nature is. Realist scientists and physical geographers will tell us it is the universe that exists outwith ourselves, an objective reality somewhere out there; social constructionists and social geographers will disagree saying Nature is subjective and what we make her: she can never be separate from ourselves.[34] As always, I do not claim to know, but I like to see her the way some Ancient Greeks saw her, Nature is everything including ourselves,[35] and from that perspective she is certainly not just out there. We are in her and if we affect her we affect ourselves.

However, whatever Nature is, it is important to note that she has provided resources—in this case bees—to many organizations and should, at the very least, be considered by those organizations as a stakeholder.

Engagement and Nature

How though do we engage with her? At this point I see claimants lining up to represent her; natural scientists of all types—geographers, biologists and geologists to name but three—social scientists, psychologists and psycho-analysts, artists, poets and philosophers even would all make their case. All and more have a valid claim and prioritizing their claims, if that were even possible, would involve arguments beyond this paper. I therefore propose to focus on what I would do if I were representing her.

Earlier I stated my belief that labour and bees should not be seen as resources. Better in my opinion would be co-producers. Following this in representing Nature I should not see bees as my resource, rather I should see them as contemporaries in a connected world, a stakeholder in their own right. A view, which not surprisingly is more akin to the deep ecology perspective I hold to. A perspective that respects life and land, and that sees inter-connectedness between them in all their forms.

I would therefore seek to understand why bees do what they do. This is no small task and will inevitably be anthropomorphic (I can after all only see things as a human). However what I propose to do in the next section is reflect on the lifestyle of *Bombus terrestris*, the buff-tailed bumblebee (also known as the large earth bumblebee), as described by Benton and summarized earlier in this chapter. I will then deconstruct this description and reconstruct it in an alternative light that challenges established attitudes to nature, competition and what is "natural".

34 See N. Castree, *Nature* (London: Routledge, 2005).

35 See C.S. Lewis, *The Discarded Image: An Introduction to Medieval and Renaissance Literature* (Cambridge, Cambridge University Press, 1964).

Bees: two perspectives

Benton's 2006 narrative[36] follows a traditional scientific analysis. It is based on a theory of evolution that interprets nest behaviour, including breeding behaviour and foraging patterns in terms of competition.

Nest behaviour is mostly explained in terms of kin selection theory, which supposes that individual bees act as they do to ensure the predominant, that is more numerous, genes within the colony are passed on to the next generation. As Benton puts it "Co-operation, such as we find in bumblebee colonies... seems to fly in the face of common understanding of the way natural selections works..." (p. 61). However "...if we take the 'gene's eye view'...so called altruistic behaviour is thus most likely to be shown towards close kin...this outcome is known as kin selection..." (p. 62).

Behaviour outside the nest is explained in terms of colony success or colony protection in the face of a hard, competitive, sometimes criminal world. *Bombus terrestris* fly as individuals to avoid the possibility of too many being ambushed by predators, pests or disease at any particular site. The colony can afford the loss of one or two individuals but would not want to risk more unnecessarily. Foragers who drill into plants whose shape precludes access to nectar are said to "rob" those plants; that is, remove the nectar without transferring pollen. Back at the nest again bees post guards to deter or even kill invaders who might try to steal the colony's supply of nectar and pollen.

Benton informs us that bumblebees have been a popular choice for research and various theories have been developed to explain their foraging behaviour. Interestingly "The main assumptions and models used in the theory closely resemble those of the most influential approach in modern economics" (p. 172). That is "...bumblebees (or any other forager) are treated as actors who try to maximise their gains while minimising the costs of achieving them" (p. 173).

Perhaps it is this choice of model that leaves Benton (and other bee observers) puzzled by apparently cooperative behaviour such as leaving chemical signals to show other bees, almost certainly from another colony, that a flower has recently been visited and its nectar harvested. Why share such useful information with potential competitors when to do so "...might entail costs in a competitive situation" (p. 92)? Further the most cooperative behaviour of all, the symbiotic partnership between bee and flower is passed over without comment. Good is taken for granted, seemingly evil is what shapes the world!

Competition and survival of the fittest are concepts frequently attributed to Charles Darwin. However his theory of evolution is precisely that, a theory. It was the culmination of a body of thinking in the mid-19th century and the most eloquent exposition of that thinking.

36 See Benton, *Bumblebees*. All the direct quotations in this section can be found in his book on the page numbers shown.

Like most theories it draws on the most prevalent conceptions of the time[37] and arguably Hobbes's dystopian view of the world figured largely in Darwin's thinking. In his perhaps most famous poem, *In Memoriam*, the soon-to-be Poet Laureate Tennyson spoke of a "Nature, red in tooth and claw" some 10 years prior to Darwin's *Origin of the Species*. Such a view was perhaps an inevitable consequence of the "Enlightenment" with its claim that Nature was a malevolent force that had to be contained and ultimately viewed as a resource. A claim beginning with Bacon's *The Great Instauration*, continuing through Hobbes's *Leviathan* and Defoe's allegorical novel *Robinson Crusoe* and bolstered by Locke's essay *Of Property*.[38] There were of course alternative views such as those of the Romantic poets and philosophers and that of Rousseau but it is probably safe to say these were seen as reactionary minority views.[39]

The influential philosopher Mill published his essay *Nature* a decade or so after the publication of Darwin's theory of evolution and the Enlightenment perspective on Nature was confirmed.[40] The idea that competition was natural was then taken up enthusiastically by the nouveau-riche industrialists gaining power and wealth at that time to justify their new status in society.[41] Thus competition and survival of the fittest were cemented as prime motivators underlying behaviour in plants, animals and life in all its forms including human life.

So now we see *Bombus terrestris* as a collection of genes fighting for eternal survival. Why? (And for that matter, why aren't humans therefore seen as a collection of genes rather than as individuals?) Surely it is just as possible to see *Bombus terrestris* as a manifestation of life, life that is more than a collection of genes, something more akin to Ch'i or Tao or some other more holistic force. As such we have no need to focus on small elements of it which we might call a species, or smaller still, an individual. Standing back from the minutia that is the analysis provided by science what we would see here on Earth, in glorious technicolour, is life expressing itself in a multitude of (potentially) ever diversifying forms. Each piece of which comes into existence from others as they give way, and which in its turn gives way to become absorbed within new pieces; an eternal kaleidoscope of wonder.

37 See M. Foucault, *The Archaeology of Knowledge* (London: Routledge, 1972) and M. Foucault, *The Order of Things* (Abingdon: Routledge, 2002) for more on how discourses shape the way we think.

38 See D.R. Keller, *Environmental Ethics: The Big Questions* (Chichester, UK: Wiley-Blackwell, 2009) on the work of Bacon, and Locke and N. Bingham *et al.*, *Contested Environments* (Chichester, UK: Wiley, 2003) on Hobbes and Defoe.

39 See The Rt Hon. Lord Quinton, "Romanticism, philosophical", in *The Oxford Companion to Philosophy*, edited by T. Honderich (Oxford, Oxford University Press, 2005) concerning the work of the Romantics, and Keller, *Environmental Ethics: The Big Questions*, and K. McPhail and D. Walters, *Accounting and Business Ethics: An Introduction* (London: Routledge, 2009) on Rousseau.

40 See Keller, *Environmental Ethics: The Big Questions*.

41 See Hinchliffe, *Geographies of Nature: Societies, Environments, Ecologies*, 28–33.

Only because we humans insist on analysing it do we lose its wonder. In developing self-awareness we developed self-interest and began the destruction of anything that stood in our way. We now claim this destruction is "natural" and like the Victorian nouveau riche interpret the behaviour of others to suit our purpose. Thus we implicitly accept that ultimately only the powerful have any right to survive and that all else, all other life, is simply there to support them.

Meanwhile *Bombus terrestris* continues to play her part in the world, contributing to the diverse array of life. I doubt she thinks of competition or even about survival, she is content to play her part. As a stakeholder I imagine all she would ask is for space to get on with her life, a little consideration before we act would be sufficient.

Does she have a lesson for us humans? Perhaps we need to live more in the moment, enjoy what we do and the sharing of life. Because of our special need, our self-awareness, we may need to consider the future. In doing so we might question where the competition concept is taking us and whether we need to change direction. Further perhaps we can again learn to enjoy the wealth in Nature that surrounds us. This is not to deny human achievement in art and literature, or even in science and technology. But what we should do is ask where these fit into the Nature of things, how do they enhance that which is different? We should not just press ahead in pursuit of personal gain.

Bibliography

Antonacopoulou, E. P. and J. Meric. "A Critique of Stake-Holder Theory: Management Science or Sophisticated Ideology of Control?" *Corporate Governance*, 5(2) (2005): 22–33.

Bebbington, J. and C. Larrinaga. "Accounting and Sustainable Development: An Exploration". *Accounting, Organizations and Society* 39 (2014): 395–413.

Bebbington, J., J. Brown, B. Frame and I. Thomson. "Theorizing Engagement: The Potential of a Critical Dialogic Approach". *Accounting, Auditing and Accountability* 20 (3) (2007): 356–81.

Belshaw, C. *Environmental Philosophy: Reason, Nature and Human Concern*. Chesham: Acumen Publishing, 2001.

Benton, T. *Bumblebees*. Fulham: Collins, 2006.

Bingham, N., A. Blowers and C. Belshaw. *Contested Environments*, Chichester, Wiley, 2003.

Bookchin, M. "Social Ecology versus Deep Ecology". *Green Perspectives, Newsletter of the Green Program Project*, 4/5 (1987).

Burritt, R. L. "Environmental Performance Accountability: Planet, People, Profits". *Accounting, Auditing and Accountability Journal*, 25(2) (2012): 370–405.

Castree, N. *Nature*. London: Routledge, 2005.

Curry, P. *Ecological Ethics: An Introduction*, Cambridge, Polity Press, 2006.

Darwin, C. *The Origin of Species*, Oxford, Oxford University Press, 1998.

Deegan, C. and J. Unerman. *Financial Accounting Theory*, Maidenhead, McGraw-Hill Education, 2006.

Desjardins, J. R. *Environmental Ethics: An Introduction to Environmental Philosphy*. Belmont: Wadsworth Publishing, 1993.

Desjardins, J. R. *Business, Ethics, and the Environment*. Upper Saddle River, New Jersey: Pearson Education, 2007.
Devall, B. and G. Sessions. *Deep Ecology: Living As If Nature Mattered*. Layton: Gibbs Smith, 2007.
Donaldson, T. and L.E. Preston. "The Stakeholder Theory of the Corporation: Concepts, Evidence and Implications". *The Academy of Management Review* 20(1) (1995): 65–91.
Drengson, A. and B. Devall. "The Deep Ecology Movement: Origins, Development & Future Prospects". *The Trumpeter* 26 (2) (2010): 48–69.
Foucault, M. *The Archaeology of Knowledge*, London: Routledge, 1972.
Foucault, M. *The Order of Things*. Abingdon: Routledge, 2002.
Fox, W. "Deep Ecology: A New Philosophy of Our Time". *The Ecologist* 14 (1984): 194–200.
Fox, W. "On Guiding Stars of Deep Ecology". *The Ecologist* 16 (1984): 203–4.
French, W. C. "Against Biospherical Egalitarianism". *Environmental Ethics* 17 (1995): 39–57.
Goodpaster, K. E. "On Being Morally Considerable". *Journal of Philosophy* 75 (6) (1978): 308–25.
Gray, R. and R. Laughlin. "It Was 20 Years Ago Today: Sgt Pepper, Green Accounting and the Blue Meanies". *Accounting, Auditing and Accountability Journal* 25(2) (2012): 228–55.
Gray, R., C. A. Adams and D. Owen. *Accountability, Social Responsibility and Sustainability: Accounting for Society and the Environment*. Harlow, Pearson Education, 2014.
Guha, R. "Radical American Environmentalism and Wilderness Preservation: A Third World Critique". *Environmental Ethics* 11(1) (1989): 71–83.
Hinchliffe, S. *Geographies of Nature: Societies, Environments, Ecologies*. London: Sage, 2007.
Jung, C. G. "Foreword". In *An Introduction to Zen Buddhism*, edited by D. T. Suzuki. London: Rider.
Keller, D. R. *Environmental Ethics: The Big Questions*. Chichester, Wiley-Blackwell, 2009.
Lewis, C. S. *The Discarded Image: An Introduction to Medieval and Renaissance Literature*, Cambridge, Cambridge University Press, 1964.
McPhail, K. and D. Walters. *Accounting and Business Ethics: An Introduction*, London, Routledge, 2009.
McIntosh, A. *Soil and Soul: People versus Corporate Power*. London: Aurum Press, 2004.
Naess, A. "The Shallow and the Deep, Long-Range Ecology Movements: A Summary". *Inquiry* 16 (1973): 95–100.
Naess, A. "Intuition, Intrinsic Value, and Deep Ecology". *The Ecologist* 14 (1984): 201–3.
Newton, L. H. *Ethics and Sustainability: Sustainable Development and the Moral Life*. Upper Saddle River, NJ: Prentice-Hall, 2003.
Pepper, D. *Modern Environmentalism: An Introduction*. London: Routledge, 1996.
Quinton, The Rt Hon. Lord. "Romanticism, philosophical". In *The Oxford Companion to Philosophy*, edited by T. Honderich. Oxford, Oxford University Press, 2005.
Singer, P. *Practical Ethics*. Cambridge: Cambridge University Press, 1993.
Sivaraksa, S. *Conflict, Culture, Change Engaged Buddhism in a Globalizing World*. Somerville: Wisdom Publications, 2005.
Suzuki, D. T. *An Introduction to Zen Buddhism*. London: Rider, 1991.
Thomson, I. and J. Bebbington. "Social and Environmental Reporting in the UK: A Pedagogic Evaluation". *Critical Perspectives on Accounting* 16 (2005): 507–33.
Tilt, C. A. "External Stakeholders' Perspectives on Sustainability Reporting". In *Sustainability Accounting and Accountability*, edited by J. Unerman, J. Bebbington and B. O'Dwyer. Abingdon: Routledge, 2007.
Unerman, J. "Stakeholder Engagement and Dialogue". In *Sustainability Accounting and Accountability*, edited by J. Unerman, J. Bebbington and B. O'Dwyer. Abingdon: Routledge, 2007.
Unerman, J., J. Bebbington and B. O'Dwyer. *Sustainability Accounting and Accountability*, Abingdon: Routledge, 2007.

Watson, R. "A Critique of Anti-Anthropocentric Biocentrism". *Environmental Ethics* 5 (1983): 245–56.
Weiming, T. "The Continuity of Being: Chinese Visions of Nature". In *Confucianism and Ecology: The Interrelation of Heaven, Earth and Humans,* edited by M. E. Tucker and J. Berthrong, 105–22. Cambridge, MA: Harvard University Press, 1998.
Witoszek, N. and A. Brennan. *Philosophical Dialogues: Arne Naess and the Progress of Ecophilosophy*. Lanham: Rowan & Littlefield Publishers, 1999.
Zimmerman, M. E. "Heidegger, Buddhism, and Deep Ecology". In *The Cambridge Companion to Heidegger,* edited by C. B. Guignon. New York: Cambridge University Press, 2006.

Part II

Investors, bees and the stock market

6

The bee and the stock market

An overview of pollinator decline and its economic and corporate significance*

Rick Stathers
Schroders, UK

Pollination plays a critical role in sustaining much of the biodiversity on Earth and, as a result, plays an essential service in maintaining the integrity and resilience of most terrestrial ecosystems and the services that these ecosystems provide.[1] There are around 150,000 flower-visiting pollinators, the majority of which are flies, butterflies, moths, bees, wasps and beetles as well as a small number of animal pollinators such as birds, bats and some non-flying mammals. One of the more common and most prolific pollinator groups is the bee, of which there are between 25,000 and 30,000 different species.[2]

* Originally published at in January 2014 by www.schroders.com (http://www.schroders.com/staticfiles/Schroders/Sites/global/pdf/The_Bee_and_the_Stockmarket.pdf, accessed 19 March 2016).
1 For an introduction to ecosystem services please see Schroders' 2009 report, *Ecosystem Credit Crunch*, accessed 19 March 2016, http://www.schroders.com/staticfiles/Schroders/Sites/global/pdf/Ecosystem_credit_crunch.pdf.
2 FAO, *Rapid Assessment of Pollinators' Status: A Contribution to the International Initiative for the Conservation and Sustainable Use of Pollinators* (Rome: FAO, 2008).

There is growing scientific concern about observed declines in bee numbers and the associated impact that this may have on agricultural production, not just from the perspective of tonnage produced but also from the impacts that this may have on the diversity and nutritional quality of the modern diet. In addition, the decline in pollinator numbers may not just be limited to the bee family, but it may actually be a reflection of a general decline in pollinator numbers, most of which are not as well studied as the bees.

This chapter provides an overview of the current rates of decline in pollinator numbers and some of the potential causes for it. It will then review the importance of pollinators to agriculture (bearing in mind that the 3,000 agricultural crops are only a small proportion of the estimated 240,000 flowering plants that depend on pollination services[3]) and the economic significance of pollinators. Finally, the chapter covers the current level of engagement and discussion on the topic by companies that have direct exposure to pollination services (e.g. soft drink manufacturers, food retailers, food producers, luxury goods and agrochemical companies).

Pollinator decline

In conducting the background research for this book chapter, it quickly became apparent that despite the large numbers of pollinator species, the most important pollinator group is that of the Apidae family (the bees). Bees are found on every continent in the world, except Antarctica, and in all habitats that contain insect-pollinated flowering plants. There are between 25,000 and 30,000 different members of the Apidae family, though the one that is probably most studied is *Apis mellifera*, or the European honey bee, because of its value as a source of honey and, more recently, as a commercial pollinator.

As mentioned in earlier chapters, declines in managed beehives (*Apis mellifera*) have been documented since the 1940s, with a 60% decline in the number of North American managed hives from 1940 to 2009.[4] Similar declines in managed colonies have also been documented in Europe where it is estimated that between 15 and 35% of honey bee colonies have been lost between 1985 and 2005.[5] Bee disappearance is predominantly reported in Europe and North America where little natural

3 UNEP, "Pollinator Key Issues", *Convention on Biological Diversity* (UNEP, 2009).

4 M.A. Aizen et al., "How Much Does Agriculture Depend on Pollinators? Lessons from Long-Term Trends in Crop Production", *Annals of Botany* 103(9) (2009): 1579–88, doi: 10.1093/aob/mcp076.

5 FAO, *Rapid Assessment of Pollinators' Status*.

habitat remains. The decline in managed stocks in these regions has been offset by increases in Asia, Latin America and Africa resulting in a 45% increase in managed colonies over the last 50 years[6] reflecting the increase in demand for pollination services as agricultural production has increased to keep up with the demands of a growing population.

The observance of declines in managed colonies in the US and Europe has increased in the last decade or so with the spread of colony collapse disorder (CCD). CCD is the rapid loss of a colony of its adult bee population; no dead bees are found inside or in close proximity to the colony. This term was originally coined in the US in 2006/2007 when US beekeepers reported an average loss of 38% of colonies during this period (significantly higher than the background rate of 10% loss). Most recently the US Department of Agriculture reported a 42.1% annual decline (Aril 2014 to April 2015), up from 34.2% for 2013–2014.[7]. In 2011, CCD was also observed in China and Japan.[8]

It is not just in honey bee populations that declines have been recorded. A study by the University of Illinois into the relative abundance of eight North American bumblebee species found that four of the sampled species showed declines of up to 96% and their geographic ranges had contracted by between 23% and 87% within the past two decades. While in the UK, 3 of 25 British species of bumblebee are extinct and 50% are in serious decline (often by up to 70%) since the 1970s.[9]

There is limited information available about pollinators other than honey bees, although it is estimated that three-quarters of pollinator species have declined by more than a third in the last decade.[10] (The Food and Agriculture Organisation [FAO] states that in the UK, half of British pollinators have disappeared from over 20% of their range, with a quarter declining by more than 50%.)[11] It should also be noted that data collection on pollinator numbers is more a reflection of the concentration of specialists to gather the data rather than a reflection of the zones of greatest concern (hence the higher number of studies in North America and Europe). Pollinator decline has been reported in at least one region or country on every continent except Antarctica. To date, however, there has been no global assessment of the changes in distribution and levels of pollination services, but the decline in numbers of key species should be cause for concern.

6 UNEP, "Global Honey Bee Colony Disorders and Other Threats to Insect Pollinators", *UNEP Emerging Issues* (2010).

7 K. Kaplan, *Bee Survey: Lower Winter Losses, Higher Summer Losses, Increased Total Annual Losses* (Washington, DC: United States Department of Agriculture, 2015).

8 *Environmental News Network*, "Honeybee End?" *Environmental News Network*, 11 March 2011.

9 *The Guardian*, "Bees in Free-Fall as Study Shows Sharp US Decline", *The Guardian*, 3 January 2011.

10 *The Guardian*, "Science Under Pressure as Pesticide Makers Face MPs over Bee Threat", *The Guardian*, 28 November 2012.

11 FAO, *Rapid Assessment of Pollinators' Status*.

Causes of decline

It is clear that pollinator decline cannot be attributed solely to one individual factor, but to a number of stressors. What's more, the exposure to one of these stressors will decrease the resilience of populations to other stressors. These stressors include habitat loss, predators and pesticide use.

Habitat loss

Pollinators require foraging, nesting, reproduction and shelter resources from their environment and the loss of one of these resources will affect pollinator numbers. Evidence shows that the diversity of wild bees has declined greatly over much of Western Europe owing mostly to natural habitat destruction.[12] An Anglo-Dutch study found that since the 1980s there has been a 70% decline in key wildflowers,[13] and there is evidence to suggest that only farms located near natural habitats are able to sustain communities of pollinators sufficient to provide the necessary pollination services.

Several of the companies we spoke to raised the concept of "green deserts". These are areas where intensification of agriculture has resulted in large areas of land being converted to extensive monoculture systems, which will provide a food bonanza for pollinators during the short period of flowering for the entire monoculture, but a desert thereafter, compounding the impacts of natural habitat loss. This can occur as the unintended consequence of national or regional policies, such as the EU's renewable fuel policies which have significantly increased the cultivated crop area given over to biofuel crops, such as oil seed rape.

Predators

Parasitic mites, fungal parasites and viral diseases are all recognized as having an impact on pollinator numbers. However, it has been the spread of one particular mite, the varroa mite, which has been blamed most for the decline of pollinator numbers, and more specifically for CCD. The varroa mite has spread from Asia to the rest of the world over the last 50 years.

Pesticide use

In April 2013 the European Commission voted to ban the use of a group of pesticides known as neonicotinoids across the whole of Europe because of concerns about their impacts on beneficial, non-target insects such as pollinators. The neonicotinoids (the world's most widely used insecticide) were banned for two years

12 Aizen et al., "How Much Does Agriculture Depend on Pollinators?"
13 *The Guardian*, "Bees in Free-fall as Study Shows Sharp US Decline".

from use on flowering crops such as corn, oil seed rape and sunflowers. Prior to this decision, neonicotinoids were banned in France, Italy, Germany and Slovenia. In addition to this ruling, a coalition of beekeepers, environmental groups and food campaigners are suing the federal Environmental Protection Agency to get the insecticides banned in the USA, and in 2014 a lawsuit was filed against Syngenta and Bayer CropScience (manufacturers of neonicotinoids) by beekeepers in Ontario alleging their pesticides have caused widespread bee deaths that have driven up costs and reduced honey production.[14] Chapter 8 analyses this case in detail.

This follows a series of scientific studies that have linked neonicotinoid use to losses in the number of queen bees (UK research found that the pesticide leads to an 85% decline in queen bee production[15]) and to increases in the incidence of "disappeared" bees. Neonicotinoids can affect navigation capacity of bees by interfering with the bee's ability to learn and remember, which subsequently impacts their ability to forage and to communicate to other hive members.

Given the financial significance of this range of pesticides to the agrochemical companies (Syngenta told us the neonicotinoid ban would reduce sales by $75 million, which is approximately 6.5% of Syngenta's worldwide sales),[16] and their global dominance as a pesticide, it is unsurprising that both advocates for the pesticide and its opponents will contest each other's findings. Disputes range from the impracticality of comparing lab tests with field tests, lack of focus on wider species range, lack of analysis of the impacts of chemicals in cocktail versus in isolation and contamination of the experimental process.

The findings from our research favour the conclusion that neonicotinoids do have an impact on bee behaviour, although they are not necessarily the sole cause of pollinator decline. Perhaps the words of Professor May Berenbaum (a leading US expert on CCD from the University of Illinois) best sums up the situation:

> There is no question that neonicotinoids are being used recklessly, for want of a better word. Fifty years of experience should have taught us that overuse of a single class of compounds is an inherently unsustainable practice, and that pre-treating seeds when pest problems might not even be present is colossally unwise. But neonicotinoids could be banned everywhere in the world, and honeybees would still have problems with pathogens, parasites, habitat degradation and overuse of just about every other class of chemical pesticide.[17]

14 *Globe and Mail*, "Beekeepers File Suit Against Pesticide Makers Syngenta and Bayer", *Globe and Mail*, 3 September 2014.

15 *BBC News*, "Pesticides Hit Queen Bee Numbers", *BBC News*, 29 March 2012.

16 Author's calculation based on Syngenta disclosure.

17 *BBC News*, "Pesticides Hit Queen Bee Numbers".

In referring to the lessons learnt over the last 50 years, Berenbaum refers back to a book, published in 1962, which was widely regarded as being instrumental in the launch of the environmental movement. As discussed in Chapter 1, Rachel Carson's *Silent Spring* documented the effects of pesticides (DDT) on the environment and especially on birds, highlighting that chemicals cannot be indiscriminately used without regard for their wider, indirect consequences on ecology and human health. As a result of her work DDT was banned from use in the US in 1972 and subsequently banned from agricultural use worldwide under the Stockholm Convention. It is therefore interesting to note that the American Bird Conservancy, perhaps in the spirit of *Silent Spring*, is claiming that neonicotinoids can also harm birds.[18]

As well as the direct impact on bees, there are also questions about the wider environmental impacts associated with the use of these pesticides. In June 2014 the conclusions of the Worldwide Integrated Assessment on the risks of neonicotinoids and fipronil (a synthesis of over 800 peer-reviewed journal articles) found that:

> the present scale of use, combined with the properties of these compounds (neonicotinoids and fipronil), has resulted in the widespread contamination of agricultural soils, freshwater resources, wetlands, non-target vegetation, estuaries and coastal marine systems which means that many organisms inhabiting these habitats are being repeatedly and chronically exposed to effective concentrations of these insecticides[19]

and that "there is a growing body of evidence that these effects pose risks to ecosystem functioning, resilience and the services and functions provided by terrestrial and aquatic ecosystems".[20] As a result exposure to neonicotinoids and fipronil can have negative effects on non-target invertebrates.

In addition to the above stressors, mention should also be made of the impacts of climate change. Climate changes at all levels (global, regional, national and local) will alter the greening, flowering and ageing cycles of plants which will impact on pollinators, multiplying the impacts of habitat loss, predators and pesticide use. Ultimately climate change could alter the natural synchronization between pollinator and plant life-cycles and over the longer term be the most serious threat to pollinator numbers.

18 *New Scientist*, "Bees to Get Day in Court", *New Scientist*, 30 March 2013.

19 J.P. van der Sluijs *et al.*, "Conclusions of the Worldwide Integrated Assessment on the Risks of Neonicotinoids and Fipronil to Biodiversity and Ecosystem Functioning, *Environmental Science and Pollution Research* 22 (2014): 148–54.

20 *Ibid.*

Importance of pollinators to agricultural production

Pollination is not essential for agricultural production and the level of dependence on pollination for different crops varies from 100% dependence (e.g. cocoa beans, kiwi fruits and melons) to no dependence (e.g. barley, rice, wheat and citrus fruits).[21] This is because some plants can self-pollinate or depend on wind pollination as opposed to animal pollination.

As a result, if we are talking about crop production in terms of quantity produced, then 60% of global food production comes from crops that do not depend on animal pollination (e.g. wheat, maize and rice) and 35% of crop production shows improvement (either in yield or quality) as a result of pollination.[22] However, within this 35% resides a far higher diversity of crops from fruits to vegetables and from oil-crops to spices which mostly have some degree of dependence on pollination. It is this diversity of crops that has a far greater importance to human nutrition as they add the proteins, vitamins and minerals to our diets, which we would not be able to get through eating cereals alone. A recent study on the relationship between nutrition and pollination found that deficiencies in micronutrients are three times as prevalent where production of micronutrients is heavily dependent on pollinators, such as sub-Saharan Africa, India and the Middle East.[23] Therefore, a focus simply on the total mass of production as dependent on pollination is misleading.

FAO states that 70% of the 115 most produced crops, including fruits and oil seeds, are animal pollinated,[24] and it is this 70% that account for 35% of agricultural production by mass mentioned in the previous paragraph. However, as we have discussed earlier, the dependence of these crops on animal pollination varies so it is misleading to say that, in the absence of pollinators, 35% of agricultural yields (by mass) would be lost. It is, though, clear that the quality and yield themselves would be affected to some degree.

Impact of pollinators on quality and yield

Pollinators have a significant influence on the quality of fruits and vegetables. For example, the more ovaries that are pollinated the higher the quality of the fruit (e.g. a perfect strawberry needs every single ovary to be pollinated).

21 FAOSTAT, "Array for the Economic Valuation of the Contribution of Insect Pollination to Agriculture", 2013, http://faostat3.fao.org/home/E.

22 FAO, *Rapid Assessment of Pollinators' Status.*

23 Rebecca Chaplin-Kramer *et al.*, "Global Malnutrition Overlaps with Pollinator-Dependant Micronutrient Production", *Proceedings of the Royal Society B* 281 (2014), doi: 10.1098/rspb.2014.1799.

24 UNEP, "Global Honey Bee Colony Disorders and Other Threats to Insect Pollinators".

One study assessed the impact of pollination on global agricultural production. It selected crops which represented around 99% of global food production (94.5% of production came from 57 leading single crops and 4.5% from five commodities which included fruit and vegetables). The study found that of the 57 single crops, 39 showed improvements as a result of pollination and 48 of the 67 commodity crops also demonstrated increased yields with pollination. The majority of crops would experience production loss owing to pollinator decline.[25] Another study found that 70% of 1,330 tropical crops showed increased fruit and seed quality and quantity as a result of animal pollination, as did 85% of 264 crops cultivated in Europe.[26] There is clearly a relationship between yield improvements and animal pollination but, as demonstrated by the table in Appendix 6.1, the significance of this link varies depending on the crop in question.

Crop dependence on pollination

As we have already mentioned, animal pollination is only one form of pollination with wind and self-pollination also playing a role. The majority of the world's agricultural production (by mass) is wind or self-pollinated (e.g. carbohydrate crops such as wheat, rice and corn), but the majority of agricultural crops by diversity are affected by animal pollination. The level of dependence will affect crop yield, with some crops being 100% reliant on animal pollination, while in others the dependence on pollination for yield improvements could be as low as 5% (please refer to Appendix 6.1 for a list of the dependence ratios of some of the world's main crops). In addition to its yield benefits (e.g. a well-pollinated flower will contain more seed which will deliver an enhanced capacity to germinate leading to bigger, better shaped fruits), improved pollination can also reduce the time between flowering and fruit set which in turn helps to reduce a crop's exposure to pests, disease, bad weather, agrochemicals and water demand.[27]

There is still debate about the impact of pollinator decline on global crop yields; if taken in terms of mass, then studies suggest that at the global level, if pollinators were to disappear, this would only reduce agricultural production by 4% to 6%. This is according to an article published in the *New Scientist*[28] which was supported by

25 A.M. Klein *et al.*, "Importance of Pollinators in Changing Landscapes for World Crops", *Proceedings of the Royal Society B* 274 (2007), accessed 7 March 2016, doi: 10.1098/rspb.2006.3721.

26 Aizen *et al.*, "How Much Does Agriculture Depend on Pollinators?"

27 UNEP, "Global Honey Bee Colony Disorders and Other Threats to Insect Pollinators".

28 *New Scientist*, "The Truth About Honeybees", *New Scientist*, 24 October 2009.

the findings of a study published in the *Annals of Botany* stating that the production deficit in the absence of pollinators would be 3% to 5% in the developed world and 8% in the developing world.[29] However, the latter article notes that this does not measure the other values, such as nutritional or economic value, that would be impacted. With regards to the economic value argument it is estimated that the production value of 1 tonne of pollinator-dependent crop is five times greater than a crop category that is not dependent on animal pollination.[30]

Global agricultural production has increased by about 140% between 1961 and 2006, though production in the developed world has levelled off since the 1980s, while production in developing countries has shown a constant rate of increase over the last 50 years. However, the area that is given over to dependent crops has been growing faster than that allocated to non-dependent crops, implying a greater demand on pollinator services.[31] This was supported by a recent piece of research which states that in Europe the demand for pollination services from increasing areas of pollinator dependent crops is rising at a rate that is 4.9 times faster than available pollinator stocks.[32] At the same time, research has indicated that yields of pollinator-dependent crops have levelled off (and this has been compensated for by expanding the cultivated area given to these crops) while yields from pollinator-independent crops have continued to increase.

The developing world accounts for around two-thirds of agricultural production and cultivated land, and supports agricultural systems that are 50% more dependent on pollinators than those of the developed world. This means that the cultivated area needed to compensate for pollinator collapse is six times larger in the developing world than in the developed world. At the same time, research has indicated that native crop pollinators seem to be lost faster in tropical agricultural landscapes than in temperate ones. The production deficit as a result of pollinator decline has increased by 50% in the developed world (despite production levelling

29 Aizen *et al.*, "How Much Does Agriculture Depend on Pollinators?"

30 UNEP, "Global Honey Bee Colony Disorders and Other Threats to Insect Pollinators".

31 L.A. Garibaldi *et al.*, "Pollinator Shortage and Global Crop Yield". *Communicative and Integrative Biology*, January/February 2009.

32 Breeze *et al.*, "Agricultural Policies Exacerbate Honeybee Pollination Service Supply-Demand Mismatches Across Europe", *PLoS ONE* 9(2) (2014): e91459, doi: 10.1371/journal.pone.0091459.

off in the developed world) and by 62% in the developing world (cultivated area has increased by 25% between 1961 and 2006).[33]

So when discussing the impact of pollinator decline on global agricultural systems one has to differentiate between a focus on the mass of global crop yield or a focus on the diversity of plants affected and their nutritional and economic value. Currently, yields in dependent crops have started to decline which may have been compensated for by an increase in cultivated area for dependent crops. While there is currently no clear evidence of negative impacts of pollinator decline on global crop yields, the increasing demands for pollination services (due to increasing cultivated area and changing diets) at a time when pollinator numbers are in decline imply that this could threaten future crop yields, with broad economic and nutritional implications. The following section will discuss the economic value of pollination services and their decline.

Economic impacts of pollinator decline

There are a number of studies which have attempted to quantify the value of pollination services to agricultural production. Typically these have focused on calculating the losses that could be accrued as a result of pollination service decline, although they could also be calculated by focusing on the costs of mitigation. The loss calculation approach (deficit approach) requires the quantification of the decrease in productivity measures (e.g. yield, quality) in the absence of pollination, while the second approach (or compensation method) requires a calculation of the inputs needed to offset pollination declines (e.g. increasing cultivated areas, increasing the use of commercial bees or the labour costs for hand pollination). The dominant methods used in the reports read as background to this paper have used the deficit approach.

Any economic calculation should calculate not only the production yield changes but also the price changes which will impact the consumer surplus, because typically if yields decrease then prices increase and less food will be traded which will cause the consumer surplus to decline.

In terms of the economic impacts of pollinator decline there are a wide range of figures presented in the research, some of which relate to the economic loss from managed bee decline, others simply from pollinator decline. The most common figures are given at a global level though there are some figures which also provide an indication of the economic impact in the UK and the USA. At the global level, predictions range from £26 billion to £141 billion per year, with most estimates clustering around £130 billion per year (approximately 10% of the economic value of agricultural production). In the UK, two media sources quote a figure of

33 Aizen *et al.*, "How Much Does Agriculture Depend on Pollinators?"

around £430 million per year at the national level and in the USA the figure ranges from £3.9 billion to £9.4 billion per year at the national level. The figure of €153 billion (or £129 billion) is used by both the FAO and the United Nations Environment Programme.[34]

As mentioned above these figures tend to only focus on the deficit calculation, and a more complete economic analysis would also include the consumer surplus figure. Table 6.1 is adapted from the website for the statistics division of the FAO. It not only demonstrates the different levels of dependence of a selection of crops but also how this affects consumer surplus based on two different elasticities of demand. FAO estimate that the global value of pollination services was €153 billion in 2005 (around £129 billion) and that the associated consumer surplus loss would be €190 billion to €310 billion (£160 billion to £261 billion).

These figures should not be considered precise, but should be taken as an indicator of the magnitude of the value of pollination services which are currently not being recognized by the market as providing a service of economic value. In addition, these figures have tended to focus only on agricultural production of crops for human consumption; they do not take into account the importance of pollination services in the production of seed used for planting (important for vegetable and forage crops) or of non-food crops—wild flowers and all services that natural flora and fauna provide to agriculture and society as a whole (e.g. the pharmaceutical, perfume and biofuel industries all have exposure to pollination services within their supply chains). Furthermore they do not consider that pollination service decline will, more than likely, impact a multitude of crops at once, further exacerbating the economic and consumer surplus losses, as opposed to individual crops.

34 Author's calculation of mean figures and UNEP, "Global Honey Bee Colony Disorders and Other Threats to Insect Pollinators".

Table 6.1 **A hypothetical example of the methodology for calculating the economic valuation of the contribution of insect pollination to agriculture and the associated impacts on consumer losses***

Sources: Adapted from FAOSTAT's "Array for the Economic Valuation of the Contribution of Insect Pollination to Agriculture and Impact on Welfare";[1] http://faostat.org and [2] Klein *et al*. "Importance of Pollinators in Changing Landscapes for World Crops", 2007.

Crop common name[1]	Crop category following FAO[1]	Dependence on animal pollination[2]	Min[2]	Max[2]	Mean (D)[2]	Producer price per metric tonne (US$/metric tonne	Production (metric tonne)	Total value of crop (TVC) Price* production (US$)	Economic value of insect pollinators (EVIP) TVC*D (US$)	Ratio of vulnerability (RV) EVIP/TVC	Consumer surplus loss (CSL) with elasticity =	
											-0.8 (US$)	-1.2 (US$)
Barley	Cereals	No increase	0	0	0	139	10	1,390	0	0%	0	0
Apples	Fruits	Great	0.4	0.9	0.65	452	10	4,520	2,938	65%	5,280	4,280
Rapeseed	Oilcrops	Modest	0.1	0.4	0.25	385	10	3,850	363	25%	1140	1076
Beans, dry	Pulse	Little	0	0.1	0.05	515	10	5,150	258	5%	266	263
Potatoes	Roots and tubers	Increase (breeding)	-	-	-	137	10	1,370	0	0%	-	-
Vanilla	Spices	Essential	0.9	1	0.95	1,003	10	10,030	3,529	95%	41,151	22,604
Cocoa beans	Stimulant crops	Essential	0.9	1	0.95	1,225	10	12,250	11,638	95%	50,260	27,607
Sugar beet	Sugar crops	No increase	0	0	0	177	10	1,770	0	0%	0	0
Almonds, with shell	Treenuts	Great	0.4	0.9	0.65	1,269	10	12,690	6,249	65%	14,624	12,017
Watermelons	Vegetables	Essential	0.9	1	0.95	468	10	4,680	4,446	95%	19,201	10,547

*We have used average prices for crop category and a generic production unit of 10 tonnes. Using the full data set it is possible to calculate the economic value at national and global levels for all crops.

A question of scale

The majority of studies have aimed to quantify the global impact of pollinator decline and have generally concluded that the impact would be around 10% of the total value of human food production. However, a global analysis disguises the impacts of pollination decline at national and local levels or on specific sectors (e.g. the production of cocoa beans is 95% dependent on pollination, and so chocolate confectionary production has a very real, material exposure to this issue).

At the national level the contribution of agriculture to GDP varies immensely. World Bank figures indicate that, in 2011, it can range from 57% in Sierra Leone to less than 0.1% in Singapore,[35] with developing economies showing a greater dependence on agriculture and hence a greater exposure to the risks of pollinator decline. In Latin America, Africa and Asia an average of 40% of the cultivated land is planted with crops with some dependence on pollinators,[36] although this is considered to be a low estimate by FAO. In addition, some countries may have a material exposure to one crop (e.g. coffee) as a significant contributor to its GDP, magnifying the impact that pollinator decline could have.

At the local level, there are a couple of examples in the literature demonstrating the importance of pollination services (and of the supporting natural habitat) to agricultural returns. A study of a Canadian apple orchard found that good pollination increased the number of seeds per apple, which in turn led to larger and better formed apples. These improved apples were estimated to provide marginal returns of about 5% to 6%, or CAN$250/hectare, compared with insufficiently pollinated orchards. The paper also noted that farmers were also willing to pay for commercial pollination services when their crops may be getting adequate pollination from wild pollinators, yet farmers didn't pay to secure these services (e.g. through habitat maintenance) though they clearly placed a value on pollination.[37]

This reference to the value of natural habitat for wild pollinators was echoed in a study of the commercial importance of pollination services to coffee production in Costa Rica. The study found that yields of coffee plants near to natural forests increased by 20.8% due to wild pollinators, giving an annual surplus of $62,000 generated by these forest patches (around 7% of the annual income of the plantation).[38] This indicates that the value of intact forests (according to the author) was far greater than the expected annual earnings from the same land if it were deforested and converted to agricultural use (not to mention the other

35 http://data.worldbank.org/indicator/NV.AGR.TOTL.ZS, accessed 7 March 2016.

36 FAO, *Rapid Assessment of Pollinators' Status.*

37 L. Hein, "The Economic Value of Pollination Service: A Review Across Scales", *The Open Ecology Journal* 2 (2009): 74–82.

38 USDA, *Bee Survey: Lower Winter Losses, Higher Summer Losses, Increased Total Annual Losses*

ecosystem services provided by the forest such as pest control, carbon sequestration, soil stabilization and water regulation).[39] A study by Michigan State University found that establishing habitat that attracts and supports wild bees can pay for itself in four years or less through increased yields in adjacent fields.[40] This supports calls for natural areas to be managed not just for the goods that they provide, but also the services they provide, and that agricultural intensification maybe having a long-term negative impact on yields through the decline in the provision of wild habitats for pollinators.

Stock level relevance of pollinator decline

So far this chapter has illustrated that there is mounting evidence of a global decline in pollinators around the world as a result of various factors. The impact of this decline depends on what is being assessed; at the global level the impact of pollinator decline is limited though at the micro level the impact can be quite significant. We have also seen that yields and quality do vary as a result of pollinator decline, affecting the productivity of the majority of crop species, which can be supplemented by increasing cultivated area at the expense of natural habitat and through the use of commercial pollination services. However, there is also evidence supporting the economic benefits of maintaining natural habitats.

I do not believe that this issue is currently having a material impact on the global economy, though its relevance for national economies will vary depending on the significance of agriculture to a nation's economy. However, I do believe that with increasing food demand (the FAO projects that food production will need to increase by 70% by 2050 to meet demand) and decreasing pollinator numbers this is an issue that needs to receive more attention. Furthermore, future academic, non-governmental, governmental and corporate analysis should broaden in geographic scope and species assessed (recognizing that declines in bee numbers may simply be a reflection of what may be happening in other pollinator species), as well as assessing the potential impacts on other non-target organisms which are essential for ecosystem function.

At the stock level we have highlighted the sectors and companies that we believe would have exposure to this issue (see Table 6.2).

39 FAO, *Rapid Assessment of Pollinators' Status.*
40 Michigan State University AgBioResearch, *Attracting Wild Bees to Farms is a Good Insurance Policy* (East Lansing, MI: Michigan State University, 2014).

Table 6.2 **Companies assessed for coverage of pollinator decline within publicly available information**

Sector	Companies	Rationale
Agrochemicals	• Syngenta • Bayer	Producers of neonicotinoids pesticides which are blamed for being one of the stressors responsible for pollinator decline. These companies also have self-interest in ensuring the long-term survival of pollinators to ensure the economic success of their main client base.
Food retailers	• Tesco • Marks & Spencer • J Sainsbury • Metro Group • Carrefour • WM Morrison • Ahold • Colruyt	Food retailers have exposure to most if not all agricultural crops, whether through the provision of fresh fruit and vegetables, processed meals, fruit juices or confectionary. This means they should be monitoring and managing any medium- to long-term risks in their supply chain.
Food producers	• Nestlé • Unilever • Kraft • Danone • ABF • Hershey	These companies have differing exposure to pollinators because of different product ranges. Some produce confectionary (cocoa exposure) and dairy products with natural fruit flavourings and others produce fruit juices. In addition some of the oil crops will be used in marinades and sauces made by these companies.
Luxury goods	• LVMH • L'Oreal • Christian Dior • Hermes	Perfumes use different parts of a plant (among other sources) for their aromatic properties. The largest and most common source is from the flowers and blossom of a plant, which evolved aromatic properties in order to attract pollinators. These would be at risk should pollinator numbers decline.
Beverages	• Coca-Cola • Dr Pepper • Snapple Group • Britvic • Pepsico	Most of the companies in this sector have exposure to the issue through some of their products, whether flavoured fizzy drinks or concentrated fruit drinks to fresh fruit juices.

This research has reviewed the publicly available information (annual reports and accounts, websites and corporate responsibility reports) of the companies shown in Table 6.2, searching for key topics such as bees, pollinators, pesticides, biodiversity and sustainable agriculture in order to develop a snapshot of the level of discussion on this topic.

Almost a third of the 25 companies made explicit reference to their long-term success being dependent on the functioning of a healthy ecosystem, with 20% also recognizing their role in having to preserve and restore ecosystem services and biodiversity. Half (50%) of the companies referred to pesticide usage, though this predominantly focused on concerns for human health from the impact of pesticide residues on food products and on sustainable pesticide use. However, recognizing that discussion on bees may be too specific, we also searched for references to sustainable agriculture and biodiversity which was covered by almost a third of the companies assessed.

There is clearly recognition of the importance of healthy, functioning ecosystem services to most of the companies within this research. Nestlé notes that: "to ensure our long-term success we must ensure that biodiversity and ecosystem services continue to flourish". Comments like this demonstrate the increasing recognition of the services that ecosystems play in the successful execution of a business strategy. Interestingly, when it came to the specific topic of pollinator decline, both the agrochemical companies covered the topic, and it was also explicitly covered by two supermarket chains and one luxury good provider.

Scientific awareness of pollinator decline has been around for some time now, so it is disappointing not to discover a higher level of discussion on the topic. However, we have to recognize that around 25% of the companies researched have very poor levels of disclosure on corporate responsibility in general. So it is therefore encouraging to see the emerging recognition of the role that healthy ecosystem services play in the long-term success of the business, and also to see that there is some recognition of the role of pollinators within this.

The following companies (from Table 6.2) responded to our request for a discussion on this topic: **LVMH**, **Syngenta**, **Britvic**, **Morrisons**, **Sainsbury's** and **Marks & Spencer**. These meetings served to reinforce the findings of our literature review, though also to provide confidence that more work is being dedicated to this area as well as to broader biodiversity concerns. Broadly speaking, no company was able to directly link pollinator decline to an impact on operating costs, though most would agree with M&S which said that there clearly was a cost but its value would only be recognized when the service was completely lost (M&S also said that the costs and quality of hand pollination made this, as an option, prohibitive). Britvic, perhaps with a touch of irony, also recognized that if quality decreases with declining pollinator numbers then this may actually be beneficial to its business as there would be larger quantities of low-value crop available in its supply chain.

In terms of recognizing the causes of pollinator decline, all companies engaged with recognized that habitat loss was a major driver and many were taking advice on pesticide use and providing guidance on this in their supply chains. Most of the specific action to tackle pollinator decline was limited to recently updated guidelines for suppliers and the establishment of pilot studies within supply chains to assess the outcomes of adopting different production practices, reflecting the relative newness of this issue.

However, LVMH provided a very clear case study of a longer-term involvement in managing wild bee habitat in recognition of its economic value to the business. LVMH uses the honey from a remote, pesticide and predator free, island population of black bees off the coast of Brittany in one of its brand's cosmetic creams. LVMH recognizes the importance of keeping this population pesticide and predator free not just for its product (where the purity of a product applied to the body is a value) but also as a reference population for future scientific study. The value it places on this is so great that it has used its legal team to help protect the pristine state of the environment in which the black bees live, while also supporting various bee conservation organizations.

Perhaps less encouraging was the finding that, because of concerns about competition rules, discussion *between* companies is limited. This means any industry responses will have to be coordinated through external bodies. It is clear, though, that this issue has numerous stakeholders in different industrial sectors and it is here that investors, as universal owners, have a role to play through the recognition that a system-wide response is needed to tackle the issue and investors will have an exposure to many of the actors involved in such a response.

Conclusion

The aim of this research was to determine the seriousness of the issue of pollinator decline and the potential ramifications for investable stocks. We have found that this is clearly an issue with economic and nutritional significance, but that we were only able to gather a few current examples of pollinator decline directly affecting yields. However, the research would suggest that should pollinator numbers continue to decline then this is surely an eventuality.

There is also little to suggest that it is visibly impacting operating costs for the companies we analysed. That said, all companies recognize it as an issue, acknowledging that pollinators provide an economic service which is only likely to be recognized should the service cease or be seriously reduced. Efforts, by corporates, to address the issue are in the early stages and predominantly focus on pilot projects with a small group of suppliers.

However, the broad scope of pollinator-dependent products—covering sectors from food retailers to cotton manufacturers—means that this is a systemic issue and one to which investors have a unique exposure as universal owners. It is too early to reflect a consideration of pollinator decline in valuations but it is not too early to engage with companies throughout the investment universe on this issue in order to develop a response to halt and reverse this decline in a valuable part of our economy before its consequences become more significant. This will cover a broad range of companies from water utilities with large land reserves, food retailers with direct exposure in their supply chain, agrochemical companies with product exposure, cosmetic firms and everything in between.

Bibliography

Aizen, M.A., L.A. Garibaldi, S.A. Cunningham and A.M. Klein. "How Much Does Agriculture Depend on Pollinators? Lessons from Long-Term Trends in Crop Production". *Annals of Botany* 103(9) (2009): 1579–88. doi: 10.1093/aob/mcp076.

BBC News. "Pesticides Hit Queen Bee Numbers". *BBC News*, 29 March 2012.

Breeze, T.D., Bernard E. Vaissière, Riccardo Bommarco, Theodora Petanidou, Nicos Seraphides, Lajos Kozák, Jeroen Scheper, Jacobus C. Biesmeijer, David Kleijn, Steen Gyldenkærne, Marco Moretti, Andrea Holzschuh, Ingolf Steffan-Dewenter, Jane C. Stout, Meelis Pärtel, Martin Zobel, and Simon G. Potts. "Agricultural Policies Exacerbate Honeybee Pollination Service Supply-Demand Mismatches Across Europe". *PLoS ONE* 9(2) (2014): e91459. doi: 10.1371/journal.pone.0091459.

Chaplin-Kramer, Rebecca, Emily Dombeck, James Gerber, Katherine A. Knuth, Nathaniel D. Mueller, Megan Mueller, Guy Ziv, and Alexandra-Maria Klein. "Global Malnutrition Overlaps with Pollinator-Dependant Micronutrient Production". *Proceedings of the Royal Society B* 281 (2014). doi: 10.1098/rspb.2014.1799.

Environmental News Network. "Honeybee End?". *Environmental News Network*, 11 March 2011.

FAO. *Rapid Assessment of Pollinators' Status. A Contribution to the International Initiative for the Conservation and Sustainable Use of Pollinators*. Rome: FAO, 2008.

FAOSTAT. "Array for the Economic Valuation of the Contribution of Insect Pollination to Agriculture", 2013. http://faostat3.fao.org/home/E.

Garibaldi, L.A., M.A. Aizen, S.A. Cunningham and A.M. Klein. "Pollinator Shortage and Global Crop Yield". *Communicative and Integrative Biology*, January/February 2009.

Globe and Mail. "Beekeepers File Suit Against Pesticide Makers Syngenta and Bayer". *Globe and Mail*, 3 September 2014.

Guardian. "Bees in Free-Fall as Study Shows Sharp US Decline". *The Guardian*, 3 January 2011.

Guardian. "Science Under Pressure as Pesticide Makers Face MPs over Bee Threat". *The Guardian*, 28 November 2012.

Hein, L. "The Economic Value of Pollination Service: A Review Across Scales". *The Open Ecology Journal* 2 (2009): 74–82.

Kaplan, K. *Bee Survey: Lower Winter Losses, Higher Summer Losses, Increased Total Annual Losses*. Washington, DC: United States Department of Agriculture, 2015.

Klein, A.M., B.E. Vaissiere, J.H. Cane, I. Steffan-Dewenter, S.A. Cunningham, C. Kremen, and T. Tscharntke. "Importance of Pollinators in Changing Landscapes for World Crops". *Proceedings of the Royal Society B* 274 (2007). Accessed 7 March 2016. doi: 10.1098/rspb.2006.3721.

Michigan State University AgBioResearch. *Attracting Wild Bees to Farms is a Good Insurance Policy*. East Lansing, MI: Michigan State University, 2014.

New Scientist. "Bees to Get Day in Court". *New Scientist*, 30 March 2013.

New Scientist. "The Truth About Honeybees". *New Scientist*, 24 October 2009.

Schroders. *Ecosystem Credit Crunch*. London: Schroders, 2009. Accessed 19 March 2016. http://www.schroders.com/staticfiles/Schroders/Sites/global/pdf/Ecosystem_credit_crunch.pdf.

van der Sluijs, J.P., V. Amaral-Rogers, L. P. Belzunces, M. F. I. J. Bijleveld van Lexmond, J-M. Bonmatin, M. Chagnon, C. A. Downs, L. Furlan, D. W. Gibbons and 21 more. Conclusions of the Worldwide Integrated Assessment on the Risks of Neonicotinoids and Fipronil to Biodiversity and Ecosystem Functioning, *Environmental Science and Pollution Research* 22 (2014): 148–54.

UNEP. "Global Honey Bee Colony Disorders and Other Threats to Insect Pollinators". *UNEP Emerging Issues* 2010.

UNEP. "Pollinator Key Issues". *Convention on Biological Diversity*. UNEP, 2009.

Appendix 6.1

Dependence of crops on insect pollination

			Dependence ratio		
Crop common name[1]	Crop category following FAO[1]	Dependence on animal pollination[2]	Min[2]	Max.[2]	Mean (D)[2]
Barley	Cereals	No increase	0	0	0
Hops	Cereals	No increase	0	0	0
Maize	Cereals	No increase	0	0	0
Oats	Cereals	No increase	0	0	0
Rice, paddy	Cereals	No increase	0	0	0
Rye	Cereals	No increase	0	0	0
Sorghum	Cereals	No increase	0	0	0
Wheat	Cereals	No increase	0	0	0
Apples	Fruits	Great	0.4	0.9	0.65
Apricots	Fruits	Great	0.4	0.9	0.65
Avocados	Fruits	Great	0.4	0.9	0.65
Bananas	Fruits	Increase - breeding	-	-	-
Cashewapple	Fruits	Great	0.4	0.9	0.65
Cherries	Fruits	Great	0.4	0.9	0.65
Cranberries	Fruits	Great	0.4	0.9	0.65
Currants	Fruits	Modest	0.1	0.4	0.25
Grapes	Fruits	No increase	0	0	0
Kiwi fruit	Fruits	Essential	0.9	1.0	0.95
Lemons and limes	Fruits	Little	0	0.1	0.05
Oranges	Fruits	Little	0	0.1	0.05
Peaches and nectarines	Fruits	Great	0.4	0.9	0.65
Pears	Fruits	Great	0.4	0.9	0.65
Pineapple	Fruits	Increase - breeding	-	-	-
Plums and sloes	Fruits	Great	0.4	0.9	0.65
Raspberries	Fruits	Great	0.4	0.9	0.65
Strawberries	Fruits	Modest	0.1	0.4	0.25
Tangerines, mandarin, clem	Fruits	Little	0	0.1	0.05

			Dependence ratio		
Crop common name[1]	**Crop category following FAO[1]**	**Dependence on animal pollination[2]**	**Min[2]**	**Max.[2]**	**Mean (D)[2]**
Coconuts	Oilcrops	Modest	0.1	0.4	0.25
Groundnuts, with shell	Oilcrops	Little	0	0.1	0.05
Linseed	Oilcrops	Little	0	0.1	0.05
Mustard seed	Oilcrops	Modest	0.1	0.4	0.25
Oil palm fruit	Oilcrops	Little	0	0.1	0.05
Olives	Oilcrops	No increase	0	0	0
Rapeseed	Oilcrops	Modest	0.1	0.4	0.25
Seed cotton	Oilcrops	Modest	0.1	0.4	0.25
Sesame seed	Oilcrops	Modest	0.1	0.4	0.25
Soybeans	Oilcrops	Modest	0.1	0.4	0.25
Sunflower seed	Oilcrops	Modest	0.1	0.4	0.25
Beans, dry	Pulse	Little	0	0.1	0.05
Chick peas	Pulse	No increase	0	0	0
Lentils	Pulse	No increase	0	0	0
Peas, dry	Pulse	No increase	0	0	0
String beans	Pulse	Little	0	0.1	0.05
Cassava	Roots and tubers	Increase – breeding	–	–	–
Potatoes	Roots and tubers	Increase – breeding	–	–	–
Sweet potatoes	Roots and tubers	Increase – breeding	–	–	–
Chillies and peppers, dry	Spices	Little	0	0.1	0.05
Cloves	Spices	Unknown	–	–	–
Ginger	Spices	Unknown	–	–	–
Pepper (Piper spp.)	Spices	No increase	0	0	0
Vanilla	Spices	Essential	0.9	1.1	0.95
Cocoa beans	Stimulant crops	Essential	0.9	1.1	0.95
Coffee, green	Stimulant crops	Modest	0.1	0.4	0.25
Tea	Stimulant crops	No increase	0	0	0
Sugar beet	Sugar crops	No increase	0	0	0
Sugar cane	Sugar crops	No increase	0	0	0

			Dependence ratio		
Crop common name[1]	**Crop category following FAO[1]**	**Dependence on animal pollination[2]**	**Min[2]**	**Max.[2]**	**Mean (D)[2]**
Almonds, with shell	Treenuts	Great	0.4	0.9	0.65
Brazil nuts, with shell	Treenuts	Essential	0.9	1.0	0.95
Cashew nuts, with shell	Treenuts	Great	0.4	0.9	0.65
Pistachios	Treenuts	No increase	0	0	0
Walnuts, with shell	Treenuts	No increase	–	–	–
Artichokes	Vegetables	Increase – seed production	–	–	–
Asparagus	Vegetables	Increase – seed production	–	–	–
Beans, green	Vegetables	Little	0	0.1	0.05
Cabbages and other brassicas	Vegetables	Increase – seed production	–	–	–
Carrots and turnips	Vegetables	Increase – seed production	–	–	–
Cauliflowers and broccoli	Vegetables	Increase – seed production	–	–	–
Chillies and peppers, green	Vegetables	Little	0	0.1	0.05
Cucumbers and gherkins	Vegetables	Great	0.4	0.9	0.65
Eggplants (aubergines)	Vegetables	Modest	0.1	0.4	0.25
Garlic	Vegetables	Increase – breeding	–	–	–
Lettuce and chicory	Vegetables	Increase seed production	–	–	–
Mushrooms and truffles	Vegetables	No increase	0	0	0
Onions (inc. shallots) green	Vegetables	Increase seed production	–	–	–
Other melons (inc. cantaloupes)	Vegetables	Essential	0.9	1.0	0.95
Peas, green	Vegetables	No increase	0	0	0
Pumpkins, squash and gourds	Vegetables	Essential	0.9	1.0	0.95
Spinach	Vegetables	No increase	0	0	0
Tomatoes	Vegetables	Little	0	0.1	0.05
Watermelons	Vegetables	Essential	0.9	1.0	0.95

Sources: Adapted from FAOSTAT's "Array for the Economic Valuation of the Contribution of Insect Pollination to Agriculture and Impact on Welfare";[1] http://faostat.org and[2] Klein *et al.* "Importance of Pollinators in Changing Landscapes for World Crops", 2007.

7

Pollinators as a portfolio risk

Making the case for investor action

Abigail Herron*
Aviva Investors, UK

This chapter explores the role and responsibilities of investors in preventing further declines in pollinators through company and policy-maker engagement and dialogue.

Responsible investment is an approach to investment that explicitly acknowledges the relevance to the investor of environmental, social and governance (ESG) factors, and the long-term health and stability of the market as a whole. It recognizes that the generation of long-term sustainable returns is dependent on stable, well-functioning and well-governed social, environmental and economic systems. It is driven by a growing recognition in the financial community that effective research, analysis and evaluation of ESG issues is a fundamental part of assessing the value and performance of an investment over the medium and longer term, and that this analysis should inform asset allocation, stock selection, portfolio construction, shareholder engagement and voting. Responsible investment requires investors and companies to take a wider view, acknowledging the full spectrum of risks and opportunities facing them, in order to allocate capital in a manner that is aligned with the short- and long-term interests of their clients and beneficiaries.[1]

In 2009, the UK Government's Chief Scientist at the time, John Beddington, warned that in 2030 we will live in a world demanding 50% more energy, 50% more food and 30% more water against a background of far greater climate variability. The challenge of this water, energy and food security nexus does not take into

* Abigail is head of engagement at Aviva Investors and writes in a personal capacity.
1 http://www.unpri.org/introducing-responsible-investment/, accessed 8 March 2016.

account pollinator loss, which has the potential to compound this looming "perfect storm".[2]

A number of sectors stand to be affected adversely by the decline of pollinators, not least agricultural commodities, food producers and the pharmaceutical sector. The global nature of many investors' assets under management means systemic issues such as a loss of pollinators, have the potential to impact heavily on the valuation models of companies and, ultimately, the value of the asset base. However, the impact of a loss of pollinators is not an established topic within the responsible investment community. This chapter explores the investor response to the decline in pollinators and makes the case for strategic action including, but not limited to, engagement.

The need for pollinators

Declines have been identified in at least one region or country on every continent (except Antarctica), including the UK and the Netherlands.[3] Bees have a unique role essential to flower and fruit pollination. While this alone is a reason to take notice, some studies place the benefits of bees and other pollinators to the global economy at US$1 trillion.[4] There are many other studies detailing the economic benefit from which to draw upon for this paper.[5] However, due to the difficulty of placing a realistic monetary worth on ecosystems, their services are not afforded adequate importance when making policy decisions.[6] It has been estimated that pollination is responsible for as much as 30% of agricultural food production.[7]

2 Professor Sir John Beddington, chief scientific adviser to HM Government, speaking at the GovNet SDUK09, accessed 18 March 2016. https://www.theguardian.com/science/2009/mar/18/perfect-storm-john-beddington-energy-food-climate

3 J. Biesmeijer *et al.*, "Parallel Declines in Pollinators and Insect-Pollinated Plants in Britain and the Netherlands", *Science* (2006): 351–4.

4 T. Juniper, *What Has Nature Ever Done For Us?: How Money Really Does Grow On Trees* (London, UK: Profile Books, 2013).

5 FAO, *Valuation of Pollination Services: Review of Methods* (Rome: FAO, 2006), accessed 8 March 2016, http://www.fao.org/fileadmin/templates/agphome/documents/Biodiversity-pollination/econvaluepoll1.pdf.

6 van Jaarsveld *et al.*, "Measuring Conditions and Trends in Ecosystem Services at Multiple Scales: The Southern African Millennium Ecosystem Assessment (SAfMA) Experience", *Philosophical Transactions of the Royal Society B-Biological Sciences* 360 (2005): 425–41.

7 S. Kluser and P. Peduzzi, "Global Pollinator Decline: A Literature Review", 2007, UNEP/GRIDEurope, accessed 18 March 2016, http://www.grid.unep.ch/products/3_Reports/Global_pollinator_decline_literature_review_2007.pdf

Accepted wisdom indicates we are in the midst of thc sixth major extinction of biological diversity.[8] The planet is losing between 2 and 10% of biodiversity per decade, due largely to habitat loss, pest invasion, pollution, over-harvesting and disease.

Specifically, between 75% and 95% of all flowering plants need help with pollination.[9] Pollinators are a keystone species; they are vital to the food chain of our planet.[10] Pollination improves the fruit or seed quantity and/or quality of 70% of the 1,330 most commercially valuable tropical crops[11] and 85% of the 264 highest yield crops cultivated in Europe.[12] This reliance is illustrated in Figure 7.1.

Figure 7.1 **Economic impact of insect pollination on agricultural production used directly for human food worldwide**

Source: United Nations Environment Programme, *Global Bee Colony Disorders and other Threats to Insect Pollinators* (2011)

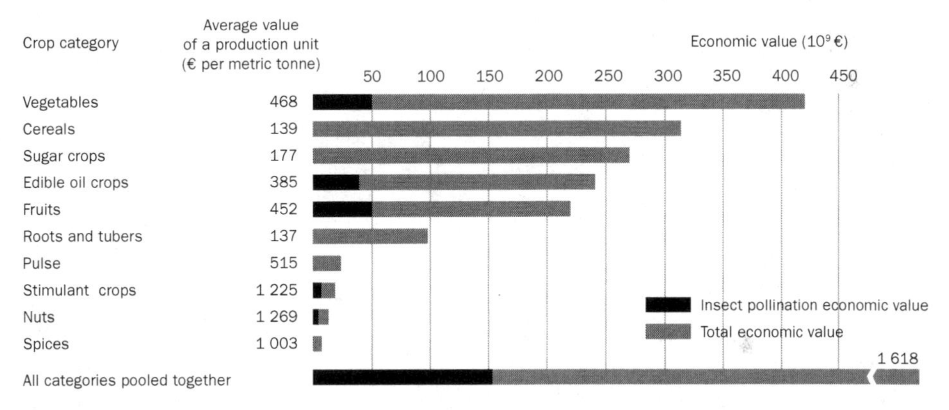

Note: The contribution of pollinators to the production of crops used directly for human food has been estimated at €153 billion globally, which is about 9.5% of the total value of human food production worldwide[13]

8 Elizabeth Kolbert, *The Sixth Extinction: An Unnatural History* (A&C Black, 2014).

9 J. Ollerton *et al.*, "How Many Flowering Plants are Pollinated by Animals?" *Oikos* 120 (2011): 321–6.

10 A.M. Klein *et al.*, "Importance of Crop Pollinators in Changing Landscapes for World Crops", *Proceedings of the Royal Society B: Biological Sciences* 274 (2007): 303–13.

11 D.W. Roubik, "Pollination of Cultivated Plants in the Tropics", *FAO Agricultural Services Bulletin* 118, Rome, 1995.

12 H. Williams, "The Dependence of Crop Production within the European Union on Pollination by Honey Bees", *Agricultural Zoology Reviews* 6 (1994): 229–57.

13 N. Gallai *et al.*, "Economic Valuation of the Vulnerability of World Agriculture Confronted with Pollinator Decline", *Ecological Economics* 68 (2009): 810–21.

The apple orchards of south-west China no longer benefit from pollinators due to excessive pesticide use and natural habitat erosion. Hand pollination is now standard for these high-value crops, but there are not enough humans on the planet to manually pollinate all crops.[14]

Shifting from crop reliance to economic impact, studies estimate the annual value of pollinator services provided in the US to be at least US$57 billion.[15] Juniper aggregated the economic impact data and concluded the annual sales dependent globally upon natural pollination was US$1 trillion.[16] He further estimated the value of services provided annually to farming by natural pollination at US$190 billion annually.

Risks of neonicotinoids

The growing news flow around this topic has the strong potential to act as a catalyst for more consideration of the issues. As discussed in earlier chapters, the EU moratorium on the use of neonicotinoids has been in place for nearly two years. The case for making this restriction permanent is growing beyond the precautionary principle rationale. Studies published in *Nature*[17] address outstanding questions around the threat that the family of pesticides pose to bees, and come as regulators around the world gear up for a fresh debate on pesticide restrictions.

Drawing parallels with other biodiversity challenges

Other investors, for instance the Commonwealth Development Commission Group (CDC), have mitigated exposure to the persistent organic pollutants listed under the Stockholm Convention. Disruption in biodiversity seldom has a positive systemic effect. We can also draw parallels with the loss of apex marine predators, for instance sharks, and the resultant population booms of prey species, such as rays and squid, and the resultant negative impacts on fish further down the food chain and even on the utility sector. Sweden's Oskarshamn nuclear power plant, which supplies 10% of the country's energy, had to shut down one of its three reactors in 2013 after a jellyfish invasion clogged the piping of its cooling system. Similarly,

14 Dave Goulson, "Decline of Bees Forces China's Apple Farmers to Pollinate by Hand", Chinadialogue, 2 October 2012, accessed 8 March 2016, https://www.chinadialogue.net/article/show/single/en/5193-Decline-of-bees-forces-China-s-apple-farmers-to-pollinate-by-hand.

15 J.E. Losey and M. Vaughan, "Conserving the Ecological Services Provided by Insects", *American Entomologist* 54 (2008): 113–5.

16 Juniper, *What Has Nature Ever Done For Us?*

17 M. Rundlöf *et al.*, "Seed Coating with a Neonicotinoid Insecticide Negatively Affects Wilds Bees", Nature 521 (2015): 77–80, doi: 10.1038/nature14420.
S.C. Kessler *et al.*, "Bees Prefer Foods Containing Neonicotinoid Pesticides", *Nature* 521 (2015): 74–76, doi: 10.1038/nature14414

a desalination plant in Oman had its intake cut by 50% when jellyfish damaged intake screens. [18]

Finally, the impact of the EU regulation banning neonicotinoids is comparable with the changes to the UK's Financial Reporting Council's regime pertaining to the mandatory disclosure of greenhouse gas emissions, and gender and diversity in the boardroom. Some challenges appear on the radar of companies only when regulation escalates them to the mainstream.

Role of the investor

The investment arena is pertinent to this challenge. The United Nations-supported Principles for Responsible Investment (PRI) Initiative is an international network of investors working together to put the six Principles for Responsible Investment into practice. Its goal is to understand the implications of sustainability for investors and support signatories to incorporate these issues into their investment decision-making and ownership practices.[19] In April 2014 the PRI announced that the total signatory assets under management was US$45 trillion. Furthermore, a Trucost, PRI and United Nations Environment Programme (UNEP) report suggests that the cost of environmental damage caused by the world's 3,000 largest publicly listed companies was US$2.15 trillion in 2008.[20]

Fiduciary duty

From the investor perspective, the United Nations Environment Programme for Financial Institutions (UNEP FI) reports on the multifaceted relationships between fiduciary law, ESG issues and institutional investment, often referred to simply as the follow-up report, "Freshfields II".[21]

Rewind back to the 2005 launch of the original ground-breaking Freshfields Report on fiduciary duty, which stated:

18 Gwynn Guilford, "Jellyfish are Taking Over the Seas and it Might be Too Late to Stop Them", qz.com, 15 October 2013, accessed 8 March 2016, http://qz.com/133251/jellyfish-are-taking-over-the-seas-and-it-might-be-too-late-to-stop-them/.

19 http://www.unpri.org

20 PRI, "Pricing Environmental Damage: US$28 trillion by 2050", accessed 8 March 2016, http://www.unpri.org/press/pricing-environmental-damage-28-trillion-by-2050-2/.

21 United Nations Environment Programme Finance Initiative, Asset Management Working Group, *Fiduciary Responsibility: Legal and Practical Aspects of Integrating Environmental, Social and Governance Issues into Institutional Investment* (UNEP, 2009), accessed 8 March 2016, http://www.unepfi.org/fileadmin/documents/fiduciaryII.pdf.

> ...in our opinion, it may be a breach of fiduciary duties to fail to take account of ESG considerations that are relevant and to give them appropriate weight, bearing in mind that some important economic analysts and leading financial institutions are satisfied that a strong link between good ESG performance and good financial performance exists.[22]

The basic premise of the report is that asset managers have a duty to raise ESG issues proactively with their clients and that failure to do so presents "a very real risk that they will be sued for negligence on the ground that they failed to discharge their professional duty of care to the client".

The European Commission recently reviewed[23] the integration of ESG issues in the context of fiduciary duty. Its preliminary findings were that the inclusion of such issues in the investment decision-making process of institutional investors is not a breach of fiduciary duties in the EU or any of its member states. Indeed, there is regulation at EU level that explicitly sets out the connection between ESG issues and fiduciary duty, namely Article 29 of the Institutions for Occupational Retirement Provision (IORP II) Directive[24] which mandates the evaluation of risks, including new or emerging risks relating to climate change, use of resources and the environment. Similarly, Article 8 (3c) of the PRIIPs KID (key information documents for packaged retail and insurance-based investment products) regulation[25] requires the inclusion of specific environmental or social objectives targeted by the product, where applicable.

In the UK, the Kay Review (2012) of UK Equity Markets and Long Term Decision Making[26] identified widespread concerns relating to the interpretation of fiduciary duties, prompting a recommendation that the Law Commission review the legal concept of fiduciary duty. The subsequent Law Commission Review[27] (2014) focused on fiduciary duties of investment intermediaries and concluded that ESG issues, believed to be financially material to an investment, should be taken into account.

22 United Nations Environment Programme Finance Initiative (UNEP FI), *A Legal Framework for the Integration of Environmental, Social and Governance Issues into Institutional Investment* 2005, UNEP FI, Nairobi [commonly referred to as the Freshfields Report] http://www.unepfi.org/fileadmin/documents/freshfields_legal_resp_20051123.pdf, 100.

23 For the call for tender of the European Commission, DG ENV and CLIMA, 'Study on resource efficiency and fiduciary duties of investors' see: https://etendering.ted.europa.eu/cft/cft-display.html?cftId=470, accessed 8 March 2016.

24 See http://eur-lex.europa.eu/legal-content/EN/TXT/?uri=celex:52014PC0167, accessed 8 March 2016.

25 See http://eur-lex.europa.eu/legal-content/EN/TXT/?uri=CELEX:32014R1286, accessed 8 March 2016.

26 John Kay, "The Kay Review of UK Equity Markets and Long-Term Decision Making", accessed 8 March 2016, https://www.gov.uk/government/uploads/system/uploads/attachment_data/file/253454/bis-12-917-kay-review-of-equity-markets-final-report.pdf.

27 See http://lawcommission.justice.gov.uk/docs/lc350_fiduciary_duties_summary.pdf, accessed 8 March 2016.

While the existence of these arrangements in the Europe Union and the UK are significant, there remain jurisdictions where the consideration of ESG issues is absent from the fiduciary duty requirement. For example, the US legal framework does not include reference to ESG issues and it could be argued that this omission in the Employee Retirement Income Security Act (ERISA) 1974, which regulates pension investments in the USA, serves to discourage the consideration of ESG issues in the decision-making process.

Universal owners

In parallel to the establishment of fiduciary duty, the Universal Owner (UO) hypothesis[28] states, "an investor benefiting from a company externalizing costs might experience a reduction in overall returns due to these externalities adversely affecting other investments in the portfolio, and hence overall market return".[29] UOs, therefore, have an incentive to reduce negative externalities (e.g. pollution and corruption) and increase positive externalities (e.g. from sound corporate governance and good human capital practices) across their investment portfolios.[30]

A number of sectors stand to be affected adversely by the decline of pollinators, not least agricultural commodities, food producers and retailers and the pharmaceutical sector. The global nature of investment means systemic issues, such as a loss of pollinators, have the potential to impact heavily on the valuation models of companies and, ultimately, the value of the asset base.

Many large, diversified funds invest in an array of companies with significant environmental impacts that undermine the environment's ability to support the economy. One company's externalities can damage the profitability of other portfolio companies, adversely affecting other investments, and hence overall market return. The Principles for Responsible Investment found that in an average equity portfolio weighted according to the MSCI All Country World Index, externalities could equate to more than half of the companies' combined earnings before

28 The hypothesis was first formulated by Robert Monks and Nell Minow, and has since been elaborated by James Hawley and Andrew Williams.

29 R. Thamotheram and H. Wildsmith, *Putting the Universal Owner Hypothesis into Action: Why Large Retirement Funds Should Want to Collectively Increase Overall Market Returns and What They Can Do About It*, 2006, accessed 8 March 2016, http://www.rijpm.com/pre_reading_files/Putting_the_Universal_Owner_Hypothesis_into_Action_Raj_Thamotheram_and_Helen_Wildsmith.pdf

30 R. Thamotheram and H. Wildsmith, *Putting the Universal Owner Hypothesis into Action: Why Large Retirement Funds Should Want to Collectively Increase Overall Market Returns and What They Can Do About It* (2006), accessed 8 March 2016, http://www.rijpm.com/pre_reading_files/Putting_the_Universal_Owner_Hypothesis_into_Action_Raj_Thamotheram_and_Helen_Wildsmith.pdf.

interest, taxation, depreciation and amortization (EBITDA), weighted according to Index constituents.[31]

Given the dominance of indices and tracker funds, a significant number of investment portfolios both institutional and retail, have exposure to companies with large market capitalization who produce the family of pesticides linked potentially with a loss of pollinators, the neonicotinoids; for instance, BASF, Bayer and Syngenta.

Reputational risk

The recent phenomenon of long-distance transportation of bee colonies across the United States (US) to pollinate almond orchards is the subject of a recent film, "More than Honey". As we saw in Chapter 2 several other film documentaries have been released along a similar theme including "Who killed the honey bee?", "Vanishing of the bees" and "Queen of the sun".[32]

Widely syndicated film documentaries are influencing the debate on company reputation and valuation. "Blackfish" premiered at the 2013 Sundance Film Festival and was syndicated by CNN who claim 21 million viewers in the US.[33] Its premise focuses on SeaWorld Entertainment Inc (SeaWorld) and their husbandry of performing killer whales. The share price of SeaWorld dropped sharply after the release of the documentary. Revenue and attendance has been down since the documentary's release and SeaWorld reported a net loss of $43.6 million, or 51 cents per diluted share, for the first quarter of 2015.[34]

In a similar vein, "Virunga" is a 2014 British documentary film focused on the conservation work of rangers within Virunga National Park that was Academy Award nominated. The Virunga National Park is protected by the World Heritage Convention that both the UK and the Democratic Republic of Congo (DRC) governments have ratified and is home to thousands of bird, plant and mammal species, including the largest concentration of hippopotami in Africa and the famous mountain gorillas.

Around the same time as the release of the documentary, Aviva Investors began engaging heavily with SOCO International plc, a FTSE250 oil and gas exploration and production company that featured in the Virunga documentary.

31 Principles for Responsible Investment, *Universal Ownership: Why Environmental Externalities Matter to Institutional Investors* (London: Principles for Responsible Investment, 2011).

32 "Vanishing of the Bees", directed by George Langworthy, Maryam Henein (UK: Hive Mentality Films, 2009).

33 Ben Beaumont-Thomas, "SeaWorld Shares Tumble 33% Following Blackfish Documentary", *The Guardian*, 14 August 2014, accessed 8 March 2016, http://www.theguardian.com/film/2014/aug/14/seaworld-shares-blackfish-documentary.

34 Sandra Pedicini, "SeaWorld Has Strong First Quarter but Warns of Challenges Ahead", *Orlando Sentinel*, 7 May 2015, accessed 8 March 2016, http://www.orlandosentinel.com/business/os-seaworld-earnings-20150507-story.html.

In addition to the film, a UK National Contact Point for the Organisation for Economic Cooperation and Development (OECD) Guidelines for Multinational Enterprises complaint that was lodged by WWF against SOCO in October 2013 and statements by the UK's Foreign & Commonwealth Office (FCO) all added to the pressure for SOCO to take action.

The nature of operations in the Virunga National Park was putting SOCO at serious risk of falling short in its stated "responsible approach to oil and gas exploration and production". This could be damaging for SOCO and, consequently, a serious concern for shareholders. A key tool in this engagement was the independent report into SOCO's activity in Block V of the DRC that Aviva Investors commissioned EIRIS to undertake.[35]

The public perspective

The impact of a loss of pollinators is a topic that has been embraced with gusto in the media and civil society arena. A 2014 survey by internet-based market research firm YouGov found the decline of bees was seen as more serious than climate change.[36] Compared with more mature topics, such as biofuels or gender and diversity, the fate of pollinators remains a fledgling topic within the responsible investment community. However, several scene-setting pieces have recently been published including:

- "The Bee and the Stockmarket" by Schroders.[37]
- The Cooperative Asset Management (now part of Royal London Asset Management) covered this topic as early as 2009 and its "Review of Sustainable Investing" made pollinators the cover story in 2014.[38]
- "Put the Bee Back in Beta"—beta being a measure of the volatility, or systematic risk, of a company or a portfolio in comparison to the market as a whole.[39]

The loss of pollinators presents a dual opportunity for the investment community to mitigate impacts to investors' long-term holdings, not least in the food

35 EIRIS, "SOCO International activities in DR Congo (Block V)" (2014), accessed 8 March 2016, http://www.eiris.org/publications/#sthash.2Rc06P8r.dpuf.

36 Will Dahlgreen, "Decline of Bees Seen as More Serious than Climate Change", YouGov, 27 June 2014, accessed 8 March 2016, https://yougov.co.uk/news/2014/06/27/bees-dying-most-serious-environmental-issue/.

37 Rick Stathers, "The Bee and the Stockmarket", Schroders, accessed 8 March 2016, http://www.schroders.com/tp/?id=a0j50000007L8FuAAK.

38 The Co-operative Asset Management, "Responsible Investment: Annual Review 2009/2010", accessed 8 March 2016, https://www.rlam.co.uk/PageFiles/9799/TCAM%20Responsible%20Investment%20Annual%20Review%2009-10.pdf.

39 R. Thamotheram, "Put the Bee Back in Beta", *IPE Magazine*, 2013, accessed 8 March 2016, www.ipe.com/magazine/put-the-bee-back-in-beta_50256.php.

producers and agrochemical sectors, and to establish themselves at the forefront of this urgent yet emergent responsible investment topic in the eyes of clients, peers, policy-makers and the media.

Tools for investors

Accounting for Sustainability was set up by HRH The Prince of Wales in 2004 "To help ensure that sustainability—considering what we do not only in terms of ourselves and today, but also of others and tomorrow—is not just talked and worried about, but becomes embedded in organizations' 'DNA'".[40]

The Accounting for Sustainability (A4S) decision-making tool is sufficiently detailed and adaptable for application to the challenge of a loss of pollinators and offers practical ways of quantifying sustainability challenges and approaching engagement. The A4S programme identified three main areas that could enable organizations to consider sustainability more effectively in their decision-making:

- **Understanding the issues**. Companies need to gain a better understanding of the main sustainability issues at a product level, including how important these are to stakeholders.
- **Prioritizing the issues**. It is important to monitor performance and prioritize the identified issues based on their environmental, social and economic impact.
- **Engaging others**. Positive change can be compounded if suppliers and other stakeholders are included in the process.

This model allows specific refinement of the challenge by substituting:

- "Sustainability" for "loss of pollinators";
- "Companies" for "investors";
- "Product-level decision-making" for "portfolio-level decision-making".

In doing so, this model expands to guide the responsible investment process.

Questions for investee companies

While the topic is systemic in nature the line of enquiry one could adopt to probe into investee companies' responses is fairly straightforward. Within a few questions it should be possible to determine the sophistication of an investee company's awareness and action on the topic. Such questions could include:

40 https://www.accountingforsustainability.org/about-us, accessed 8 March 2016.

- How does your supply chain, both upstream and downstream, stand to be affected by a loss of pollinators?
- Have any studies been commissioned to quantify this impact?
- What contingency measures and scenarios have been discussed around this issue?
- What initiatives and incentives have been put in place to mitigate the impact?
- How do you keep abreast of the latest science around pollinator decline?

Given the emergent but pertinent nature of this issue the investee company's response can be an excellent proxy for their overall ESG risk and opportunity management.

Engagement is a natural starting point for determining the calibre of investee companies' responses to this threat.

Challenges: a lack of research

Sell-side brokers are typically part of an investment bank. Such brokers provide investment analysis and recommendations to fund managers. Research on pollinators is notable by its absence.

It is not clear how the challenge interlinks with other ESG topics; e.g. will reduced crop yields intensify demand for agricultural land and how does this fit with the controversy associated with emerging market land grabs and the use of agricultural land for biofuels?

The reliance on sponsored research which may be conflicted is a concern in identifying causality. However, this is a concern not just for the investment community.

Macro impacts could see decreased crop reliability, increased food insecurity, commodity price fluctuation, to the detriment of pharmaceutical research and development; all of which are difficult to capture using conventional valuation models.

Risks to investee companies

As discussed in Chapter 6, investee companies producing neonicotinoids are starting to become the subjects of class-action lawsuits. A live example is between Syngenta and Bayer and the Ontario Beekeepers Association with the latter seeking $400 million in damages.[41]

41 CBC News, "Canadian Beekeepers Sue Bayer and Syngenta over Neonicotinoid Pesticides", 3 September, 2014, accessed 8 March 2016, http://www.cbc.ca/news/technology/canadian-beekeepers-sue-bayer-and-syngenta-over-neonicotinoid-pesticides-1.2754441.

The OECD's Guidelines for Multinational Enterprises (the Guidelines) cover several themes, including the Environment. Within this section the guidance states:

> Enterprises should, within the framework of laws, regulations and administrative practices in the countries in which they operate, and in consideration of relevant international agreements, principles, objectives, and standards, take due account of the need to protect the environment, public health and safety, and generally to conduct their activities in a manner contributing to the wider goal of sustainable development. In particular, enterprises should...
>
> ...4) Consistent with the scientific and technical understanding of the risks, where there are threats of serious damage to the environment, taking also into account human health and safety, not use the lack of full scientific certainty as a reason for postponing cost-effective measures to prevent or minimise such damage.[42]

These Guidelines are supported by a unique implementation mechanism known as National Contact Points (NCPs). NCPs are agencies established by adhering governments to promote and implement the Guidelines as well as assisting companies and their stakeholders to take appropriate measures to further the observance of the Guidelines. A mediation and conciliation platform is available for resolving practical issues that may arise with the implementation of the Guidelines and has been used by many NGOs since inception.[43]

Recent studies have highlighted a relationship between a decline in insect pollinators and neonicotinoids, albeit not sole responsibility.[44] Given the number of NGOs becoming active on the topic of pollinator loss there is a distinct possibility that a company associated with neonicotinoids may be referred by an NGO to a NCP.

Recommendations for investors

Using the corporate access enjoyed by institutional investors, asset managers raising the topic at board level may escalate the topic up the agenda, especially if done in collaboration with other investors with significant combined assets under management and hierarchy on the share register.

42 OECD, *OECD Guidelines for Multinational Enterprises* (Paris: OECD), accessed 31 March 2016, https://www.oecd.org/corporate/mne/1922428.pdf.

43 OECD, "National Contact Points for the OECD Guidelines for Multinational Enterprises", accessed 8 March 2016, http://www.oecd.org/investment/mne/ncps.htm.

44 C. Moffat *et al.*, "Chronic Exposure to Neonicotinoids Increases Neuronal Vulnerability to Mitochondrial Dysfunction in the Bumblebee (*Bombus terrestris*)", *The FASEB Journal* 29.5 (2015): 2112–9.

Presently, the divestment approach is enjoying record levels of consideration buoyed in the UK by *The Guardian*'s "Keep it in the Ground" campaign that focuses on fossil fuels.[45]

Divestment is one half of the two approaches that can be couched as "Voice and Exit".[46] However, due to the rights and influence afforded to investors, the "voice" option is an arguably more influential position to adopt at least in the first instance.[47]

The following suggestions are designed as a solid foundation on which to extend and build resilience based on an average level of portfolio exposure to this challenge. The suggestions are non-hierarchical in nature, offered on a pick and mix basis, and sufficiently broad in scope to capture the varying approaches adopted by different investment houses.

- **Background research**. Commissioning sell-side and/or independent research that estimates the impact for specific companies in the portfolio as well as recommending a course of remedial action.
- **Target identification**. Private meetings are the norm for most investor engagement with investee companies.[48] Such private meetings should be requested with companies identified as high risk and scheduled for the next three months to indicate urgency, rather than around the quarterly reporting cycle. High risk can be defined as those companies that are involved in the production or use of neonicotinoids or those with heavy reliance on pollinators such as food producers. Collaboration with other investors, via established methods such as the PRI Clearinghouse may be explored to complement this. In the case of Bayer and Syngenta, who appear to be suing the EU over its moratorium,[49] steps should be taken to discuss forthwith.
- **Monitor**. Extend and enhance the current internal processes to encompass monitoring of investee companies' progress on the pollinator challenge.
- **File shareholder resolutions.** If progress is insufficient in the six months following engagement, initiate the initial stages of filing a shareholder resolution as a nuclear option.

45 *The Guardian*, "Keep it in the ground", accessed 8 March 2016, http://www.theguardian.com/environment/series/keep-it-in-the-ground.

46 A.O. Hirschman, *Exit, Voice, and Loyalty: Responses to Decline in Firms, Organizations, and States* (Cambridge, MA: Harvard University Press, 1970).

47 S. Waygood, "NGOs and Equity Investment: Capital Market as a Campaign Device" (University of Surrey, 2004).

48 J. Holland, "A Conceptual Framework for Changes in Fund Management and Accountability Relative to ESG Issues", *Journal of Sustainable Finance and Investment* 1(2) (2011): 159–77.

49 Matthew Dalton, "EU to Revisit Question of Insecticides' Responsibility for Bee Die-Offs", *Wall Street Journal*, 26 May 2015, accessed 31 March 2016, http://www.wsj.com/articles/eu-to-revisit-question-of-insecticides-responsibility-for-bee-die-offs-1432646742.

- **Alternatives**. Commission a different sell-side broker to identify companies able to capitalize on the challenge, for instance, chemical companies developing less harmful alternatives to neonicotinoids and biotech firms researching self-pollinating seed stock. This may help to guide asset allocation decisions.
- **Valuation**. Establish a working party to define how pollinator loss could be captured in discount cash flow and company valuation models.
- **Build expertise and buy in**. Draft an internal briefing paper designed for fund managers and analysts delivered at a workshop, building on the literature reviewed for this paper. Arrange complementary teach-in sessions to obtain buy in and gain feedback.
- **External positioning**. Drafting of an external position paper to inform stakeholders of activity in this area. Specific webinars and events could be scheduled to raise awareness of the topic within the investment community and beyond.
- **External fund managers**. Revise Service Level Agreements with all third party fund managers to ensure loss of pollinators is captured in asset allocation and engagement activities. If appropriate, invite fund managers to relevant fora and working groups.
- **Sharing knowledge**. Creation of a coalition of experts, suppliers and investors to share the latest research under Chatham House rules. Participants should be selected from issuers, academia, global policy-makers, NGOs and industry. Barriers, solutions and current work to be turned into a wiki-document, open for contributions.
- **Public policy**. Opportunities to contribute to consultations, for instance the Defra Consultation on National Pollinator Strategy and regulatory policy documents should be sought. The potential for influence with regulators is strong and timely given that, for instance, the moratorium imposed by the European Commission (EC) on the use of neonicotinoids in its member states for two years, expired on 1 December 2015. Monaghan and Monaghan illustrate how advocacy and public policy intervention can accelerate the delivery of a more sustainable economy.[50]

50 P. Monaghan and P. Monaghan, *Lobbying for Good: How Business Advocacy can Accelerate the Delivery of a Sustainable Economy* (London: DoSustainability), accessed 18 March 2016, http://www.dosustainability.com/shop/lobbying-for-good-how-business-advocacy-can-accelerate-the-delivery-of-a-sustainable-economy-p-51.html

Concluding thoughts

This chapter shows how investors can use their influence over companies to motivate change in corporate behaviour. Three important issues are highlighted in this chapter. The first is the growth in the number of investors who have published policy statements regarding the extent to which they are responsible owners. The pertinence, urgency and impact of this challenge is clear as is the link with responsible ownership and systemic implications.

The second issue is that the benefits of working in association with informed people cannot be understated, especially as responsible investment practitioners are unlikely to have first-hand scientific expertise. However, the investment community must be mindful not to neglect other sources of bee decline, for instance the varroa mite, in the search for a single smoking gun.

The third is that the presence of new means of highlighting responsible investment issues has disrupted and increased the methods by which an investee company is exposed to brand risk. These new methods include, but are not limited to film documentaries and the NCP. The media and NGOs are increasingly savvy in making the connection between corporate behaviour and the owners of those companies. This increases the onus on investors to demonstrate their stewardship.

Bibliography

Accounting for Sustainability. *Sustainability Decision-Making Model: A Methodology for Improving the Sustainability of your Organisation's Products and Service*. Accounting for Sustainability, 2011. Accessed 8 March 2016. www.accountingforsustainability.org/wp-content/uploads/2011/10/Decision-Making-Tutorial.pdf.

Andrews, K.R. *The Concept of Corporate Strategy*. Homewood, IL: Dow Jones Irwin, 1971.

Barbour, R.S. "Checklists for Improving Rigour in Qualitative Research: A Case of the Tail Wagging the Dog?" *British Medical Journal* 322.7294 (2001): 1115–23.

BBC News. "Bee Deaths: EU to Ban Neonicotinoid Pesticides". BBC News. Last updated 13 April 2013. http://www.bbc.co.uk/news/world-europe-22335520.

Beaumont-Thomas, B. "SeaWorld Shares Tumble 33% Following Blackfish Documentary". *The Guardian*, 14 August 2014. Accessed 8 March 2016. http://www.theguardian.com/film/2014/aug/14/seaworld-shares-blackfish-documentary.

Benford, R.D. and D.A. Snow. "Framing Processes and Social Movements: An Overview and Assessment". *Annual Review of Sociology* (2000): 611–39.

Benjamin, A. and B. McCallum. *A World Without Bees*. New York: Pegasus Books, 2009.

Biesmeijer, J., P. M. Roberts, M. Reemer, R. Ohlemüller, M. Edwards, T. Peeters, A. P. Schaffers, S. G. Potts, R. Kleukers, C. D. Thomas, J. Settele, and W. E. Kunin. "Parallel Declines in Pollinators and Insect-Pollinated Plants in Britain and the Netherlands". *Science* (2006): 351–4.

Buchmann, S. L. and G.P. Nabhan. *The Forgotten Pollinators*. Washington, DC: Island Press, 1997.

Burnes, B. "Kurt Lewin and the Planned Approach to Change: A Re appraisal". *Journal of Management Studies* 41(6) (2004): 977–1002.

Carpenter, S. R., Harold A. Mooneyb, John Agardc, Doris Capistranod, Ruth S. DeFriese, Sandra Díazf, Thomas Dietzg, Anantha K. Duraiappahh, Alfred Oteng-Yeboahi, Henrique Miguel Pereiraj, Charles Perringsk, Walter V. Reidl, Jose Sarukhanm, Robert J. Scholesn, and Anne Whyteo. "Science for Managing Ecosystem Services: Beyond the Millennium Ecosystem Assessment". *Proceedings of the National Academy of Sciences* 106.5 (2009): 1305–12.

Carrington, D. "Loss of Wild Pollinators Serious Threat to Crop Yields, Study Finds". *The Guardian*, 2013. Accessed 8 March 2016. www.guardian.co.uk/environment/2013/feb/28/wild-bees-pollinators-crop-yields.

Casson, P. D. and D. Russell. *Universities Superannuation Scheme: Implementing Responsible Investment*. Sheffield, UK: Greenleaf Publishing, 2006.

CBC News. "Canadian Beekeepers Sue Bayer and Syngenta over Neonicotinoid Pesticides". 3 September 2014. Accessed 8 March 2016. http://www.cbc.ca/news/technology/canadian-beekeepers-sue-bayer-and-syngenta-over-neonicotinoid-pesticides-1.2754441.

Chandler, A.D. *Strategy and Structure*. Cambridge, MA: MIT Press, 1962.

Crompton, T. *Common Cause: The Case for Working with our Cultural Values*. World Wildlife Fund, 2010. Accessed 8 March 2016. http://assets.wwf.org.uk/downloads/common_cause_report.pdf.

Dahlgreen, W. "Decline of Bees Seen as More Serious than Climate Change". YouGov, 27 June 2014. Accessed 8 March 2016. https://yougov.co.uk/news/2014/06/27/bees-dying-most-serious-environmental-issue/.

Dalton, M. "EU to Revisit Question of Insecticides' Responsibility for Bee Die-Offs". *Wall Street Journal*, 26 May 2015. Accessed 31 March 2016. http://www.wsj.com/articles/eu-to-revisit-question-of-insecticides-responsibility-for-bee-die-offs-1432646742.

Davidson, J. "Sustainable Development: Business as Usual or a New Way of Living?" *Environmental Ethics* 22(1) (2000): 25–42.

Davis, F.C. *Implementing your Strategic Plan: How to Turn "Intent" into Effective Action for Sustainable Change*. New York: Amacom, 1999.

Delmas, M. A. and N.S. Nairn-Birch. *Is the Tail Wagging the Dog? An Empirical Analysis of Corporate Carbon Footprints and Financial Performance*. UC Los Angeles: Institute of the Environment and Sustainability, 2011. Accessed 8 March 2016. http://escholarship.org/uc/item/3k89n5b7.

Dent, E.B. and S.G. Goldberg. "Challenging 'Resistance to Change'". *The Journal of Applied Behavioral Science* 35(1) (1999): 25–41.

Department of Economic and Social Affairs Statistics Division United Nations. *Wealth Accounting and the Valuation of Ecosystem Services (WAVES): A Global Partnership*, 2011. Accessed 8 March 2016. http://unstats.un.org/unsd/envaccounting/ceea/meetings/UNCEEA-6-7.pdf.

Dias, B.S.F. A. Raw and V.L. Imperatriz-Fonseca. "International Pollinators Initiative: The Sao Paulo Declaration on Pollinators". In *Report on the Recommendations of the Workshop on the Conservation and Sustainable Use of Pollinators in Agriculture with Emphasis on Bees*. Brazilian Ministry of the Environment, 1999.

Druckman, P. "Accounting for Sustainability". Presented at the 11th Annual Environmental and Sustainability Management Accounting Network Conference. Budapest, Hungary, 6–7 October 2008.

Dunphy, D., S. Benn and A. Griffiths. *Organizational Change for Corporate Sustainability*. London: Routledge, 2003.

EIRIS. "SOCO International activities in DR Congo (Block V)", 2014. Accessed 8 March 2016. http://www.eiris.org/publications/#sthash.2Rc06P8r.dpuf.

Elkington, J. "Towards the Suitable Corporation: Win-Win-Win Business Strategies for Sustainable Development". *California Management Review* 36.2 (2004).
FAO. *Valuation of Pollination Services: Review of Methods*. Rome: FAO, 2006. Accessed 8 March 2016. http://www.fao.org/fileadmin/templates/agphome/documents/Biodiversity-pollination/econvaluepoll1.pdf.
Fiedler, H. "National PCDD/PCDF Release Inventories under the Stockholm Convention on Persistent Organic Pollutants". *Chemosphere* 67.9 (2007): S96–S108.
Foreign & Commonwealth Office. *Human Rights and Democracy: The 2012 Foreign & Commonwealth Office Report. Democratic Republic of the Congo*, 15 April 2013. Accessed 18 March 2016. http://www.refworld.org/docid/516fb7cbf.html
Gallai, N., Jean-Michel Salles, Josef Settele, and Bernard E. Vaissière. "Economic Valuation of the Vulnerability of World Agriculture Confronted with Pollinator Decline". *Ecological Economics* 68 (2009): 810–21.
Ghazoul, J. "Buzziness as Usual? Questioning the Global Pollination Crisis". *Trends in Ecology and Evolution* 20 (2005): 7.
Gond, J. and E. Boxenbaum. "The Globalization of Responsible Investment: Contextualization Work in France and Quebec". *Journal of Business Ethics* (2013): 1–15.
Goulson, D. "Decline of Bees Forces China's Apple Farmers to Pollinate by Hand". Chinadialogue, 2 October 2012. Accessed 8 March 2016. https://www.chinadialogue.net/article/show/single/en/5193-Decline-of-bees-forces-China-s-apple-farmers-to-pollinate-by-hand.
Gray, R. "Is Accounting for Sustainability Actually Accounting For Sustainability…and How Would We Know? An Exploration of Narratives of Organisations and the Planet". *Accounting, Organizations and Society* 35.1 (2010): 47–62.
Grundy, A.N. "Rejuvenating Strategic Management: The Strategic Option Grid". *Strategic Change* 13(3) (2004): 111–23.
Guilford, G. "Jellyfish are Taking Over the Seas and it Might be Too Late to Stop Them". qz.com, 15 October 2013. Accessed 8 March 2016. http://qz.com/133251/jellyfish-are-taking-over-the-seas-and-it-might-be-too-late-to-stop-them/.
Hill, J.W. and G. Jones. *Strategic Management Theory: An Integrated Approach*, 2nd edn. Boston, MA: Houghton Mifflin, 1995.
Hirschman, A.O. *Exit, Voice, and Loyalty: Responses to Decline in Firms, Organizations, and States*. Cambridge, MA: Harvard University Press, 1970.
Holland, J. "A Conceptual Framework for Changes in Fund Management and Accountability Relative to ESG Issues". *Journal of Sustainable Finance and Investment* 1(2) (2011): 159–77.
Hollender, J., B. Breen and P. Senge. *The Responsibility Revolution: How the Next Generation of Businesses Will Win*. John Wiley & Sons, 2010.
Hopwood, A., J. Unerman and J. Fries. *Accounting for Sustainability: Practical Insights*. London, UK: Routledge, 2010.
Hubbard, G. and P. Beamish. *Strategic Management: Thinking, Analysis, Action*, 4th edn. Sydney: Pearson Education Australia, 2011.
Juniper, T. *What Has Nature Ever Done For Us?: How Money Really Does Grow On Trees*. London, UK: Profile Books, 2013.
Juravle, C. and A. Lewis. "Identifying Impediments to SRI in Europe: A Review of the Practitioner and Academic Literature". *Business Ethics: A European Review* 17:3 (2008): 285–310.
Kay, J. "The Kay Review of UK Equity Markets and Long-Term Decision Making". Accessed 8 March 2016. https://www.gov.uk/government/uploads/system/uploads/attachment_data/file/253454/bis-12-917-kay-review-of-equity-markets-final-report.pdf.
Kessler, S. C., E.J. Tiedeken, K.L. Simcock, S. Derveau, J. Mitchell, S. Softley, A. Radcliffe, J.C. Stout, and G.A. Wright. "Bees Prefer Foods Containing Neonicotinoid Pesticides". *Nature* 521 (2015): 74–76. doi: 10.1038/nature14414

Klein, A.M., Ingolf Steffan-Dewenter, Damayanti Buchori and Teja Tscharntke. "Effects of Land-Use Intensity in Tropical Agroforestry Systems on Coffee Flower-Visiting and Trap-Nesting Bees and Wasps". *Conservation Biology* 16 (2002): 1003–14.

Klein, A.M., Bernard E Vaissière, James H Cane, Ingolf Steffan-Dewenter, Saul A Cunningham, Claire Kremen, and Teja Tscharntke. "Importance of Crop Pollinators in Changing Landscapes for World Crops". *Proceedings of the Royal Society B: Biological Sciences* 274 (2007): 303–13.

Kluser, S. and P. Peduzzi. "Global Pollinator Decline: A Literature Review"..UNEP/ GRI-DEurope, 2007. Accessed 18 March 2016. http://www.grid.unep.ch/products/3_Reports/Global_pollinator_decline_literature_review_2007.pdf

Kolbert, Elizabeth. *The Sixth Extinction: An Unnatural History*. A&C Black, 2014.

Kotter, J.P. "Leading Change: Why Transformation Efforts Fail". *Harvard Business Review* 73(2) (1995): 59–67.

Ligthart, J.J. "NGO Initiatives in the EU: Identifying Substances of Very High Concern (SVHCs) and Driving Safer Chemical Substitutes in Response to REACH". In: *Chemical Alternatives Assessments*. London: Royal Society of Chemistry Press, 2013.

Losey, J. E. and M. Vaughan. "Conserving the Ecological Services Provided by Insects". *American Entomologist* 54 (2008): 113–5.

Meyer, C., and J. Kirby. *The Big Idea: Leadership in the Age of Transparency*. Harvard Business, 2010.

Moffat, C., Joao Goncalves Pacheco, Sheila Sharp, Andrew J Samson, Karen A Bollan, Jeffrey Huang, Stephen T Buckland, and Christopher N Connolly. "Chronic Exposure to Neonicotinoids Increases Neuronal Vulnerability to Mitochondrial Dysfunction in the Bumblebee (*Bombus terrestris*)". *The FASEB Journal* 29.5 (2015): 2112–9.

Monaghan, P. and P. Monaghan. *Lobbying for Good: How Business Advocacy can Accelerate the Delivery of a Sustainable Economy*. London: DoSustainability. Accessed 18 March 2016. http://www.dosustainability.com/shop/lobbying-for-good-how-business-advocacy-can-accelerate-the-delivery-of-a-sustainable-economy-p-51.html

Nilsson, J. "Investment with a Conscience: Examining the Impact of Pro-Social Attitudes and Perceived Financial Performance on Socially Responsible Investment Behaviour". *Journal of Business Ethics* 83(2) (2008): 307–25.

Nilsson, J. "Consumer Decision Making in a Complex Environment: Examining the Decision Making Process of Socially Responsible Mutual Fund Investors". Diss., Umeå University, 2010.

Nilsson, J., A.C. Nordvall and S. Isberg. "The Information Search Process of Socially Responsible Investors". *Journal of Financial Services Marketing* 15(1) (2010): 5–18.

OECD. *OECD Guidelines for Multinational Enterprises*. Paris: OECD. Accessed 31 March 2016. https://www.oecd.org/corporate/mne/1922428.pdf.

OECD. "National Contact Points for the OECD Guidelines for Multinational Enterprises". Accessed 8 March 2016. http://www.oecd.org/investment/mne/ncps.htm.

Ollerton, J., R. Winfree and S. Tarrant "How Many Flowering Plants are Pollinated by Animals?" *Oikos* 120 (2011): 321–6.

Pedicini, S. "SeaWorld Has Strong First Quarter but Warns of Challenges Ahead". *Orlando Sentinel*, 7 May 2015. Accessed 8 March 2016. http://www.orlandosentinel.com/business/os-seaworld-earnings-20150507-story.html.

Porritt, J. *Capitalism as if the World Matters*. London, UK: Routledge, 2012.

Principles for Responsible Investment (PRI). *Universal Ownership: Why Environmental Externalities Matter to Institutional Investors*. London: Principles for Responsible Investment, 2011.

PRI. "Pricing Environmental Damage: US$28 trillion by 2050". Accessed 8 March 2016. http://www.unpri.org/press/pricing-environmental-damage-28-trillion-by-2050-2/.

Roubik, D.W. "Pollination of Cultivated Plants in the Tropics". *FAO Agricultural Services Bulletin* 118. Rome, 1995.

Rundlöf, M., G.K.S. Andersson, R. Bommarco, I. Fries, V. Hederstrom, L. Herbertsson, O. Jonsson, B.K. Klatt, T.R. Pedersen, J. Yourstone and H.G. Smith. "Seed Coating with a Neonicotinoid Insecticide Negatively Affects Wilds Bees". *Nature* 521 (2015): 77–80. doi: 10.1038/nature14420.

Senge, P.M. and B. Smith. *The Necessary Revolution: Working Together to Create a Sustainable World*, Broadway Books, 2010.

Stathers, R. "The Bee and the Stockmarket", Schroders. Accessed 8 March 2016. http://www.schroders.com/tp/?id=a0j50000007L8FuAAK.

Sullivan, R. *Valuing Corporate Responsibility: How Do Investors Really Use Corporate Responsibility Information?* Sheffield, UK: Greenleaf Publishing, 2011.

Sullivan, R., C. Mackenzie, and S. Waygood. *Does a Focus on Social, Ethical and Environmental Issues Enhance Investment Performance. Responsible Investment.* Sheffield, UK: Greenleaf Publishing, 2006.

TEEB (The Economics of Ecosystems and Biodiversity for National and International Policy Makers). *Summary: Responding to the Value of Nature.* TEEB, 2009. Accessed 8 March 2016. http://www.teebweb.org/publication/teeb-for-policy-makers-summary-responding-to-the-value-of-nature/.

Thamotheram, R. "Put the Bee Back in Beta". *IPE Magazine*, 2013. Accessed 8 March 2016. www.ipe.com/magazine/put-the-bee-back-in-beta_50256.php.

Thamotheram, R. and H. Wildsmith. *Putting the Universal Owner Hypothesis into Action: Why Large Retirement Funds Should Want to Collectively Increase Overall Market Returns and What They Can Do About It.* 2006. Accessed 8 March 2016. http://www.rijpm.com/pre_reading_files/Putting_the_Universal_Owner_Hypothesis_into_Action_Raj_Thamotheram_and_Helen_Wildsmith.pdf.

The Co-operative Asset Management. "Responsible Investment: Annual Review 2009/2010". Accessed 8 March 2016. https://www.rlam.co.uk/PageFiles/9799/TCAM%20Responsible%20Investment%20Annual%20Review%2009-10.pdf.

The Guardian. "Keep it in the ground". Accessed 8 March 2016. http://www.theguardian.com/environment/series/keep-it-in-the-ground.

United Nations Environment Programme. *Global Bee Colony Disorders and other Threats to Insect Pollinators*. UNEP, 2011. Accessed 8 March 2016. www.unep.org/dewa/Portals/67/pdf/Global_Bee_Colony_Disorder_and_Threats_insect_pollinators.pdf.

United Nations Environment Programme Finance Initiative (UNEP FI). *A Legal Framework for the Integration of Environmental, Social and Governance Issues into Institutional Investment*, 2005. UNEP FI, Nairobi. Accessed 18 March 2016. http://www.unepfi.org/fileadmin/documents/freshfields_legal_resp_20051123.pdf.

United Nations Environment Programme Finance Initiative, Asset Management Working Group. *Fiduciary Responsibility: Legal and Practical Aspects of Integrating Environmental, Social and Governance Issues into Institutional Investment.* UNEP, 2009. Accessed 8 March 2016. http://www.unepfi.org/fileadmin/documents/fiduciaryII.pdf.

United Nations Environment Programme Financial Initiative and the United Nations Principles for Responsible Investment. *Universal Ownership: Why Environmental Externalities Matter to Institutional Investors.* UNEP, 2010. Accessed 8 March 2016. www.unepfi.org/fileadmin/documents/universal_ownership_full.pdf.

Van der Lugt, C., S. Gilbert and W. Evison. *Measuring and Reporting Biodiversity and Ecosystem Impacts and Dependence.* The Economics of Ecosystems and Biodiversity, 2010. Accessed 8 March 2016. http://www.teebweb.org/media/2012/01/TEEB-For-Business.pdf

van Jaarsveld, A.S., R. Biggs, R.J. Scholes, E. Bohensky, B. Reyers, T. Lynam, C. Musvoto, and C. Fabricius. "Measuring Conditions and Trends in Ecosystem Services at Multiple Scales: The Southern African Millennium Ecosystem Assessment (SAfMA) Experience". *Philosophical Transactions of the Royal Society B-Biological Sciences* 360 (2005): 425–41.

Waterfield, B. "EU Overrules Britain to Ban Pesticides Linked to Bee Deaths". *The Telegraph*, 2013. Accessed 8 March 2016. www.telegraph.co.uk/news/worldnews/europe/eu/10025667/EU-overrules-Britain-to-ban-pesticides-linked-to-bee-deaths.html.

Waygood, S. "NGOs and Equity Investment: Capital Market as a Campaign Device". Doctoral thesis, University of Surrey, 2004.

Waygood, S. "Measuring the Effectiveness of Investor Engagement". *Responsible Investment* 1(77) (2006): 206–13.

Waygood, S. and W. Walter. "A Critical Assessment of How Non-Governmental Organizations Use the Capital Markets to Achieve their Aims: A UK Study". *Business Strategy and the Environment* 12(6) (2003): 372–85.

Williams, H. "The Dependence of Crop Production within the European Union on Pollination by Honey Bees". *Agricultural Zoology Reviews* 6 (1994): 229–57.

Wittmer, H. *The Economics of Ecosystems and Biodiversity: Mainstreaming the Economics of Nature: A Synthesis of the Approach, Conclusions and Recommendations of TEEB*. The Economics of Ecosystems and Biodiversity, 2010. Accessed 8 March 2016. http://www.teebweb.org/publication/mainstreaming-the-economics-of-nature-a-synthesis-of-the-approach-conclusions-and-recommendations-of-teeb/.

World Economic Forum *Accelerating the Transition towards Sustainable Investing: Strategic Options for Investors, Corporations and other Key Shareholders*. World Economic Forum, 2011.

8

Bees and pesticides

The Ontario controversy

Margaret Clappison and Aris Solomon
Athabasca University, Faculty of Business, Canada

Ontario has suffered a major decimation of its bee populations since 2007. Although only 15% of the Canadian bee yards are in Ontario, they represent 50% of the beekeepers in Canada.[1] As a result, Ontario is in the midst of a controversy between the environmental groups on one side and the pesticide companies on the other. This has polarized beekeepers into two camps. Both sides agree bees are very important and any substance that kills bees in significant doses is detrimental to the farmers and the commercial honey crop. What they do not agree about is which, if any, pesticides should be used, the positive and negative effects of the pesticides, and what level of dosages are acceptable, if any. Those who take the pesticide companies' position claim other factors are causing the bees to die and that some Ontario beekeepers are just causing problems by agitating to ban a type of pesticide known as neonicotinoids.[2] Even some beekeepers are against any ban on these pesticides and do not believe they are harmful. This is indicated by suggestions that "hobby

1 Claire Brownell, "Bees, Bans and Bungling: How an Anti-Pesticide Campaign May Spell Serious Trouble", *Financial Post*, November 7, 2014, accessed 8 March 2016, para. 13, http://business.financialpost.com/news/economy/bees-bans-and-bungling-how-an-anti-pesticide-campaign-may-spell-serious-trouble.

2 Brownell, "Bees, Bans and Bungling".

beekeepers" in Ontario are less diligent and less knowledgeable so are responsible for the increased decline of bees.[3]

Currently some Ontario beekeepers are involved in a class action lawsuit against agrochemical companies claiming the pesticides are killing bees and jeopardizing their livelihood as beekeepers.[4] In particular, they are concerned about neonicotinoids used to prevent insect damage and increase the yields of crops such as corn, canola and soybeans. Neonicotinoids are a synthetic form of nicotine that is systemic and there seems to be evidence that the declining bee population is due to the impact of these pesticides on crops such as corn.[5] Ontario doctors and nurses are also encouraging a ban on neonicotinoids since they consider neonicotinoids have "a direct and indirect relation on health".[6] Those agreeing with the environmentalists are mainly from Ontario where the overwinter bee losses in the spring of 2014 were devastating. To add to the controversy studies have implicated these pesticides in the decline of the bee population in Europe, and the European Union has banned them for a period of two years. However, the pesticide companies Bayer and Syngenta have been fighting Europe's temporary ban on neonicotinoids, which came into effect in 2014.[7] Bayer and Syngenta earned $2.79 billion in sales in 2011, representing 40% of the insecticide market.[8] A pesticide ban may harm the companies' sales.

This chapter begins with a literature review discussing current research on the effect neonicotinoids may or may not have on bees and continues with an overview of the existing controversy. Then a discussion of the pesticide companies' and supporters' viewpoints is followed by a section on the viewpoints of the beekeepers and their supporters. Lastly, there is a discussion of the possible implications and the potential knowledge gaps.

3 Paul Driessen, "Beekeepers Blaming Pesticides for Bee Losses Could Face Bigger Losses in Court", *Financial Post*, October 27, 2014, accessed 8 March 2016, http://business.financialpost.com/fp-comment/beekeeepers-blaming-pesticides-for-bee-losses-could-face-bigger-losses-in-court.

4 Driessen, "Beekeepers Blaming Pesticides for Bee Losses".

5 Driessen, "Beekeepers Blaming Pesticides for Bee Losses".

6 CBC/Radio-Canada, "Doctors, Nurses Urge Ontario to Ban Neonicotinoids", November 17, 2014, accessed 8 March 2016, http://www.cbc.ca/news/business/doctors-nurses-urge-ontario-to-ban-neonicotinoids-1.2837919.

7 Brownell, "Bees, Bans and Bungling".

8 Aleksandra Sagan, "Pesticides Linked to Bee Deaths Must Be Banned Scientists Say: Neonicotinoids, Dipromil Linked to Ecosystem Damage in New Study". *CBC/Radio-Canada*, June 26, 2014. Accessed 8 March 2016. http://www.cbc.ca/news/technology/pesticides-linked-to-bee-deaths-must-be-banned-scientists-say-1.2685492.

Other significant factors

A further complication was the discovery of colony collapse disorder (CCD) in 2006–2007 on the Eastern side of the United States where beekeepers "…witnessed large-scale losses of managed honey bee … colonies".[9] CCD is a bee disorder where the worker bee population continues declining and the queen does not begin laying eggs to ensure new bees will emerge to re-populate the hive for the coming spring and instead there is a decline in the population in winter as the worker bees start to leave the hive.[10] Furthermore, a review of the literature on neonicotinoids in bees showed honey bees exposed to sub-lethal levels of neonicotinoids experience difficulties with flying and navigation, reduced taste sensitivity, reproduction and slower learning of new tasks, which all impact foraging ability.[11] A new study confirmed that the "queen ['s] health is critical to colony survival of social bees".[12] Researchers discovered neonicotinoids increased the size of queen honey bee ovaries and decreased sperm quality and quantity thereby adversely affecting queen health and colony survival.[13] In addition, research to determine the effects of neonicotinoid pesticides (mainly imidacloprid and clothianidin) discovered a higher percentage of overwinter bee losses where hives affected by the pesticides were almost empty of bees, although the hives seemed healthy and well populated at the start of winter. In contrast, those hives that died from infection were full of dead bees.[14] Beekeeper Tibor Szabo experienced the phenomenon of almost deserted hives typical of colony collapse disorder in his apiary in Ontario 2012 where strong bee colonies that had plenty of food had piles of dead bees outside the hives and the few remaining were expelling the remaining dead bees. This loss of hives occurred in 242 bee yards within Ontario and Quebec; the next year there were 322 bee yards affected.[15] The result of an investigation was that "…80% of bee yards tested had residues from neonic pesticides [and furthermore] most of the hives were located near cornfields".[16] This suggests the neonicotinoids may be involved in causing or enhancing colony collapse disorder.

9 Dennis van Engelsdorp *et al.*, "Colony Collapse Disorder: A Descriptive Study". *PLoS ONE* 4 (8) (2009): doi: 10.1371/journal.pone.0006481.
10 van Engelsdorp *et al.*, "Colony Collapse Disorder".
11 Chensheng Lu *et al.*, "Sub-Lethal Exposure to Neonicotinoids Impaired Honey Bees Winterization before Proceeding to Colony Collapse Disorder", *Bulletin of Insectology* 67 (1) (2014): 125–30.
12 Geoffrey R. Williams *et al.*, "Neonicotinoid Pesticides Severely Affect Honey Bee Queens", *Scientific Reports* 5 (2015): 14621, doi: 10.1038/srep14621, p. 1.
13 Williams *et al.*, "Neonicotinoid Pesticides Severely Affect Honey Bee Queens".
14 Lu *et al.*, "Sub-Lethal Exposure to Neonicotinoids Impaired Honey Bees Winterization".
15 Brownell, "Bees, Bans and Bungling", para. 19
16 *Ibid.*, para. 20

Integrated pest management is a method to control the amount of crop damage from pests by using a set of techniques such as providing a habitat for predators of the insects and monitoring the pest population. When using conventional agriculture practices the farmer would use pesticides if insect damage reaches high levels. In integrated pest management if the population of pests does increase, there are less toxic ways to control an infestation. Alternative methods include trap crops that prevent the insect from attacking the commercial crop or pheromone mating disruption where ties coated with the female pheromone are affixed throughout the crop causing confusion for the male because it cannot easily find a female with which to mate.[17]

Sub-lethal exposure levels can disrupt cognitive abilities (learning, navigation), mobility, communication, various behaviours and physiology, and the ability to collect and store food, all of which depend on coordination and communication.[18] Since adult bees feed on nectar and take nectar and pollen back to the hives to feed their young, the larvae would be developing on raw pollen and undiluted nectar laced with neonicotinoids.[19] "The economic importance of pollination, and its aesthetic and ethical values, makes it clear that the conservation of pollination systems is an important priority".[20] So far, most of the concern is for bees killed by toxic levels of the neonicotinoids but not as much thought is given to the bees that survive. However, there are noticeable unusual behaviours the bees seem to acquire such as the inability to fly and social problems such as displacing queens or abnormal communication dances.[21] The other problem is that foraging honey bees take all the water, pollen and nectar back to the hive so if one bee becomes contaminated with pesticides so will the rest of the hive.[22] Some Ontario beekeepers are also concerned that neonicotinoid residues found in pollen and nectar in flowers may be at lethal levels, and pollinators such as bees could consume them when gathering nectar and pollen. They mention "neonics aren't like other pesticides: since they're applied to the seed, they're absorbed throughout the plant, distributed into every cell as it grows, and poison the neurological systems of insects that attempt to feed on them".[23]

The resulting controversy prompted three associations to commission The Ontario Omnibus Survey Report to conduct a survey of residents. The Ontario Omnibus Survey Report in December 2014 found 76% of those surveyed were concerned about the future of honey bees and wild bees; the figures were higher in the two main agricultural areas with 80% in the south-western region and 83% in

17 Jennifer Hopwood *et al.*, *Are Neonicotinoids Killing Bees?*
18 Hopwood *et al.*, *Are Neonicotinoids Killing Bees?*
19 Hopwood *et al.*, *Are Neonicotinoids Killing Bees?*
20 Carol Kearns *et al.*, "Endangered Mutualisms: The Conservation of Plant-Pollinator Interactions", *Annual Review of Ecology and Systematics* 29 (1998), Annual Reviews: 83–112, http://www.jstor.org/stable/221703, p. 84
21 Kearns *et al.*, "Endangered Mutualisms", p. 91.
22 *Ibid.*
23 Brownell, "Bees, Bans and Bungling".

the Niagara region.[24] In addition, 77% of those surveyed agreed with the Ontario Government's intent to limit neonics use by 80%.[25] This suggests that people in Ontario appear to be extremely concerned by the decreasing bee population. The public debate caused the Ontario provincial government to put a neonicotinoid pesticide ban in place starting on 1 July 2015.[26] As the first jurisdiction in North America to ban neonicotinoids, the government planned to decrease the overwintering death rate to 15% from 34%.[27] Furthermore, the target is an 80% expected decline in neonicotinoid usage by 2017.[28] Although the current Ontario Minister of the Environment Glen Murray believes the target is reasonable, the regulations have increased the controversy since the grain growers in Ontario immediately challenged the Ontario Government's decision by filing a notice to the court.[29] Despite grain farmers' hope of a quick and favourable outcome, in the judge's decision from the Ontario Supreme Court given on 23 October 2015, "the court rejected an attempt by the Grain Farmers of Ontario to enact a stay on the seed treatment regulations passed into law in July".[30]

The current controversy in Ontario

In 2015, a class action lawsuit was brought by some Ontario beekeepers against Bayer and Syngenta pesticide companies for $400 million in damages because they claim that the neonicotinoid pesticides (neonics) are killing their bees.[31] This has

24 Environmental Communications Options Orcalepoll Research Limited, *Ontario Omnibus Survey Report* (Canada, 2014). The three organizations Friends of the Earth, Canadian Association of Physicians for the Environment, and the Ontario Beekeeper's Association commissioned the report to discover the level of support there was for the government positon on neonicotinoids.

25 Environmental Communications Options Orcalepoll Research Limited, *Ontario Omnibus Survey Report*

26 R. Benzie, "Ontario First in North America to Curb Bee-Killing Neonicotinoids Pesticides", *The Star*, June 9, 2015, accessed 8 March 2016, http://www.thestar.com/news/queenspark/2015/06/09/ontario-first-in-north-america-to-ban-bee-killing-neonicotinoid-pesticides.html.

27 E. Arkins, "Ontario Restricts Use of Pesticides Blamed for Decline of Bee Populations", *Globe and Mail*, June 9, 2015, accessed 8 March 2016, http://www.theglobeandmail.com/report-on-business/ontario-unveils-first-restrictions-on-class-of-pesticides/article24874268/, para. 4.

28 Arkins, "Ontario Restricts Use of Pesticides Blamed for Decline of Bee Populations", para. 2.

29 Arkins, "Ontario Restricts Use of Pesticides Blamed for Decline of Bee Populations".

30 CBC News, "Grain Farmers of Ontario Lost Bid to Block Law That Limits Use of Neonicotinoids on Corn, Soybeans", October 26, 2015, accessed 8 March 2016, http://www.cbc.ca/news/business/grain-farmers-neonics-1.3289326, para. 2.

31 Driessen, "Beekeepers Blaming Pesticides for Bee Losses".

caused deep divisions among Ontario beekeepers on how dangerous they believe the neonic pesticides are. Those suing the pesticide companies have called for a ban on neonicotinoids while the other side claims the neonicotinoid pesticides are vital to the survival of the commercial beekeeping business. Furthermore, the pesticide companies claim the neonicotinoid pesticides target insects that eat the plants that threaten the crop yields.[32] Art Schaafsma, a pest management researcher at the University of Guelph, claims the Ontario Government is severely restricting neonicotinoids for political reasons and further suggests the government "…is anti-modern agriculture".[33] He continues, saying it is "a delusion that the world can be fed with small-holdings organic agriculture".[34] The assumption is that pesticides are essential in order to support agriculture even though it was only towards the end of the 20th century that some agriculture practices actually began to use pesticides. Thus, Schaafsma's remarks may have the effect of scaring consumers about food shortages so that they lobby the Ontario Government to relax the restrictions on neonicotinoids.

Due to the polarization between pesticide producers and anti-pesticide lobbies, the pesticide companies and their supporters have begun a politicized advertising campaign to inform the public of their side of the story and let them know many farmers believe neonics are safe.[35] According to an article in the *Toronto Star*[36] and a news item on WBFO radio,[37] the pesticide lobby paid $350,000 for an advertisement in the form of an open letter in all the major newspapers in Ontario claiming that bee losses have been exaggerated. The open letter stated "…real-world field level research consistently demonstrates that response of neonicotinoid

32 Driessen, "Beekeepers Blaming Pesticides for Bee Losses".

33 Janet Thomson and Manmeet Ahluwaili, "Bee-Killing Pesticides: The Fight Ramps Up", *CBC/Radio-Canada*, May 16, 2015, accessed 8 March 2016, http://www.cbc.ca/news/canada/bee-killing-pesticides-the-fight-ramps-up-1.3075620, para. 10.

34 *Ibid.*, para 12.

35 Dan Karpenchuk, "Ontario Beekeepers Say Farm Pesticide Is Killing Millions of Drones", *WBFO*, June 8, 2015, accessed 8 March 2016, http://news.wbfo.org/post/ontario-beekeepers-say-farm-pesticide-killing-millions-drones.

36 Catherine Porter, "Safety of neonicotinoids for bees and other creatures unclear: Porter", *Toronto Star*, 6 February 2015, accessed 31 March 2016, http://www.thestar.com/news/world/2015/02/06/safety-of-neonicotinoids-for-bees-and-other-creatures-unclear-porter.html.

37 Karpenchuk, "Ontario Beekeepers Say Farm Pesticide Is Killing Millions of Drones".

seed treatments does not result in honey bee colony health issues".[38] In contrast, the Task Force on Systemic Pesticides discovered that bees are being killed by the neonicotinoid pesticides.[39] Adult bees feed on nectar and bring nectar and pollen back to the hives to feed their young; therefore, raw pollen and undiluted nectar on which the larvae feed is laced with neonicotinoids.[40] Furthermore, no complete copy of the advertisement entitled "An Open Letter to Ontarians: Getting the Facts Straight on Honey Bees" is available online (the middle section is missing) and the ad is no longer available on the Bayer website.[41] The pesticide companies have been accused of cherry-picking the facts that support the continued use of the pesticides[42] and the open letter prompted written rebuttals against the pesticide companies. For example, Environmental Defence Canada refutes the open letter's claim that honey bee colonies are up almost 60% since 2003 and notes how the advertisement neglects to mention there was a 58% overwinter bee loss in Ontario in 2014.[43] Another claim in the open letter was that honey production increased by 29% in 2014; Environmental Defence Canada contends that this omits the decrease in production of 32% in 2013 thus leaving an overall decrease of 3%.[44] These two viewpoints represent two different extremes and neither side seems willing to listen to the other. Those who agree with the pesticide companies have started a new beekeeping association called the Independent Commercial Beekeepers while the Canadian Honey Council is also critical of those lobbying for a ban on the neonicotinoids.[45] Ted Menzies of CropLife Canada, which represents pesticide companies such as Monsanto, Syngenta, Dow and Bayer, commented "...neonics are the best thing that ever happened for farmers" because they only target invertebrates

38 Bayer CropScience, "An Open Letter to Ontarians: Getting the Facts Straight on Honey Bees", January 31, 2015, accessed 8 March 2016, https://www.google.com/search?q=an+open+letter+to+ontarians+bees&source=lnms&tbm=isch&sa=X&ved=0CAgQ_AUoAmoVChMIu9PQ9tGWxgIVhi-sCh2DkwC-&biw=1246&bih=648#imgrc=ukNSfFjYCE-5qM%3A;gnaA233HcdwPtM;http%2.

39 J.P. van der Sluijs *et al.*, "Conclusions of the Worldwide Integrated Assessment on the Risks of Neonicotinoids and Fipronil to Biodiversity and Ecosystem Functioning", *Environmental Science Pollution Resources* (2014): 148–54, doi: 10.1007/s11356-014-3229-5.

40 Hopwood *et al.*, *Are Neonicotinoids Killing Bees?*

41 Bayer CropScience, "An Open Letter to Ontarians".

42 Porter, "Safety of Neonicotinoids for Bees and Other Creatures Unclear".

43 Mariah Griffin-Angus, "Getting Down to the Facts: What's Really Happening to Ontario's Bees?" 2014, Accessed 8 March 2016, http://environmentaldefence.ca/blog/getting-down-facts-what%E2%80%99s-really-happening-ontario%E2%80%99s-bees.

44 *Ibid.*

45 Brownell, "Bees, Bans and Bungling".

and are safe for animals and humans.[46] The organization further suggests pesticide use has no effect on bees so there is no need for a ban on neonicotinoids and instead blames the varroa mite as the primary cause of bee death.[47] While it may be that these pesticides are less harmful to humans, restrictions on pesticides and particularly neonicotinoids would be likely to decrease sales. There seems to be an attitude that maintaining high commercial crop yields is a top priority and, therefore, it is essential to eliminate the offending insect pests. This seems to suggest that there is either no measurable effect on the wider environment from eliminating the crop eating-insects or that any effect is insignificant since they are not considered "beneficial" insects. The method used to apply the neonicotinoids makes the pesticide water soluble so that insects are poisoned by the very food they must eat to survive. Those against neonics say the systemic nature of the neonicotinoids contrasts sharply with the consumer trend towards natural and organic foods.[48]

Professional apiculturists consider the "long term acceptable level" of winter bee deaths to be 15% of the population, but in Ontario[49] there has been a far higher level of bee deaths during the past eight winters (over 25% in six winters and only one year at the expected level of 15%).[50] A comparison of bee losses in the spring of 2014 found a low of 15% in British Columbia, while Alberta, Saskatchewan, Prince Edward Island and Quebec were between 18 and 19.1%, Nova Scotia was at 22.7%, Manitoba was at 24% and New Brunswick was at 26.3%; Ontario's bee loss was much higher, at 58% (see Fig. 8.1).[51] This suggests that Ontario had two to three times the losses in other provinces.[52]

46 Janet Thomson and Manmeet Ahluwaili, "Bee-Killing Pesticides: The Fight Ramps Up", *CBC/Radio-Canada*, May 16, 2015, accessed 8 March 2016, http://www.cbc.ca/news/canada/bee-killing-pesticides-the-fight-ramps-up-1.3075620, para. 13.

47 Sagan, "Pesticides Linked to Bee Deaths Must Be Banned Scientists Say".

48 Brownell, "Bees, Bans and Bungling".

49 Sagan, "Pesticides Linked to Bee Deaths Must Be Banned Scientists Say".

50 *Ibid.*

51 CAPA National Survey Committee and Provincial Apiarist *et al.*, "CAPA Statement on Honey Bee Wintering Losses in Canada", 2014.

52 CAPA National Survey Committee and Provincial Apiarist *et al.*, "CAPA Statement on Honey Bee Wintering Losses".

Figure 8.1 **Map of Canada with the percentage of overwinter bee losses in 2014**

Sources: Statistics Canada, Canadian Association of Professional Apiculturists. Map modified from: Canada (polygons) (JPG 37 kb): Brock University Map, Data & GIS Library. Available: Brock University Map, Data & GIS Library Controlled Access http://www.brocku.ca/maplibrary/maps/outline/North_America/can_colour.jpg (accessed February 2, 2016).

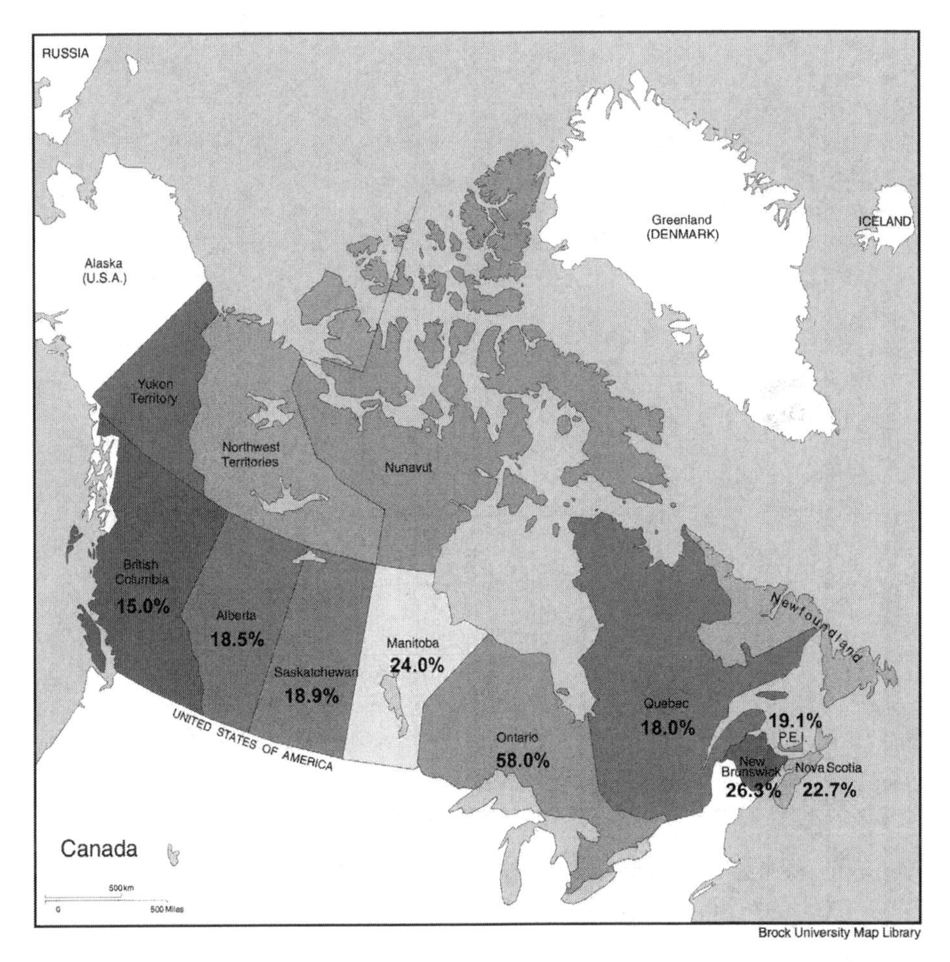

Ontario also has fewer commercial beekeeping operations and more small-scale beekeepers compared with the other provinces and, as outlined above, had much higher beehive losses in 2014.[53] This has caused some commercial beekeepers to blame the small-scale beekeepers for the high overwinter bee losses.[54] It was inferred that the hobbyists are less current and proficient because it is not their main source of income.[55] This suggests a disdain for small-scale beekeepers as only

53 Driessen, "Beekeepers Blaming Pesticides for Bee Losses".
54 *Ibid.*
55 *Ibid.*

hobbyists by those with larger bee yards. The assumption is that those beekeepers in large commercial bee yards have greater skills. However, small-scale beekeepers may be more likely to research all sides of an issue exactly because they know they are not experts and so it is possible they would have a greater understanding of the issues than some who make their living from beekeeping. An article by Paul Driessen[56] provides details on the current controversy while appearing to favour the pesticide companies.

In Ontario, the corn seeds coated with neonics tend to generate more pesticide-laced dust than those of other crops therefore may be a possible cause of the province's bee losses.[57] Some beekeepers point out that Alberta uses neonicotinoids on the canola crop but that this does not result in the same effect on bees; Alberta's 2014 bee loss from overwintering was one of the lowest in Canada thereby indicating that neonics cannot be the cause of the increased overwinter bee losses in Ontario.[58] Alberta farmers seem to need less of the pesticide since they grow canola rather than corn and the canola crop tends to generate less pesticide dust which may explain Alberta's relatively lower overwinter bee losses.[59]

It may also be possible that the concentration of neonicotinoids in the soil and water sources have simply not yet reached a critical level where an increase in the overwinter bee losses would be noticeable in Alberta. In fact, Tibor Szabo, a beekeeper "…thinks it's only a matter of time before repeated exposure from planting neonics treated seeds year after year starts to kill Alberta's bees too".[60] British Columbian beekeepers have not generally suffered such significant overwinter bee losses; however they were concerned because part of the province had a 32% overwinter bee loss in 2013 and colony collapse disorder seems to them to be related to some extent to the neonicotinoids.[61] The president of the British Columbia Honey Producer Association, Catherine Culley, mentioned that they were concerned about the ramifications of a potential local large-scale overwinter bee loss due to neonicotinoid pesticides.[62] While there is major controversy in Ontario, British Columbia honey producers and farmers have learned from Ontario and seem to be willing to work together on this problem.[63] British Columbia does not grow any significant amount of corn, canola and soybeans, suggesting that pesticide-laced dust may not be a problem in that province.

56 *Ibid.*
57 Brownell, "Bees, Bans and Bungling".
58 *Ibid.*
59 *Ibid.*
60 *Ibid.*, para. 32
61 Glenda Luymes, "Large-scale Pesticide Use Threatens Hives B.C. Beekeepers Say", *Press Reader British Columbia*, 15 March 2015, NewspaperDirect Inc. http://www.pressreader.com/canada/times-colonist/20150315/textview
62 *Ibid.*
63 *Ibid.*

Viewpoint of the pesticide companies and supporters

Some independent commercial beekeepers are against lobbying Ottawa to ban or restrict neonicotinoids. One beekeeper, Mr Simpson, complains the Ontario Beekeepers Association's "…agenda has really become overwhelmed by banning neonicotinoids, neonicotinoid advocacy, [and] lawsuits".[64] He continues, saying he has "no tolerance for highly politicized [discussion that is] mostly focused on anti-agriculture, anti-crop protection, [and] anti-science [points of view]".[65] He is an example of those farmers who are vehemently opposed to banning neonicotinoids. However, Pride Seeds Canada suggests that the Ontario Government's actions "effectively [are] saying that it doesn't trust farmers".[66] There seems to be no question that neonics are not a healthy food for bees nor that they kill bees if the dosage is high enough.[67] Some say the pesticide is useful and there is no proof that the deaths are directly due to the pesticide while others may concede that a loss of bees is acceptable to ensure a good crop.[68] Interestingly, the concern for those aligned with the pesticide companies seems focused on how to keep the crop yields high if the neonics are not used. They believe banning the neonicotinoids would be more harmful to the Canadian agriculture business than continuing their use.[69]

There was a change in application methods in 2014 to decrease dust exposure, which was followed by a 70% decline in overwinter bee losses that it is hoped may correct the problem of the diminishing bee population.[70] The alternative proposed seems to be returning to the more toxic pesticides that negatively affect humans, bees and the environment,[71] and supporters of neonicotinoids appear to be rejecting certified organic agriculture as another viable option. Part of the argument is that, although Ontario has half of all beekeepers in Canada, many are smaller honey producers which means that they account for only 15% of bees countywide.[72] Rod Scarlett of the Canadian Honey Council suggests, "Most of Ontario's beekeeping operations are run by keepers with a lot less to lose in antagonizing relationships with farmers and pesticide makers".[73] However, it also makes those beekeepers more independent and willing to voice their concerns. With the high overwinter losses of bees in Ontario, the tendency would be that a smaller bee apiary with less hives would be likely to suffer more and lose a greater percentage of their hives.

64 Driessen, "Beekeepers Blaming Pesticides for Bee Losses", para. 7
65 *Ibid.*
66 Thomson and Manmeet Ahluwaili, "Bee-Killing Pesticides", para. 9.
67 Brownell, "Bees, Bans and Bungling", para. 15
68 Brownell, "Bees, Bans and Bungling".
69 *Ibid.*
70 *Ibid.*
71 *Ibid.*
72 *Ibid.*
73 *Ibid.*, para. 15

The viewpoint of the beekeepers and supporters

Although proponents of the use of pesticides maintain that planting practices are the source of the problem (including over-use of pesticides in an application, creating dust clouds when planting the treated seed and forgetting to clean the equipment after planting), many beekeepers want to follow the European example. In 2013 the EU banned neonicotinoids for two years, starting in 2014, following the observation of a global bee population decline.[74] The Ontario Government approved the neonicotinoids ban in the spring of 2015 despite the advertising campaign to prevent it. Even the Canadian Association of Physicians for the Environment and the Registered Nurses of Ontario jointly entered the controversy by advertising on the Toronto transit system about needing to ban neonicotinoids for a healthy population.[75]

Comments such as those of Steve Denys (a grain farmer and vice-president at Pride Seeds) suggest that the use of coated seed technology is less expensive (it involves less tilling and uses less fuel) and uses a lower level of pesticides for the same amount of plant protection.[76] This comment inadvertently illustrates at least part of the problem because it implies that the use of pesticides for protection is essential. However, there is also the option to use active pest management. The former approach uses pesticides just in case there is an insect infestation (which is a fast and easy solution) while the latter monitors the pest levels and spot-sprays when and where there is an infestation (which generally requires more time and knowledge of the insects and their life-cycle). A new lubricant devised by Bayer was able to decrease the neonicotinoid dust concentration by 21%.[77] This is a decrease of close to one-quarter of the previous pesticide dust level and could be a significant improvement. However, to assess the lubricant's ability to decrease the neonicotinoids levels by generating less pesticide dust it is important to compare both the new and the original levels of pesticide dust on a per unit basis. In fact, Tibor Szabo (a beekeeper) suggested "A 21% per cent reduction of something that's very, very toxic isn't going to make me feel better" and does not believe it will save the bees.[78] In fact, Health Canada stated that, "Current agricultural practices related to the use of neonicotinoid-treated corn and soybean seed are not sustainable due to their impacts on bees and other pollinators".[79] This statement was the result of testing conducted by Health Canada on dead bees where they discovered 70% of

74 Janet Thomson, "Bee Researchers Raise More Warning Flags about Neonicotinoid Pesticides", *CBC/Radio-Canada*, May 27, 2014, accessed 8 March 2016, http://www.cbc.ca/news/technology/bee-researchers-raise-more-warning-flags-about-neonicotinoid-pesticides-1.2644354.

75 CBC/Radio-Canada, "Doctors, Nurses Urge Ontario to Ban Neonicotinoids".

76 Thomson, "Bee Researchers Raise More Warning Flags about Neonicotinoid Pesticides".

77 *Ibid.*, para. 14

78 *Ibid.*, para. 13

79 *Ibid.*, para. 8

the dead bees tested in the spring of 2013 had neonicotinoids on them because of the dust clouds created when planting the treated seeds.[80]

Madeleine Chagnon, a Canadian environmental scientist, discovered bees come in contact with neonicotinoids even if not exposed to planting dust, thus suggesting the bees "were exposed to the pesticide while collecting pollen from maturing plants".[81] She adds, "They're taking in something that they will ultimately die from, and [furthermore] they're taking this into the hive and feeding on it all winter, [and] then we wonder why we have winter mortality".[82] Other scientists corroborate her findings. One scientist, Henk Tennekes, remarked, "There is little doubt that neonicotinoids are implicated in bee decline".[83]

Research on neonicotinoids for and against their use

In the mid-1990s neonicotinoids started to be used in commercial farms and domestic gardens.[84] At the time, these pesticides were a major improvement since they were targeted at the insects eating the crop thus increasing crop yields and were much less harmful to humans.[85] Neonicotinoids are systemic because they are soluble and so can be absorbed into the plant easily thereby protecting the crop by causing harm to those insects that pollinate or eat the plant.[86] Since neonicotinoids application methods are by injection, soil drench, spray or even as a seed coating they are relatively easy to use.[87] These pesticides may affect birds, bees and other insects eating pollen or nectar because they are designed to attack the nervous system by blocking nerve impulses, which effectively paralyses the organism.[88]

The relationship between pollinators and plants is very important; many plants need pollinators such as bees in order to reproduce and the plants contain pollen and nectar which allow the insect or bird to survive.[89] This symbiotic relationship between the pollinator animal and the plant is ecologically sensitive because the decrease in plants would cause a decline in pollinator animals and similarly if the pollinator animals such as bees and hummingbirds started to decline this could threaten the plant population. In fact, current research states neonicotinoid pesticide "provides a direct threat to bees and other flower-visitors".[90] This could become

80 Thomson, "Bee Researchers Raise More Warning Flags about Neonicotinoid Pesticides".
81 *Ibid.*, para. 17
82 *Ibid.*, para. 18
83 *Ibid.*, para. 19
84 Hopwood *et al.*, *Are Neonicotinoids Killing Bees?*
85 *Ibid.*
86 *Ibid.*
87 *Ibid.*
88 *Ibid.*
89 *Ibid.*
90 *Ibid.*, p. 1

a problem since globally 85% of flowering plants rely on pollinators that include bees, butterflies, moths and wasps.[91] With the decline of pollinators such as honey bees, this could decrease the biodiversity of plants and significantly decrease crop yields thus causing a possible worldwide food shortage in the future. "Pollinators are keystone species in most terrestrial systems: they pollinate the seeds and fruits that feed everything from songbirds to Grizzly bears".[92] Further, honey is a commercial crop that relies on having healthy hives. Thus, this is not just about the commercial honey bees but possible disruption to the temperate biospheres. Furthermore, cities continue to grow larger while the need for agricultural land in production increases which has resulted in diminishing forests, increased pollution and increased use of pesticides and herbicides. This is causing the destruction of natural habitats or creating small pockets rather than the large, continuous natural habitats needed for sustainability. These changes are affecting wildlife including pollinators such as bees and moths.

However, habitat is only part of the issue—water is also a critical concern. A study of surface water showed that honey bees prefer warmer water temperatures, and standing water which provides them with minerals makes puddles attractive to the bees.[93] However since the neonicotinoids are water soluble, these pesticides have contaminated the puddles honey bees use as a water source which results in the bees receiving sub-lethal doses.[94] The possible dosages found in the puddles will naturally increase as the evaporation of water occurs thus increasing the problem. This exposure to the neonicotinoids has resulted in effects such as "reduced consumption…impaired foraging behaviour and reduced colony growth and queen production".[95] Thus, this suggests the neonicotinoid pesticides leaching into water and especially groundwater may be causing a disruption in the bee food chain.

Neonicotinoids can persist in soil for months or even years after a single application and measurable amounts of residues were found in woody plants up to six years after application.[96] The most frequently used neonicotinoids are imidacloprid and clothianidin. These are also the most persistent in the soil (with the longest half-life); therefore the likelihood of increased exposure is greater since imidacloprid has a half-life of 40–997days while clothianidin has a half-life in the soil of 148–1,155 days.[97] This suggests that continued neonicotinoid use may increase bees'

91 Jeff Ollerton *et al.*, "How Many Flowering Plants Are Pollinated by Animals?" *Oikos* 120(3) (2011): 321–26, doi:10.1111/j.1600-0706.2010.18644.x.

92 Hopwood *et al.*, *Are Neonicotinoids Killing Bees?* p. 2.

93 Olivier Samson-Robert *et al.*, "Neonicotinoid-Contaminated Puddles of Water Represent a Risk of Intoxication for Honey Bees", *PLoS ONE* 9(12) (2014): e108443, doi:10.1371/journal.pone.0108443.

94 *Ibid.*

95 *Ibid.*, p. 9

96 Hopwood *et al.*, *Are Neonicotinoids Killing Bees?*

97 *Ibid.*

and other pollinators' exposure to sub-lethal concentrations since farmers may be applying neonicotinoids while there is still a residue in the soil from the previous application.

Discussion and implications

This fight between the two groups of beekeepers, pesticide companies and environmentalists does not seem to have an end in sight. The pesticide companies wish to convince the public that those against the pesticides are short-sighted, are hurting the farmers and that pesticides are essential for a good crop. In addition, they state their belief that "bees matter" and that they are working hard for the honey bee. Simultaneously, these companies maintain that the neonicotinoids are not adversely affecting the bees and that they are being misrepresented by those who oppose the use of such pesticides. The idea that farmers cannot grow crops without pesticides is one that has become entrenched in the agriculture business through the expectation of a perfect, pest-free crop. However, humans have practised agriculture for centuries without pesticides and only after the Second World War did the practice of using pesticides begin in order to produce a perfect crop. Perhaps this expectation of less work and high yields is causing ecosystems to become more unbalanced thus increasing the pest infestations seen in the monoculture agricultural environment in Ontario.

Knowledge gaps

One of the difficulties is that pesticide company research may not be in the public domain, so scientists do not have all the available information on the pesticides. Health Canada was pressured by environmental groups to make pesticide companies release their studies on bee deaths related to use of pesticides but no studies had been released up to May 2014.[98]

A large and growing body of research demonstrates that neonicotinoid insecticides harm multiple bee species, yet substantial knowledge gaps remain. Based on its findings, the Xerces Society for Invertebrate Conservation makes some major recommendations that mostly relate to pesticides. The first recommendation suggests reassessing agricultural products previously considered safe for bees such as those containing neonicotinoid insecticides to ensure that bees are not at risk from these pesticides and that beekeepers are able to keep their bees healthy

98 Sagan, "Pesticides Linked to Bee Deaths Must Be Banned Scientists Say".

and safe.[99] This suggests there needs to be more oversight of patented products to ensure complete transparency of the possible immediate and long-term effects of the each product. In particular, there may be a cumulative effect of repeated neonicotinoid applications and synergistic effects when using neonicotinoid with multiple products such as herbicides and fungicides.[100] Second, it is important to understand how the pesticides affect not only honey bees but also other beneficial and "pest" insects.[101] Since bee larvae eat the pollen and honey there is a possibility they could be exposed to neonicotinoids in their food and the effects on the larvae are not known.[102] Could this effect prevent larvae from emerging or affect the future reproduction or development of bee larvae? The ecosystem may be dramatically affected if the insects considered as pests have been eliminated, as they are food for other animals. The approved uses of all products should be either conditional registrations, or re-examined and/or suspended until we understand how to manage the risk to bees. The risk from exposure to neonicotinoid insecticides then needs to be re-evaluated so there is a greater understanding of how they affect different soil types, plants and insects and what level of residue is associated with the different rates of application and resulting effects on groundwater.[103] Before registration for a specific crop or ornamental plant species, research facilities should investigate the influences of application rate, application method, target plant species and environmental conditions on levels of neonicotinoid residues in pollen and nectar.[104] Since residues can persist for long periods, it is important to determine the levels of the neonicotinoids in perennial and ornamental plants to see if they may also have accumulated concentrations that could be or become sub-lethal or lethal. The effects on plants, insects, other animals and humans will be very diverse; it is likely that the effects on insects will be relatively direct and immediate while the effect on other animals and humans could develop over a long period. This means that developing a methodology to contain and correct effects of the pesticides is crucial.

Since there is a high risk for bees and other pollinators to be affected adversely by neonicotinoids, alternative control methods are recommended, such as targeted pest treatment and naturalized habitat zones for animals and insects to allow for natural predators to help control the pest insects and increase the diversity of wildflowers and other plants. Canada needs increased scrutiny for new pesticides and continual monitoring of existing pesticides. There has not been sufficient research on solitary bees to see how the neonicotinoids may affect them. And yet, usage has now gone up for some time—is it too late to reverse the damage?

99 Hopwood *et al.*, *Are Neonicotinoids Killing Bees?*
100 *Ibid.*
101 *Ibid.*
102 *Ibid.*
103 *Ibid.*
104 *Ibid.*

Furthermore, comprehensive and ongoing independent research is important to counteract any biased research. Health Canada should adopt a more cautious approach to approving all new pesticides, using a comprehensive assessment process that adequately addresses the risks to honey bees, bumblebees and solitary bees in all life stages. Beyond bees, there is a need to understand how neonicotinoids affect other pollinators, such as butterflies, moths, beetles, flies and wasps. These insects make minor contributions to crop pollination, but serve important roles within crop systems and other ecosystems (e.g. as larvae, hover flies may be predators).[105]

Research conducted on topics such as the maximum lethal or sub-lethal dosages may prevent the bees and other pollinators from dying but will it allow them to thrive? Since the half-lives of the neonicotinoids are known, research based on maintaining a heathy ecosystem is important. This is because it is not possible to narrow the scope of research and study each component separately when every part of the ecosystem is interrelated. Therefore, the pesticides used will affect every aspect of the ecosystem—the effect on the bees and other pollinators including nectar-feeding birds is an immediate concern, as would be the microorganisms in the soil and water. However, these are at the bottom of the food chain and if they are affected those higher up in the food chain will also be affected, although the effects will not be as noticeable in the short term. There is a saying that "you are what you eat" for a very good reason—everything an animal or person takes into their body has an effect on them at macro or micro level—even the pesticides in the honey. Could pesticides and other chemicals cause new illnesses in the 21st century?

Bibliography

Arkins, E. "Ontario Restricts Use of Pesticides Blamed for Decline of Bee Populations". *Globe and Mail*, June 9, 2015. Accessed 8 March 2016. http://www.theglobeandmail.com/report-on-business/ontario-unveils-first-restrictions-on-class-of-pesticides/article24874268/.

Bayer CropScience. "An Open Letter to Ontarians: Getting the Facts Straight on Honey Bees", January 31, 2015. Accessed 8 March 2016. https://www.google.com/search?q=an+open+letter+to+ontarians+bees&source=lnms&tbm=isch&sa=X&ved=0CAgQ_AUoAmoVChMIu9PQ9tGWxgIVhi-sCh2DkwC-&biw=1246&bih=648#imgrc=ukNSfFjYCE-5qM%3A;gnaA233HcdwPtM;http%2.

Benzie, R. "Ontario First in North America to Curb Bee-Killing Neonicotinoids Pesticides". *The Star*, June 9, 2015. Accessed 8 March 2016. http://www.thestar.com/news/queenspark/2015/06/09/ontario-first-in-north-america-to-ban-bee-killing-neonicotinoid-pesticides.html.

105 *Ibid.*, para.25

Brownell, Claire. “Bees, Bans and Bungling: How an Anti-Pesticide Campaign May Spell Serious Trouble”. *Financial Post*, November 7, 2014. Accessed 8 March 2016. http://business.financialpost.com/news/economy/bees-bans-and-bungling-how-an-anti-pesticide-campaign-may-spell-serious-trouble.

CAPA National Survey Committee and Provincial Apiarist, Paul Kozak, Steve Pernal, Melanie Kempers, Rheal Lafreniere, Anne Leboeuf, Medhat Nasr, Geoff Wilson, Jessica Morris, Paul Van Westerndorp, Cris Maund, Cris Jordan, Steve Tatrie, and David Ostermann. “CAPA Statement on Honey Bee Wintering Losses in Canada”, 2014.

CBC News. “Grain Farmers of Ontario Lost Bid to Block Law That Limits Use of Neonicotinoids on Corn, Soybeans”. October 26, 2015. Accessed 8 March 2016. http://www.cbc.ca/news/business/grain-farmers-neonics-1.3289326.

CBC/Radio-Canada. “Doctors, Nurses Urge Ontario to Ban Neonicotinoids”, November 17, 2014. Accessed 8 March 2016. http://www.cbc.ca/news/business/doctors-nurses-urge-ontario-to-ban-neonicotinoids-1.2837919.

Driessen, Paul. “Beekeepers Blaming Pesticides for Bee Losses Could Face Bigger Losses in Court”. *Financial Post*, October 27, 2014. Accessed 8 March 2016. http://business.financialpost.com/fp-comment/beekeepers-blaming-pesticides-for-bee-losses-could-face-bigger-losses-in-court.

Environmental Communications Options Orcalepoll Research Limited. *Ontario Omnibus Survey Report*. Canada, 2014.

Griffin-Angus, Mariah. “Getting Down to the Facts: What’s Really Happening to Ontario’s Bees?” 2014. Accessed 8 March 2016. http://environmentaldefence.ca/blog/getting-down-facts-what%E2%80%99s-really-happening-ontario%E2%80%99s-bees.

Hopwood, Jennifer, Mace Vaughan, Matthew Shepherd, David Biddinger, Eric Mader, Scott Hoffman Black, and Celeste Mazzacano. *Are Neonicotinoids Killing Bees ?* Portland, OR: Xerces Society for Invertebrate Conservation, 2013.

Karpenchuk, Dan. “Ontario Beekeepers Say Farm Pesticide Is Killing Millions of Drones”. *WBFO*, June 8, 2015. Accessed 8 March 2016. http://news.wbfo.org/post/ontario-beekeepers-say-farm-pesticide-killing-millions-drones.

Kearns, Carol A, David W Inouye, and Nickolas M Waser. “Endangered Mutualisms: The Conservation of Plant-Pollinator Interactions”. *Annual Review of Ecology and Systematics* 29 (1998). Annual Reviews: 83–112, http://www.jstor.org/stable/221703.

Lu, Chensheng, Kenneth M. Warchol, and Richard a. Callahan. “Sub-Lethal Exposure to Neonicotinoids Impaired Honey Bees Winterization before Proceeding to Colony Collapse Disorder”. *Bulletin of Insectology* 67 (1) (2014): 125–30.

Luymes, Glenda. “Large-scale Pesticide Use Threatens Hives B.C. Beekeepers Say”. *Press Reader British Columbia*, 15 March 2015. NewspaperDirect Inc. http://www.pressreader.com/canada/times-colonist/20150315/textview

Ollerton, Jeff, Rachael Winfree, and Sam Tarrant. “How Many Flowering Plants Are Pollinated by Animals?” *Oikos* 120 (3) (2011): 321–26, doi:10.1111/j.1600-0706.2010.18644.x.

Porter, Catherine. “Safety of neonicotinoids for bees and other creatures unclear: Porter”. *Toronto Star*, 6 February 2015. Accessed 31 March 2016. http://www.thestar.com/news/world/2015/02/06/safety-of-neonicotinoids-for-bees-and-other-creatures-unclear-porter.html.

Sagan, Aleksandra. “Pesticides Linked to Bee Deaths Must Be Banned Scientists Say: Neonicotinoids, Dipromil Linked to Ecosystem Damage in New Study”. *CBC/Radio-Canada*, June 26, 2014. Accessed 8 March 2016. http://www.cbc.ca/news/technology/pesticides-linked-to-bee-deaths-must-be-banned-scientists-say-1.2685492.

Samson-Robert, Olivier, Geneviève Labrie, Madeleine Chagnon, and Valérie Fournier. “Neonicotinoid-Contaminated Puddles of Water Represent a Risk of Intoxication for Honey Bees”. *PLoS ONE* 9 (12) (2014): e108443, doi:10.1371/journal.pone.0108443.

Thomson, Janet. "Bee Researchers Raise More Warning Flags about Neonicotinoid Pesticides". *CBC/Radio-Canada,* May 27, 2014. Accessed 8 March 2016. http://www.cbc.ca/news/technology/bee-researchers-raise-more-warning-flags-about-neonicotinoid-pesticides-1.2644354.

Thomson, Janet, and Manmeet Ahluwaili. "Bee-Killing Pesticides: The Fight Ramps Up". *CBC/Radio-Canada,* May 16, 2015. Accessed 8 March 2016. http://www.cbc.ca/news/canada/bee-killing-pesticides-the-fight-ramps-up-1.3075620.

van der Sluijs, J. P., V Amaral-Rogers, L P Belzunces, M F I J Bijleveld van Lexmond, J M Bonmatin, M Chagnon, C.A. Downs, L. Furlan, D.W. Gibbons, C. Giorio, V.Girolami, D. Goulson, D. P. Kreutzweiser, C. Krupke, M. Liess, E. Long, M. McField, P Mineau, E. A. D. Mitchell, C. A. Morrissey, D. A. Noom, L Pisa, J. Settele, N. Simon-Delso, J. D. Stark, A. Tapparo, H. Van Dyck, J. van Praagh, P. R. Whitehorn, and M.Wiemers. "Conclusions of the Worldwide Integrated Assessment on the Risks of Neonicotinoids and Fipronil to Biodiversity and Ecosystem Functioning". *Environmental Science Pollution Resources* (2014): 148–54, doi: 10.1007/s11356-014-3229-5.

van Engelsdorp, Dennis, Jay D. Evans, Claude Saegerman, Chris Mullin, Eric Haubruge, Bach Kim Nguyen, Maryann Frazier, Jim Frazier, Diana Cox-Foster, Yanping Chen, Robyn Underwood, Dacis R Tarpy and Jeffery S Pettis. "Colony Collapse Disorder: A Descriptive Study". *PLoS ONE* 4 (8) (2009): doi: 10.1371/journal.pone.0006481.

Williams, Geoffrey R., Aline Troxler, Gina Retschnig, Kaspar Roth, Orlando Yañez, Dave Shutler, Peter Neumann, and Laurent Gauthier. "Neonicotinoid Pesticides Severely Affect Honey Bee Queens". *Scientific Reports* 5 (2015): 14621, doi: 10.1038/srep14621.

9

Bee colony and food supply collapse

Could investors be the cavalry?*

Raj Thamotheram and Olivia Stewart
Preventable Surprises, UK

What is the price of a bee? And more generally, where does the extinction of bee populations and the disruptive impact this would have on agriculture and global food supply ecosystems fit into discounted cash flow and other traditional risk decision-making tools used by investors?

The simple answer is that bees don't fit.[1] This is not about some, or even many, investment analysts or firms being stupid or irresponsible. Rather, it is because the investment system has a risk management system which is fundamentally unfit for purpose when it comes to managing sector-wide and systemic risks.[2] And given that much of the food supply chain requires pollination for plants to create plant-based products, bee decline and colony collapse disorder (CCD) have the potential

* This chapter is based on an earlier article co-authored by Raj Thamotheram and Aidan Ward for *Investment & Pensions Europe* entitled "Put the bee back in beta", IPE, March 2013, accessed 31 March 2016, http://www.ipe.com/put-the-bee-back-in-beta/50256.fullarticle.

1 This was confirmed in a personal communication (February 2013) with an asset owner client who contacted their fund manager who is one of the chief investors in a major pesticide company.

2 The failure of investors to address systemic risks is the focus of the Investment Integration Project: http://www.investmentintegrationproject.com/

to cause systemic risk.[3] Unless investor action on bee decline builds from this understanding of the challenge, the action will have very limited value.

This chapter focuses on a class of pesticide called neonicotinoids ("neonics"), which are derived from nicotine and target insects' nervous systems. The use of neonics has exploded over the past decade, thanks to a perception that they are both safer and more effective than the pesticides they replaced. The chapter—which assumes, in line with the other chapters in this book, that neonics are one factor in bee decline—begins by mapping out the kind of investment system that *would* be fit for purpose in terms of managing the systemic risk entailed in the loss of bee populations, how this compares with the way the investment system works today and what needs to change for this different approach to be implemented.

Fiduciary capitalism

The term "fiduciary capitalism" describes a financial system in which the investment decisions of institutional investors are based on considerations of intergenerational equity, in which negative externalities are minimized and positive externalities maximized to benefit beneficiaries across time.

As John Rogers, former CEO of the CFA Institute explains, the institutional asset owners that are the dominant players in capital formation "are legally bound to a duty of care and loyalty and must place the needs of their beneficiaries above all other considerations".[4]

When these asset owners (such as pension funds, endowments, foundations, and sovereign wealth funds) place considerations of intergenerational equity at the heart of their investment processes they will (re)shape behaviour in the financial markets and the broader economy—and encourage long-term thinking. "As 'universal owners', fiduciaries foster a deeper engagement with companies' management teams and public policy-makers on governance and strategy", continues Rogers.

The main problem with fiduciary capitalism is that it is not here yet. This means that the capacity of institutional asset owners to rise above being passive

3 According to the US Department of Agriculture, about one-third of the food humans ingest benefits directly or indirectly from honey bee pollination; US Department of Agriculture, "USA and EPA Release New Report on Honey Bee Health", accessed 23 June 2013, http://www.usda.gov/wps/portal/usda/usdamediafb?contentid=2013/05/0086.xml&printable=true23.

4 J. Rogers, "A New Era of Fiduciary Capitalism? Let's Hope So", CFA Institute Blog, 2014, accessed 13 September 2015, https://blogs.cfainstitute.org/investor/2014/04/28/a-new-era-of-fiduciary-capitalism-lets-hope-so/.

"consumers" of market trends and become "future makers"[5]—with the potential to collectively influence systemic, market-wide outcomes by assertively engaging with policy-makers and investee companies—is very much work in progress.

Indeed, it is because of the focus on portfolio metrics narrowly defined that systemic risks are inevitably ignored. What does "portfolio metrics narrowly defined" mean? It means defining the "success" or the goal of investment as portfolio returns relative to a benchmark (i.e. beating a benchmark for an active investor or matching a benchmark for a passive investor). It also mean achieving an "absolute return" goal (the benchmark here being investments with "zero" return). What this goal, this definition of success inevitably loses out on (i.e. ignores) is any assessment of investments' ability to "create value"—that is, to understand how funds have been put to "productive" use. The former approach (benchmark oriented) takes price (or more precisely returns as measured in stock price) as a proxy for value creation. (This is why the efficient market hypothesis is so crucial to modern portfolio theory, because if markets are not correctly priced, then price cannot be considered a proxy for value.) So investors concerned only with price-based performance in effect can ignore whether value is "really" created by price appreciation—they just assume it is. Systemic risk is therefore irrelevant to the equation, because it is (theoretically) accounted for in efficient prices already.[6]

The good news is that there is every reason for thinking that this transition in mental models could be accelerated (see "Who needs to do what", below) with the result that asset owners, now thinking as fiduciary capitalists, decide to shape the corporate purpose of a few companies to take account of the science of bee decline and to do so in order to safeguard the beta of the portfolio as a whole. This change in mind-set will require concerted action by, in particular, academics, consultants, regulators and specialist NGOs since the existing investment beliefs and worldviews are deeply, if unconsciously, held. It will also require pressure on the asset owners to execute actions which are easy to take: actions reinforce behaviour and beliefs, which in turn affect future actions.[7]

5 The term "future maker" has been adapted from Mercer's "Investing in a Time of Climate Change", Mercer LLC, 2015, accessed 7 April 2016, http://www.mercer.com/our-thinking/investing-in-a-time-of-climate-change.html, as an individual who builds on their fiduciary duty to influence systemic, market-wide outcomes. Mercer focus on climate risk but the principles apply to CCD. This explicitly involves engaging with policy-makers, corporates and other key stakeholders in order to seek to fulfil one's fiduciary duty of loyalty and care.

6 Steve Lydenberg, personal communication, 10 March 2016.

7 C. Heath and D. Heath, *Switch: How to Change Things When Change is Hard* (London: Random House Business Books, 2010).

Requirement for a fundamentally different risk strategy

Fiduciary capitalists are inherently more sensitive to the concept of systemic risk, which can be described as the exposure of a given node in the system to defaults by other firms.[8]

Systemic risk captures the risk of collapse inherent in an entire financial system or entire market segment. One consequence is "stranded assets", defined as assets that suffer from unanticipated or premature write-offs, downward revaluations or are converted to liability.[9] Due to their very nature, systemic risks tend to build up over time, reach a tipping point and then collapse. The effects can be widespread and (very) costly.

The global financial crisis (GFC), for example, wiped $5.4 trillion off global pension assets in 2008[10] and would have resulted in much greater stranding of assets had governments not acted to bail out many financial institutions.

And the speed with which damage can spread is hard to predict. Just one week after the VW vehicle emissions crisis broke, the whole auto industry appeared to be facing downgrades, the German economy started to show signs of loss and producers of diesel fuel began to face negative market sentiment.

A troublesome irony of catastrophic systemic collapses is that few people seem to see them coming and even fewer do something about it. Behavioural finance is good at explaining the former: why there are such events.[11] However, what remains relatively unidentified is how to deal with systemic risks *before* they reach a tipping point and crisis ensues. It is important to note that while the phenomenon of ESG (environmental, social and governance) investing is arguably a reaction to these events, there is little evidence that ESG investing as currently practised is able to meet this particular need.[12]

8 L. Eisenberg and T.H. Noe, "Systemic Risk in Financial Systems", *Management Science* 47(2) (2001): 236–9.

9 A. Ansar *et al.*, *Stranded Assets and the Fossil Fuel Divestment Campaign: What Does Divestment Mean for the Valuation of Fossil Fuel Assets?* (Stranded Assets Programme, SSEE, University of Oxford, 2013).

10 J. Yermo and C. Severinson, "The Impact of the Financial Crisis on Defined Benefit Plans and the Need for Counter-Cyclical Funding Regulations", *OECD Working Papers on Finance, Insurance and Private Pensions*, No. 3 (OECD Publishing, 2010), accessed 7 April 2016, http://www.oecd.org/finance/private-pensions/45694491.pdf.

11 D. Ariely, *Predictably Irrational* (New York: HarperCollins, 2008); and D. Rook and B. Caldecott, *Cognitive Biases and Stranded Assets: Detecting Psychological Vulnerabilities within International Oil Companies* (Smith School of Enterprise and the Environment, University of Oxford, 2015) for example.

12 Few investors who have committed to encouraging good corporate governance (e.g. members of ICGN) or ESG investing (e.g. members of PRI) have publicly indicated that there was anything they could have done differently about, or indeed, that there is anything that they need to learn, from cases such as BP, Tepco, Tesco or most recently VW.

It is of course very encouraging that at least a few investors are exploring how to correct mis-pricing of assets related to bee decline.[13] These fund managers are, indeed, well ahead of their peer group and should be recognized as such. And this approach may even have some partial value with regard to a particular company, assuming these firms are brave enough to be as contrarian as the evidence warrants.[14]

The use of better integration strategies, such as revised pricing or better diversification, should address company risk and may even manage sector risk. But as the much bigger risk is in this case systemic, these actions are not fit for purpose. Put simply, the risks associated with the failure of agriculture are so diversified across asset classes and sectors that it renders traditional risk management strategies ineffective. Because this risk is systemic, it cannot be hedged or sold.

The only means of mitigating the risks associated with bee decline is to prevent it. Whatever this means for the profitability of two or three chemical companies—and in reality it is unlikely to be as catastrophic for these companies as they imply[15]—the rapid phasing out of neonics, if these are the cause of bee decline, will almost certainly be cost-neutral or more likely positive over the long term for diversified investors.

While the risks associated with bee decline require a major shift in risk management culture, it is both important and encouraging to note that these considerations fit well with the legal debate on the purpose of the corporation and fiduciary duties of directors. Section 172 of the Companies Act 2006 obliges the directors to act in good faith so as to "promote the success of the company for the benefit of its members as a whole". Members include present as well as future shareholders. Section 172 specifically refers to the consequences of any decision in the long term and impact on the environment.

Corporate law in other jurisdictions typically also allows directors to take account of other interests that further the interests of the company.[16] The findings of the 2009 UN Corporate Law Tools Project confirmed this assertion in over 40 jurisdictions in the context of corporate responsibility to respect human rights.[17]

13 See Chapter 6.

14 Despite the strong case for being seriously underweight on banks, most institutional investors today are sector neutral. The reasons are well understood and relate to the fact that investment firms are more focused on managing the risks to themselves than their clients. But as many investment professionals found out with the tech bubble and then the GFC, being right too far ahead is not career enhancing (see R. Thamotheram and A. Ward, "Whose Risk Counts?" in *Cambridge Handbook of Institutional Investment and Fiduciary Duty*, edited by James Hawley *et al.* (Cambridge University Press, July 2015).

15 There have been repeated occasions when companies have claimed that the costs of acting responsibly would be catastrophic. Examples include the oil industry over lead free petrol and the chemical sector over the ban of CFCs.

16 The Modern Corporation Statement on Company Law available online at http://themoderncorporation.wordpress.com/company-law-memo/, accessed 9 March 2016.

17 http://business-humanrights.org/en/reports-on-corporate-law-tools-0, accessed 9 March 2016.

In the case of large institutional investors, reflecting the long-term purpose of such organizations, the discretion to take bee decline into account is arguably becoming a mandatory fiduciary duty.[18]

What should investors do in the face of irreducible uncertainty?

The above argument would, of course, be very powerful if it had been proven beyond reasonable doubt that bee decline and CCD are systemic risks and that neonics play a significant causative role.

However, as is mentioned at many points throughout this book, there is ongoing debate in the scientific literature about whether neonics are particularly dangerous to bees, especially in low doses. To summarize for non-specialists, there is growing evidence of a connection between bee decline and neonics, but the debate is by no means over. However, the fact that there is a growing possibility of a scientific connection between bee decline and pesticide use means that ignoring a link would likely be a mistake for bees, humans and stock markets! Fiduciaries can choose to override the consensus scientific opinion but if they are taking this route, they should also be willing to explain why they know better than scientists.

What will be the economic impact of bee decline? On this dimension of the challenge, we are an even longer way from consensus—in this case among economists—about the macro implications. Indeed, there are actually only a handful of studies on this question.

But this lack of consensus is neither unexpected nor should it be used to justify delay. The easy answer is we need more research, more data, but there are strong grounds for concluding that this would be a case of paralysis by analysis.

Even with climate change, more conclusive evidence has not (yet) triggered assertive leadership by investors. Because financial models about the future are normally inconclusive, the way in which data and outputs from models are interpreted and what decisions are made based on the outputs is highly dependent on culture.[19] In January 2015 one of us co-authored a paper highlighting the portfolio value at risk due to climate change.[20] This finding was largely confirmed by a more detailed study by the Economist Intelligence Unit,[21] and then again by an even more in-depth study led by Simon Dietz, a senior environmental economist at

18 F. Gregor, personal communication, 14 September 2015.

19 A. Tsanakas *et al.*, "Model Risk and Culture", *Actuary Magazine*, December 2014.

20 H. Covington and R. Thamotheram, "The Case for Forceful Stewardship (Part 1): The Financial Risk from Global Warming", accessed 15 May 2016, http://papers.ssrn.com/sol3/Papers.cfm?abstract_id=2551478.

21 The Economist Intelligence Unit, "The Cost of Inaction: Recognising the Value at Risk from Climate Change", 2015.

the London School of Economics.[22] Despite this and other evidence of significant financial risk, many investors have failed to support even moderately worded AGM resolutions at US companies that call for better disclosure of 2°C stress tests.[23]

What is noteworthy is that these resolutions have been endorsed by the two main voting advisory services (Glass Lewis and ISS) and these investors have supported essentially the same resolutions at European oil and gas and mining companies and also Suncor, a Canadian oil sands company. The conclusion? Investors assert they are "evidence-based", but in this and other cases[24] this assertion is not supported by empirical evidence. To pursue the analogy with climate change, as the dominant financial community culture and its intellectual framework do not acknowledge climate risk,[25] it is, by definition, not yet a risk.[26] Investors are blind to it or, put differently, the sector chooses to see an endogenous risk as an exogenous one.[27] The same is true for bee decline.

And on climate, despite all the data, economists are still deeply divided about the ultimate costs.[28] Again, the same is the case with bee decline. What is important to acknowledge from this comparison is that the lack of consensus about bee decline and CCD—including with regulators—is unlikely to be resolved until it is too late. Thus in practice, what this reduces to is a fundamental choice: taking intellectual refuge in portfolio risk management metrics that institutional investors

22 S. Dietz, A. Bowen, C. Dixon and P. Gradwell, "'Climate value at risk' of global financial assets", Nature Climate Change, April 2016.

23 At the time of writing, none of the four resolutions that have been voted at US AGMs this year—AES, Anadarko, Noble Energy and Occidental—have received majority support. This contrasts markedly with investor support of 96%-plus for essentially the same resolutions at European companies. The remaining five resolutions—Chevron, Devon, Exxon, First Energy and Southern Company—will define exactly how deep-rooted the problems in the investment community are.

24 R. Thamotheram, "Body of Evidence", *Investments & Pensions Europe (IPE)*, May 2014.

25 In a poll of its 44,000 members worldwide, the CFA Institute found that climate change comes bottom of the list of ESG issues, even lower than board diversity; see Raj Thamotheram, "How Investors Should Monitor their Real World Impact on Climate Related Systemic Risk", Preventable Surprises, 24 January 2016, accessed 28 March 2016, https://preventablesurprises.com/blog/how-investors-should-monitor-their-real-world-impact-on-climate-related-systemic-risk/.

26 N. Silver, "Blindness to Risk: Why Institutional Investors Ignore the Risks of Stranded Assets", 1st Global Conference on Stranded Assets and the Environment 2015, accessed 7 April 2016, http://www.strandedassets2015.org/presentations--videos.html; http://www.strandedassets2015.org/uploads/2/6/9/5/26954337/session_ii_presenter_i_nicksilver.pdf.

27 TIFF, *Exogenous Risk vs. Endogenous Risk* (The Investment Fund for Foundations, 1997), accessed 9 March 2016, http://www.tiff.org/Reports/Education/Exogenous_Risk_vs_Endogenous_Risk.pdf.

28 M. Weitzman, "Fat-Tailed Uncertainty in the Economics of Catastrophic Climate Change", *Review of Environmental Economics and Policy* 5 (Summer 2011); and S. Pindyck, *Climate Change Policy: What do the Models Tell Us?* (National Bureau of Economic Research, 2013).

find reassuring (for example volatility and tracking error) or engaging in real world risk management. And a fundamental question: if what matters is the failure to act when a crisis was preventable and approaching, rather than when it is imminent and unstoppable, should investors have a risk management strategy for dealing with a plausible worst case or not? It is obvious but worth stating that this is a type of decision that can only be taken by the trustees or board of directors of the investment organization and cannot simply be left to an ESG professional, however committed and skilled they may be, to do the best they can.

Are investors part of the solution or part of the problem?

Investors who want to play a proactive role and be part of the solution to the problem of bee decline would do the following:

Support remuneration plans that focus the executives on diversifying into more responsible chemical products. One practical way to do this would be to ensure executives know that they would have investor support should there be any underperformance in the short term related to making the transition. This might mean, for example, having an approach to executive remuneration, pay for performance and focusing management on return on capital instead of (short-term) revenue, EPS or share price metrics.[29] This recognizes that the return on capital in the short term may decline due to the removal of pesticides that are materially harmful to the food supply ecosystem and because of their contribution to bee decline and CCD. Incentives aligned to increasing R&D could be required to drive both innovation to address pesticide contribution to the phenomenon of bee decline and innovation for new pesticides that do not create harmful and systemic effects in the food supply system. As important as the incentives "carrot" is the "stick". Investors could thus helpfully encourage appropriate risk management by, for example, making it clear that they would pursue executives for malus and clawback and also hold executives personally liable for litigation costs if profits turned out to have been based on food chain risks which had gone unreported but which executives were—or should have been—aware of.

Challenge companies about their strategies for influencing regulators. The legal scrutiny that Exxon and other major oil and gas companies are facing, about what

29 Mark van Clieaf, personal communication, 29 March 2016. See also Investor Responsibility Research Center Institute and Organizational Capital Partners, "The Alignment Gap between Creating Value, Performance Measurement, and Long-Term Incentive Design", November 2014, accessed 1 April 2016, http://lerner.udel.edu/sites/default/files/WCCG/PDFs/events/final-irrc-webinar-ppt-value_webinar-nov-24.pdf.

they knew about climate risk versus what they said and what they lobbied for, is an important precedent. And VW is perhaps the best recent reminder that one of the six drivers of corporate "preventable surprises" is regulatory capture,[30] which some commentators assert is a factor in bee decline and CCD as well.[31] Given an association between corporate regulatory influence and fat tail risks,[32] investors have strong reasons for wanting greater disclosure of specific corporate positions on key scientific studies and legislation and also corporate involvement in the policy process.[33] This will ensure stronger board oversight of executive actions.

Require that their sell-side and credit rating analysts focus on this extra-financial risk in order to improve corporate performance on ESG matters.

Engage with respected scientists and NGOs—as has happened, for example, in the debate about affordable access to HIV treatment—to sensitize their members (i.e. the general public) and to influence public policy.

The authors have been unable to find any evidence of the above. Instead what we have found is that most investors have endorsed standard remuneration plans which focus on short-term share price growth for companies who are involved in pesticide production.[34] We have been unable to find any public comment from mainstream investors about concerns over the influence these corporations have on policy decisions. And there are very few reports from sell-side and credit rating agencies that cover the neonicotinoid issue in any depth. Given that these are highly commercial organizations, this lack of coverage is a very good indicator of lack of expressed client interest.

In summary, while there is no evidence that investors are *actively* encouraging neonics producers to be a part of the problem, the aggregated result of investor

30 InfluenceMap, "Is the Volkswagen Scandal the Tip of the Iceberg? Climate Policy Engagement and the Automotive Sector", November 2015, accessed 31 March 2016, http://influencemap.org/site/data/000/124/IM_Report_Automotive_Oct_2015.pdf.

31 S. Helman, "The Bee Keepers", *Boston Globe*, 23 June 2013.

32 Traditional strategies of asset pricing often rely on a normal bell curve to make market assumptions, and under a normal distribution, the majority of asset variation falls within three standard deviations of its mean. But financial markets do not always act this way and can exhibit fatter tails than traditionally predicted; i.e. there is a greater likelihood of extreme events. And as with the global financial crisis, the greater the political capture of regulators and legislators, the greater the risk (and cost) of such fat tails.

33 For further information about tools and tactics that investors could use, see: http://influencemap.org/page/About#Relevance-for-Investors

34 In an analogous case, fossil fuel company executive pay systems are coming increasingly under scrutiny as they reward short-term actions with disastrous results for the world's climate (C. Collins, "How Our Screwed-Up CEO Pay System Makes Climate Change Worse", Grist, 2015, accessed 13 September 2015, http://grist.org/climate-energy/how-our-screwed-up-ceo-pay-system-makes-climate-change-worse/?utm_content=buffer8d52b&utm_medium=social&utm_source=twitter.com&utm_campaign=buffer). The return on invested capital of oil and gas companies has fallen over 8 years and exploration and production has become more challenging and riskier. Despite this, business as usual remains.

disinterest amounts to the same. Unintentionally and probably unconsciously as well, investors are again enabling dysfunctional corporate and market behaviour.

Systematically undervaluing beta risks is nothing new and undervaluing ESG beta is probably an even bigger issue. As Jeremy Grantham has said about climate change, "we should not unnecessarily ruin a pleasant and currently very serviceable planet just to maximize the short-term profits of energy companies and others".[35] In this case, investors are enabling the short-term profits of neonicotinoids producers and putting at risk the system of agriculture today.

In defence of investors: the current tools don't work!

The authors do not blame investment professionals for this failure to do what end beneficiaries need. It is not, in general, what they were asked to do. More specifically, it is not how they are monitored, rewarded (or sacked). And it is not what their professional training equips them to do: the absence of serious ESG content in the *core* chartered financial analyst (CFA) training has been noted by many authors.[36]

Exacerbating this dilemma is the fact that the individuals who are best able to reduce the risks of bee decline are the senior executives and board directors of companies. Yet these are the individuals who have the strongest personal financial incentives to continue with "business as usual". This conflict of interest is one that only regulators and/or diversified investors can resolve.

Given the capture of corporate political and regulatory decision-making in many markets, most obviously in the USA,[37] there are strong reasons for thinking that if change is to happen, it will need a long horizon and diversified investors to be more proactive in their stewardship role. Specifically, they will need to take an approach which is more strategic (i.e. focused on the core business model) and more collaborative (since combined investor pressure will be needed to encourage companies to undertake fundamental change).

The encouraging news is that this proposal is closely aligned with new thinking about the purpose of the investment industry. Professor John Kay has evidenced in his review for the UK Government, how the stock market is now not the primary

35 Jeremy Grantham, "On the Road to Zero Growth", *GMO Quarterly Letter*, November 2012, accessed 28 March 2016, http://www.zerohedge.com/sites/default/files/images/user5/imageroot/2012/11/Grantham%20letter%20Nov%2020.pdf.

36 J. Confino, "Powerful Global Finance Institute Fails to Train Future Leaders on Sustainability", *The Guardian*, 4 March 2014. The Institute has added ESG modules but these are voluntary and not yet part of the core training for all newly qualified CFAs.

37 R. Monks, *Citizens DisUnited: Passive Investors, Drone CEOs and the Corporate Capture of the American Dream* (Billerica, MA: Miniver Press, 2013).

source of new capital but rather a secondary trading market.[38] The very important implication—and one that the system has largely sidestepped—is that the primary role for investors today is not stock picking but rather stewardship. This includes informed oversight of the strategy for capital expenditure and successful implementation of strategy to deliver sustainable cash flows. This sort of thinking is also now having traction in North American markets. For example, Martin Lipton, who has been described as "the ultimate consigliere to Wall Street and Fortune 500 companies",[39] now says in his latest memo: "A company should recognize that ESG and CSR issues and how they are managed are important to these [long-term] investors".[40]

Two investor strategies that appear to have the greatest risk management value are divestment and shareholder engagement.

Divestment

Divestment has been advocated for and used by investors on a range of issues including democracy in South Africa, human rights in Sudan, blacklisted countries (e.g. Iran) and most recently with regard to factory farming.[41] The highest profile divestment campaign has focused on the fossil fuel sector, helping to stigmatize the sector as a whole (and so reducing its political influence) and contributing to splits within the sector (coal versus oil and gas; European versus North American oil and gas majors), thus creating the opportunity to encourage "laggards" to catch up with "leaders". It is unclear, however, how much "real world" value a divestment campaign would have if it affected share price negatively and resulted in the company being owned by investors who have little interest in being responsible owners: for example some mutual funds or private equity investors. It is for this reason that the authors favour robust shareholder engagement. Clearly, however, divestment—with a public announcement—is preferable to "business as usual" engagement, which others have called "tea and biscuits" engagement given its lack of real world impact.[42]

38 John Kay, *The Kay Review of UK Equity Markets and Long-Term Decision Making, July 2012*, accessed 30 March 2016, https://www.gov.uk/government/uploads/system/uploads/attachment_data/file/253454/bis-12-917-kay-review-of-equity-markets-final-report.pdf.

39 Andrew Ross Sorkin, "Questioning an Adviser's Advice", *New York Times*, 8 January 2008.

40 Martin Lipton, "The New Paradigm for Corporate Governance", Harvard Law School Forum on Corporate Governance and Financial Regulation, 3 February 2016, accessed 7 April 2016, https://corpgov.law.harvard.edu/2016/02/03/the-new-paradigm-for-corporate-governance/.

41 Tom Levitt, "Factory Farming Divestment: What You Need To Know", *The Guardian*, 3 March 2016.

42 For a discussion of the lack of value of business as usual engagement on the climate issue, see: Raj Thamotheram, "Why We Need Forceful Stewardship (and Why BAU Engagement Isn't Fit For Purpose)", Preventable Surprises blog, 9 December 2015,

Shareholder engagement

Shareholder engagement uses long established corporate governance practices between investors and investee companies, but extends this to deal with sustainability issues. This strategy has had good success with regard to oil and gas companies such as BP, ENI, Shell, Statoil, Suncor and Total where management have taken a progressive policy role and also disclosed their risk assessments assuming a transition to a low carbon world. In general, however, engagement on sustainability issues is either well supported (including by mainstream investors) but focused on "weak" asks or has focused on "strong" sustainability asks but with only niche (i.e. sustainability, religious, local authority pension funds) investor support. Arguably, a different kind of request, which is both more robust and more systemic, is required.

Robust engagement

As with climate change, the decline in bee numbers threatens portfolio value. But unlike climate change, there is a lower level of consensus about the nature of the risk. Thus, we propose "Robust Engagement" as the optimal investor strategy for this issue at this moment in time, with the possibility that more forceful actions may be needed if there is greater scientific consensus on the scale of the risk.

Specifically, we propose that investors ask the neonics producers that they are invested in to:

- Disclose their positions on key scientific studies and relevant legislation and also corporate involvement in the policy-making process.
- Ensure there are fully independent and effective whistle-blower systems in place.[43]
- Propose remuneration plans that support executives to focus on the long term and hence diversify into more responsible chemical products. This includes a 5-year rolling average approach to executive remuneration and focusing on return on capital instead of (short-term) revenue and share price metrics.

Make clear to executives that if there is share price loss related to information which senior management did know or should have known about, investors will use all legal means to recoup losses from the executives and board directors in their

accessed 7 April 2016, https://preventablesurprises.com/blog/why-we-need-forceful-stewardship-and-why-bau-engagement-isnt-fit-for-purpose/.

43 R. Thamotheram, "Hug a Whistle Blower Today", *Investments & Pensions Europe (IPE)*, May 2016 accessed 15 May 2016, http://www.ipe.com/analysis/long-term-matters/long-term-matters-hug-a-whistle-blower/10013041.article.

individual capacity. The goal here, as in robust engagement on climate risk,[44] is to harness the power of fiduciary capitalists to encourage better systemic risk management actions by the companies they invest in. As part of this strategy, investors should also redouble efforts to engage with credible and well-informed scientists, regulators and civil society experts to avoid groupthink resulting from talking only to corporate experts.

Who needs to do what

To implement the above, the following stakeholders need to take action:

- **Concerned fund managers**. A group of investors who are willing to take the first-mover role—and if necessary to propose AGM resolutions—and lobby other investors to provide support. These investors will, in all probability, need to have few conflicts of interest and a strong leadership commitment to long-termism and the public good.[45]
- **Scientists and academics** to verify the technical and scientific legitimacy of any action. Specifically, we recommend forming an expert panel that can provide independent assessments of scientific and regulatory developments: the Intergovernmental Panel on Climate Change and the Systemic Risk Council are two models that have relevance.
- **Concerned asset owners**. Starting with asset owners, all players in the investment chain will need to appoint subcontractors (i.e. fund managers, investment consultants, research agencies and voting advisers) who have upskilled to integrate ESG analysis in a meaningful manner, and specifically give those ESG issues which have potential systemic implications the importance they deserve. NGOs, foundations and (ultra) high net worth individuals who have a concern about the loss of bees and CCD are ideally placed to play this role.
- **Environmental litigators**. Legal action for breach of fiduciary obligation should pension funds fail to instruct their fund managers to adopt this kind of approach would raise the salience of this issue considerably. Regulators

44 A good example of this is the work undertaken by the Aiming for A coalition of investors, which has tabled resolutions at oil & gas and mining companies calling for enhanced disclosure of climate risks (http://investorsonclimatechange.org/portfolio/aiming-for-a/).

45 A good example of this is the work being done by two specialist investment firms, the Sustainability Group at Loring, Wolcott & Coolidge and Trillium Asset Management. This started in 2013 with a letter sent to approximately 20 food producers, food retailers and home improvement retailers. They are currently engaging a number of companies on the issue and have a resolution pending with PepsiCo (http://www.trilliuminvest.com/shareholder-proposal/pepsico-pollinator-protection-2016/).

could also do much to contribute here. Even if individual litigation efforts do not "work", they could trigger a shift in the social norm. This has happened with fossil fuel divestment at least within some parts of the philanthropic, endowment and religious investor community, and some sovereign wealth funds that have a commitment to the public good.

- **NGOs and the media**. Underpinning all the above, concerned NGOs and the media will need to raise awareness of the link between bee decline, food supply risk and fiduciary capitalism, collectively promoting the investment beliefs and actions needed. There is a major capacity building challenge here. The good news is that there are people who are passionate about protecting bees. What they don't realize is that their pensions are working against their strongly held values and beliefs. Just as individuals who care about climate change have learned to change their light bulbs and just as individuals who care about factory farming have learned to modify their diet, so the individuals who care about bees could choose to become "citizen investors" as has been advocated by many, including Lukomnik *et al.*[46]

The good news is that mainstream investors, having had several wake-up calls, most recently from VW, now understand what they need to do to ensure that companies address the cultural roots of weak risk management in a preventative manner.[47] But this understanding is at the level of some senior practitioners and is not yet a matter of common organizational behaviour. That said, it is noteworthy that the International Corporate Governance Network's new global stewardship principles place a significant emphasis on the *internal* governance of investment institutions.[48] For this good practice to be implemented in this particular case—and before the bee crisis really hits—the stakeholders listed above will need to take the action outlined.

Bibliography

Anderson, S., S. Pizzigati and C. Collins. "Executive Excess 2015: Money to Burn". Institute for Policy Studies, 2015. Accessed 9 March 2016. http://www.ips-dc.org/executive-excess-2015/

46 J. Lukomnik, S. Davis and D. Pitt-Watson, *What They Do With Your Money: How the Financial System Fails Us and How to Fix It* (New Haven, CT: Yale University Press, 2016).

47 International Corporate Governance Network (ICGN), Institute of Business Ethics (IBE) and Institute of Chartered Secretaries and Administrators (ICSA), "Report of a Senior Practitioners' Workshop on Identifying the Indicators of Corporate Culture", 17 December 2015, accessed 7 April 2016, https://www.icgn.org/sites/default/files/redflagsfinal.pdf.

48 ICGN, *ICGN Global Stewardship Code: Member Consultation, November 2015*, accessed 8 April 2016, https://www.icgn.org/sites/default/files/ICGN%20Global%20Stewardship%20Code%20Consultation%20FINAL%20November%202016.pdf.

Ansar, A., Caldecott, B. and Tilbury, J. *Stranded Assets and the Fossil Fuel Divestment Campaign: What Does Divestment Mean for the Valuation of Fossil Fuel Assets?* Stranded Assets Programme, SSEE, University of Oxford, 2013.

Ariely, D. *Predictably Irrational.* New York: HarperCollins, 2008.

Collins, C. "How Our Screwed-Up CEO Pay System Makes Climate Change Worse". Grist, 2015. Accessed 13 September 2015. http://grist.org/climate-energy/how-our-screwed-up-ceo-pay-system-makes-climate-change-worse/?utm_content=buffer8d52b&utm_medium=social&utm_source=twitter.com&utm_campaign=buffer.

Confino, J. "Powerful Global Finance Institute Fails to Train Future Leaders on Sustainability". *The Guardian,* 4 March 2014. The Institute has added ESG modules but these are voluntary and not yet part of the core training for all newly qualified CFAs.

Eisenberg, L. and T.H. Noe. "Systemic Risk in Financial Systems". *Management Science* 47(2) (2001): 236–9.

Grantham, J. "On the Road to Zero Growth". *GMO Quarterly Letter,* November 2012. Accessed 28 March 2016. http://www.zerohedge.com/sites/default/files/images/user5/imageroot/2012/11/Grantham%20letter%20Nov%2020.pdf.

Heath, C. and D. Heath. *Switch: How to Change Things When Change is Hard.* London: Random House Business Books, 2010.

Helman, S. "The Bee Keepers". *Boston Globe,* 23 June 2013.

InfluenceMap. "Is the Volkswagen Scandal the Tip of the Iceberg? Climate Policy Engagement and the Automotive Sector". November 2015. Accessed 31 March 2016. http://influencemap.org/site/data/000/124/IM_Report_Automotive_Oct_2015.pdf.

International Corporate Governance Network (ICGN). *ICGN Global Stewardship Code: Member Consultation, November 2015.* Accessed 8 April 2016. https://www.icgn.org/sites/default/files/ICGN%20Global%20Stewardship%20Code%20Consultation%20FINAL%20November%202016.pdf.

ICGN, Institute of Business Ethics (IBE) and Institute of Chartered Secretaries and Administrators (ICSA). "Report of a Senior Practitioners' Workshop on Identifying the Indicators of Corporate Culture", 17 December 2015. Accessed 7 April 2016. https://www.icgn.org/sites/default/files/redflagsfinal.pdf.

Investor Responsibility Research Center Institute and Organizational Capital Partners. "The Alignment Gap Between Creating Value, Performance Measurement, and Long-Term Incentive Design", November 2014. Accessed 1 April 2016. http://lerner.udel.edu/sites/default/files/WCCG/PDFs/events/final-irrc-webinar-ppt-value_webinar-nov-24.pdf.

IPCC. *Climate Change 2013, The Physical Science Basis. Contribution of Working Group 1 to the Fifth Assessment Report of the Intergovernmental Panel on Climate Change.* Cambridge, UK: Cambridge University Press, 2014.

Kay, J. *The Kay Review of UK Equity Markets and Long-Term Decision Making, July 2012.* Accessed 30 March 2016. https://www.gov.uk/government/uploads/system/uploads/attachment_data/file/253454/bis-12-917-kay-review-of-equity-markets-final-report.pdf.

Levitt, T. "Factory Farming Divestment: What You Need To Know". *The Guardian,* 3 March 2016.

Lipton, M. "The New Paradigm for Corporate Governance". Harvard Law School Forum on Corporate Governance and Financial Regulation, 3 February 2016. Accessed 7 April 2016. https://corpgov.law.harvard.edu/2016/02/03/the-new-paradigm-for-corporate-governance/.

Lukomnik, J., S. Davis and D. Pitt-Watson. *What They Do With Your Money: How the Financial System Fails Us and How to Fix It.* New Haven, CT: Yale University Press, 2016.

Mercer. "Investing in a Time of Climate Change", Mercer LLC, 2015. Accessed 7 April 2016. http://www.mercer.com/our-thinking/investing-in-a-time-of-climate-change.html.

Monks, R. *Citizens DisUnited: Passive Investors, Drone CEOs and the Corporate Capture of the American Dream.* Billerica, MA: Miniver Press, 2013.

Pindyck, S. *Climate Change Policy: What do the Models Tell Us?* National Bureau of Economic Research, 2013.

Preventable Surprises. "Institutional Investors and Climate Related Systemic Risk", 26 October 2015. Accessed 7 April 2016. https://preventablesurprises.com/wp-content/uploads/2011/03/Preventable-Surprises-October-report_FINAL.pdf.

Rogers, J. "A New Era of Fiduciary Capitalism? Let's Hope So". CFA Institute Blog, 2014. Accessed 13 September 2015. https://blogs.cfainstitute.org/investor/2014/04/28/a-new-era-of-fiduciary-capitalism-lets-hope-so/.

Rook, D., and B. Caldecott. *Cognitive Biases and Stranded Assets: Detecting Psychological Vulnerabilities within International Oil Companies.* Smith School of Enterprise and the Environment, University of Oxford, 2015.

Silver, N. "Blindness to Risk: Why Institutional Investors Ignore the Risks of Stranded Assets", 1st Global Conference on Stranded Assets and the Environment 2015. Accessed 7 April 2016. http://www.strandedassets2015.org/presentations--videos.html; http://www.strandedassets2015.org/uploads/2/6/9/5/26954337/session_ii_presenter_i_nicksilver.pdf.

Sorkin, A.R. "Questioning an Adviser's Advice". *New York Times,* 8 January 2008.

Thamotheram, R. "Why We Need Forceful Stewardship (and Why BAU Engagement Isn't Fit For Purpose)". Preventable Surprises blog, 9 December 2015. Accessed 7 April 2016. https://preventablesurprises.com/blog/why-we-need-forceful-stewardship-and-why-bau-engagement-isnt-fit-for-purpose/.

Thamotheram, R. "How Investors Should Monitor their Real World Impact on Climate Related Systemic Risk". Preventable Surprises, 24 January 2016. Accessed 28 March 2016. https://preventablesurprises.com/blog/how-investors-should-monitor-their-real-world-impact-on-climate-related-systemic-risk/.

Thamotheram, R. "Hug a Whistle Blower Today". *Investments & Pensions Europe (IPE),* May 2016 (forthcoming).

Thamotheram, R. and A. Ward. "Put the bee back in beta". *Investment & Pensions Europe,* March 2013. Accessed 31 March 2016. http://www.ipe.com/put-the-bee-back-in-beta/50256.fullarticle.

Thamotheram, R. and A. Ward, "Whose Risk Counts?" in *Cambridge Handbook of Institutional Investment and Fiduciary Duty,* edited by James Hawley, Andreas Hoepner, Keith Johnson, Joakim Sandberg and Edward Waitzer. Cambridge University Press, July 2015.

TIFF. *Exogenous Risk vs. Endogenous Risk.* The Investment Fund for Foundations, 1997. Accessed 9 March 2016. http://www.tiff.org/Reports/Education/Exogenous_Risk_vs_Endogenous_Risk.pdf.

Tsanakas, A., M. B. Beck, T. Ford, M. Thompson, and I. Ye. "Model Risk and Culture". *Actuary Magazine,* December 2014.

US Department of Agriculture. "USA and EPA Release New Report on Honey Bee Health". Accessed 23 June 2013. http://www.usda.gov/wps/portal/usda/usdamediafb?contentid=2013/05/0086.xml&printable=true23.

Weitzman, M. "Fat-Tailed Uncertainty in the Economics of Catastrophic Climate Change". *Review of Environmental Economics and Policy* 5(2) (Summer 2011): 275–92.

World Bank. *Turn Down the Heat: Why a 4°C Warmer World Must Be Avoided.* Washington, DC: International Bank for Reconstruction and Development/World Bank, 2012.

Yermo, J., and C. Severinson. "The Impact of the Financial Crisis on Defined Benefit Plans and the Need for Counter-Cyclical Funding Regulations". *OECD Working Papers on Finance, Insurance and Private Pensions,* No. 3. OECD Publishing, 2010. Accessed 7 April 2016. http://www.oecd.org/finance/private-pensions/45694491.pdf.

Part III

Accounting for bees and bee decline

10

How to account for bees and pollinators?

Joël Houdet

African Centre for Technology Studies (ACTS), Albert Luthuli Centre for Responsible Leadership, Faculty of Economic and Management Sciences at the University of Pretoria, Synergiz, South Africa

Ruan Veldtman

South African National Biodiversity Institute (SANBI), South Africa

This chapter introduces the various approaches which may be used by businesses to account for their dependencies and impacts on pollinators and their services.

Business accounting systems are diverse as they are designed to satisfy the needs of various stakeholders, which may have different interests and goals regarding a company's activities and performance. From an environmental perspective, we can distinguish internal and external accounting approaches, namely environmental management accounting which aims to support internal decision-making (e.g. product pricing, cost control, budgeting) and external environmental reporting (i.e. financial, sustainability and/or integrated reporting) which targets external stakeholders for accountability purposes.

Similarly, means of crop pollination and pollination of wild plants are very diverse. Crop pollination by insects can be performed either by managed pollinators (e.g. honey bee hives that beekeepers rent to growers) or by wild pollinators resident in natural habitats or untransformed landscapes near these crops (thereby providing an ecosystem service of pollination).[1] In turn, the wild flora dependent on insects for the setting of seed (and thus also food for fauna) can be pollinated

1 A.-M. Klein *et al.*, "Importance of Pollinators in Changing Landscapes for World Crops", *Proceedings of the Royal Society B* 274 (2007): 303–13.

by a few generalist pollinator species for some generalist plants, or this interaction can be specialized and species specific. In the most extreme case, a plant species can only be pollinated by a single pollinator species. In most cases, however, bees are the most important group of animal pollinators.

First, we provide a short explanation of the links between bees, their pollination services and business operations, including how various industries are directly and indirectly dependent on the services they provide. Second, we introduce how various business activities impact on bee species, their populations and their services. Then, we discuss accounting for bees from an environmental management accounting perspective before exploring how external reporting approaches may deal with business dependencies and impacts on bees and their services.

Business dependencies on wild and domestic pollinators and their services

Insect pollination is a vital ecosystem service for both natural and agricultural systems. Without insect pollinators, roughly a third of the world's crops would fail to produce flower. For instance, in the province of Western Cape of South Africa, where the winter rainfall makes it ideal for citrus and soft fruit, wild bees and other pollinators generate between 400 million and 2,500 million South African rand worth of services to the deciduous fruit industry every year (2008 figures).[2] A report which aimed to estimate the economic value of the pollination services of 32 insect-pollinated crops in South Africa revealed that, in 2012, 10 billion rand could be attributed to insects.[3]

In other words, many businesses involved in growing, transporting, transforming and selling pollination-dependent fruit and seed crop products and by-products (e.g. fruit jams, oils, agro-fuels) depend, directly or indirectly, on wild and/or domesticated pollinators. In addition, the tourism sector also depends on the services of wild pollinators which enable the reproduction of countless plant species making up the ecosystems which tourists visit or enjoy (e.g. eco-tourism activities).[4] For instance, in the Namaqualand region of South Africa, where floral

2 DEA–SANBI, *National Biodiversity Assessment 2011: An Assessment of South Africa's Biodiversity and Ecosystems. Synthesis Report* (Pretoria: South African National Biodiversity Institute and Department of Environmental Affairs, 2012).

3 www.sanbi.org/news/sanbi-wraps-studies-pollination-and-pollinators, accessed 9 March 2016.

4 J.E. Losey and M. Vaughan, "The Economic Value of Ecological Services Provided by Insects", *BioScience* 56 (2006): 311–23.

displays are enjoyed by thousands of seasonal tourists, these annual plants depend on pollinators in order to generate seeds for future displays.

From a management perspective, pollination services of wild pollinators (including wild honey bees) and domesticated honey bees need to be distinguished. While the latter are closely managed by beekeepers (e.g. by trucking honey bee hives to targeted farm fields or orchards), the pollination benefits of wild pollinators depend on species diversity, population density and proximity to pollinated crops and wild plants).[5] Besides, each species would have specific habitat requirements within agro-ecosystems and wild ecosystems.

To assess business dependencies on pollination services we need to understand the combined benefits of wild and domestic pollinators to both crop and wild plants (Table 10.1). According to Garibaldi *et al.*[6,7] wild pollinators are often more effective than honey bees in producing seeds and fruit on some crops (e.g. oilseed rape, coffee, tomatoes, strawberries, mangoes), as managed honey bee hives often do not fully replace (lost) wild pollination services. However, in most intensive agricultural areas, pollination service demand far outstrips the supply offered by wild pollinators. In these cases, growers have no other option but to use managed pollination, 95% of the time being performed by a single honey bee species, *Apis mellifera*.[8,9]

Table 10.1 **Examples of industries depending on pollination services**

	Directly dependent industries	Indirectly dependent industries
Wild pollinators (ecosystem service)	Seed and fruit cropping; Eco-tourism; ecosystem resilience	All industries using, transporting, transforming and selling seed/fruit crops; Supporting industries such as banking and assurance
Managed pollinators (managed service)	Seed and fruit cropping; honey production	All industries using, transporting, transforming and selling seed/fruit crops; Supporting industries such as banking and assurance

5 L. Pfiffner and A. Müller, *Wild Bees and Pollination: Fact Sheet* (Research Institute of Organic Agriculture, 2014).

6 L.A. Garibaldi *et al.*, "Stability of Pollination Services Decreases with Isolation from Natural Areas Despite Honey Bee Visits", *Ecology Letters* 14(10) (2011): 1062– 72.

7 L.A. Garibaldi *et al.*, "Wild Pollinators Enhance Fruits Set of Crops Regardless of Honey Bee Abundance", *Science* 339 (6127) (2013): 1608–11.

8 J. Ghazoul, "Buzziness as Usual? Questioning the Global Pollination Crisis", *Trends in Ecology & Evolution* 20(7) 2005: 367–73, doi: 10.1016/j. tree.2005.04.026.

9 N.W. Calderone, "Insect Pollinated Crops, Insect Pollinators and US Agriculture: Trend Analysis of Aggregate Data for the Period 1992–2009", *PLoS ONE* 7(5) (2012): e37235, doi:10.1371/journal.pone.0037235.

Business impacts on pollinators and their services

While some economic sectors rely heavily on pollination services, many other industries have an impact directly and indirectly on wild and managed pollinators, which can further affect their respective services delivery to wild and managed ecosystems. Yet, this manifests itself at different spatial and temporal scales. Table 10.2 summarizes key concepts and indicators for estimating the size and status of pollinators and their pollination services.

On the one hand, wild pollinators are vulnerable to on-farm management practices in the case of their ecosystem service provision (e.g. pesticides' and herbicides' impact on non-crop forage plants). In nature, however, the scale of business-induced impacts is much larger (e.g. land conversion due to agriculture and urbanization, unsustainable rangeland management), which is sometimes amplified when acting in synergy, especially in the context of climate change.

Managed pollinators can also be affected by on-farm practices during the crop pollination period when hives are in close proximity (e.g. use of pesticides via spraying or systemic treatments of seeds, e.g. neonicotinoids). For most of the year, however, beekeepers need to support their hives by finding adequate forage sites so that bees fulfil their need for pollen and nectar.[10] They are thus predominantly influenced by off-farm impacts where, in contrast to the relatively short period of crop pollination (sometimes as little as 5 days), for a much longer period and a bigger area, they depend on a particular region's forage resources (depending how far hives are transported by the beekeeper). Populations of managed pollinators can be thus be subject to changes in forage source diversity and availability, hive vandalism and theft, transmission of diseases between apiaries, and increased level of colony stress.[11] Globally, they are impacted by global trade, climate change altering forage availability, and between continental exchange of diseases and genetic material.[12]

10 W.J. De Lange *et al.*, "Valuation of Pollinator Forage Services Provided by *Eucalyptus cladocalyx*", *Journal of Environmental Management* 125 (2013):12–18, doi: 10.1016/j.jenvman.2013.03.027.

11 M.F. Johannsmeier, *Beeplants of the South-Western Cape: Nectar and Pollen Sources of Honeybees*, 2nd edn (Pretoria: Agricultural Research Council – Plant Protection Research Institute, 2005).

12 D. vanEngelsdorp and M.D. Meixner, "A Historic Review of Managed Honey Bee Populations in Europe and the United States and the Factors that May Affect Them", *Journal of Invertebrate Pathology* 103 (2010): 580–95.

Table 10.2 **Assessing the size and status of pollinators and pollination services**

Pollinator type	Service magnitude and significance	Indicators of health and vulnerability
Wild pollinators (ecosystem service)	Crops: fruit and seed set; yield Wild plants: plant-pollinator networks; species evenness (Shannon Winer Index)	Crops: pollination deficits; % habitat loss; on farm pollinator diversity Wild plants: % fragmentation of natural habitat
Managed pollinators (managed service)	Crops: Seed and fruit cropping; honey production Wild plants: forage species richness	% colony losses; forage species diversity (pollen and nectar); parasite loads; hive rental price; transport frequency and distance

Accounting for pollinators for improved decision-making

Management accounting constitutes the central tool for internal management decisions, such as budget control. It is an internal information system which deals with questions typically pertaining to monitoring expected and actual production costs for different products and hence being able to track the efficiency of the organization and/or its employees. The main stakeholders in management accounting are thus members of different management positions within a company. It is not regulated by law but follows broadly accepted principles.

Yet, there is a growing consensus that conventional management accounting practices do not provide adequate information about the environmental issues facing a business. Environmental management accounting (EMA) has thus been developed to address this concern. In other words, it aims to support improved decision-making of companies which are faced with environmental problems or opportunities. EMA is broadly defined to be the identification, collection, analysis and use of two types of information for internal decision-making,[13,14] namely:

- Monetary information on environment-related costs, earnings and savings;
- Physical information on the use, flows and destinies of energy, water and materials (including waste), though the use of input–output models.

In terms of dependencies on pollination services, both wild and managed pollinators can be understood as inputs into crop and honey production systems,

13 UNDSD, *Environmental Management Accounting Procedures and Principles* (New York: United Nations, 2001).

14 D. Savage and C. Jasch, *International Guidance Document. Environmental Management Accounting* (New York: IFAC – International Federation of Accountants, 2005).

both in quantitative (see examples of indicators in Table 10.2) and monetary terms (e.g. replacement costs of lost and/or managed wild pollination services).[15] Environmental management accounting can thus help farmers better understand and monitor their pollination-related costs for specific crops, towards assessing the true costs of a tonne of crop, typically by taking into account the non-product outputs costs due to lost pollination services (which would be hidden in the case of wild pollinators, see Table 10.3). This would enable farmers to better cost specific pollination-maximization strategies, and thus adapt their farm management practices according to the "pollinator-needs" of their crops and their business objectives (e.g. profit maximization).

In terms of the management of impacts on pollinators and their pollination services, environmental management accounting can enable any business to track the impact mitigation costs of achieving specific pollinator-related targets (e.g. species diversity, population size, fruit and seed set, crop yield). This may be particularly relevant to projects subject to socioeconomic and/or environmental impact assessments. Here, depending on the legislation or through stakeholder engagement, the resulting environmental management plans may require the implementation of the impact mitigation hierarchy for impacted pollinators and the associated pollination services. This would lead to specific impact avoidance, minimization, restoration and, potentially, offset measures. This could then lead to changes in the project design or implementation, which may save or add to its overall cost.

In that context, although the focus would essentially be on private cost minimization, the company may also embed externalities in its model; that is, assessing, in economic terms, the negative and positive impacts on society (or specific stakeholders) of its planned project and the associated impact mitigation measures. This would allow the company to explore the costs and benefits of various scenarios regarding the sustainable management and conservation of pollinators and their pollination services.

Table 10.3 **Key questions that EMA can help answer regarding pollinators and pollination services**

Pollinator type	Business dependencies	Business impacts
Wild pollinators (ecosystem service)	Share of crop production value/non-product output, costs savings (compared to managed pollinators), share of total product costs (if specific measures taken)	Costs of implementation of the impact mitigation hierarchy per unit of product/service value, costs to society per unit of product/service value
Managed pollinators (managed service)	Share of crop production value/non-product output, share of total product costs	Costs of implementation of the impact mitigation hierarchy per unit of product/service value, costs to beekeepers per unit of product/service value

15 M.H. Allsopp *et al.*, "Valuing Insect Pollination Services with the Cost of Replacement", *PLOS One* 3 (2008): e3128.

Which reporting approaches for pollinators and their services?

External reporting aims to satisfy the information requirement of various external stakeholders and includes financial, sustainability and integrated reporting practices. While financial reporting is based on strict mandatory accounting standards (International Financial Reporting Standards, Generally Accepted Accounting Principles), sustainability and integrated reporting are based on voluntary guidelines (e.g. International Integrated Reporting Framework[16], Global Reporting Initiative guidelines).

To date, three main distinct environmental reporting methods have been used to disclose ecosystem dependencies and/or impacts to external stakeholders,[17] namely environmental financial reporting, extra-financial environmental reporting (as part of conventional sustainability reporting), and the disclosure of environmental externalities in ad hoc reports (see Table 10.4).

In environmental financial accounting and reporting, the primary focus for reporting organizations should be on providing additional information on pollinators/pollination-related risks for the financial value of assets (e.g. farm land) or sales (crop/food prices in agribusinesses and potentially for food retailers). Disclosures on contingent liabilities to impact third parties (e.g. farmers, wild flower-related tourism) might also be explored for companies which may have significant impacts on pollinator populations through the sale of their products (e.g. chemical companies selling pesticides).

In extra-financial environmental reporting, disclosure approaches are more flexible. The reporting organization may use various indicators. For instance, disclosing a company's reliance on pollination services, in quantitative and/or in economic metrics, has been proposed by Gilbert et al. (2011).[18] It is also possible for a reporting organization to report on the impacts of its operations, products and/or services on pollinator species, especially if there are any legally protected species concerned, as well as disclose the impact mitigation measures put in place in space

16 IIRC, *International Integrated Reporting Framework* (London: IIRC, 2013), accessed July 23 http://integratedreporting.org/wp-content/uploads/2013/12/13-12-08-THE-INTERNATIONAL-IR-FRAMEWORK-2-1.pdf.

17 J. Houdet *et al.*, *Promoting Business Reporting Standards for Biodiversity and Ecosystem Services. The Biodiversity Accountability Framework* (Orée - FRB, 2010).

18 S. Gilbert *et al.*, *Approach for Reporting on Ecosystem Services: Incorporating Ecosystem Services into an Organization's Performance Disclosure*, CREM–GRI Research and Development Series (Amsterdam: GRI, UNEP-WCMC, 2011).

and time. Because of the lack of widely agreed metrics though, this type of reporting is likely to remain at the level of commentaries in the foreseeable future.

An interesting approach relates to disclosing the costs to society of the company's impacts on pollinators and pollination services.[19] This has recently been strongly advocated by some stakeholders (consultancies, forward thinking companies such as Kering, independent associations and lobby groups) because both environmental financial and extra-financial reporting fail to disclose the full economic dimensions of business dependencies and impacts on natural capital (including pollination services), both for their own sustainability and that of its stakeholders. This would involve disclosing the negative (and positive, if any) environmental externalities of the reporting entity: for example, impacts on pollinators and their pollination services in economic (monetary) terms.

Finally, in integrated reporting, the challenge would be to disclose how pollinators and their pollination services affect how a business generates value through the interactions between ecological, social, governance and financial performance. For instance, how do pollinators contribute to the financial capital of an agribusiness or a food retailer (i.e. direct economic contributions)? What does it mean for environmental and stakeholder management at the farm level (i.e. ecological and social performance)? What's more, how do governance mechanisms take into account pollination risks for the company, in terms of both input costs and its social licence to operate? Here again, clear guidelines on how to do so remain to be developed and tested.

Table 10.4 **Various external reporting approaches on pollinators and pollination services**

Pollinator type	Business dependencies	Business impacts
Wild and managed pollinators	Financial reporting: risks for asset value, sales Sustainability reporting: qualitative and quantitative information on reliance on pollinators and their services Positive externalities disclosure (i.e. contribution of pollinators to business value creation)	Financial reporting: Potential contingent liabilities, any material expenses Sustainability reporting: qualitative and quantitative information on impacts on pollinators and their services, description of impact mitigation measures Negative externalities disclosure (i.e. costs to society of impacts on pollinators)

19 J. Houdet *et al.*, *What Natural Capital Disclosure for Integrated Reporting? Designing and Modelling an Integrated Financial – Natural Capital Accounting and Reporting Framework* (Synergiz–ACTS, Working Paper 2014-01, 2014).

Bibliography

Allsopp, M.H., W.J. De Lange, and R. Veldtman. "Valuing Insect Pollination Services with the Cost of Replacement". *PLOS One* 3 (2008): e3128.

Calderone, N.W. "Insect Pollinated Crops, Insect Pollinators and US Agriculture: Trend Analysis of Aggregate Data for the Period 1992–2009". *PLoS ONE* 7(5) (2012): e37235, doi:10.1371/journal.pone.0037235.

DEA–SANBI. *National Biodiversity Assessment 2011: An Assessment of South Africa's Biodiversity and Ecosystems. Synthesis Report.* Pretoria: South African National Biodiversity Institute and Department of Environmental Affairs, 2012.

De Lange, W.J., R. Veldtman and M.H. Allsopp. "Valuation of Pollinator Forage Services Provided by *Eucalyptus cladocalyx*". *Journal of Environmental Management* 125 (2013):12–18, doi: 10.1016/j.jenvman.2013.03.027.

Garibaldi, Lucas A., Ingolf Steffan-Dewenter, Claire Kremen, Juan M. Morales, Riccardo Bommarco, Saul A. Cunningham, Luísa G. Carvalheiro, Natacha P. Chacoff, Jan H. Dudenhöffer, Sarah S. Greenleaf, Andrea Holzschuh, Rufus Isaacs, Kristin Krewenka, Yael Mandelik, Margaret M. Mayfield, Lora A. Morandin, Simon G. Potts, Taylor H. Ricketts, Hajnalka Szentgyörgyi, Blandina F. Viana, Catrin Westphal, Rachael Winfree, and Alexandra M. Klein. "Stability of Pollination Services Decreases with Isolation from Natural Areas Despite Honey Bee Visits". *Ecology Letters* 14(10) (2011): 1062–72.

Garibaldi, L.A., I. Steffan-Dewenter, R. Winfree, M.A. Aizen, R. Bommarco, S.A. Cunningham, C. Kremen, L.G. Carvalheiro, L.D. Harder, O. Afik, I. Bartomeus, F. Benjamin, V. Boreux, D. Cariveau, N.P. Chacoff, J.H. Dudenhöffer, and 34 more. "Wild Pollinators Enhance Fruits Set of Crops Regardless of Honey Bee Abundance". *Science* 339 (6127) (2013): 1608–11.

Ghazoul, J. "Buzziness as Usual? Questioning the Global Pollination Crisis". *Trends in Ecology & Evolution* 20(7) 2005: 367–73, doi: 10.1016/j. tree.2005.04.026.

Gilbert, S., M. Fleur, M. Barcellos-Harris, S. Brooks, T. Tyrrell, W. Broer, and J. van Schaik. *Approach for Reporting on Ecosystem Services: Incorporating Ecosystem Services into an Organization's Performance Disclosure.* CREM–GRI Research and Development Series. Amsterdam: GRI, UNEP-WCMC, 2011.

Houdet, J., M. Trommetter, and J. Weber. *Promoting Business Reporting Standards for Biodiversity and Ecosystem Services. The Biodiversity Accountability Framework.* Orée - FRB, 2010.

Houdet, J., R. Burritt, K.N. Farrell, J. Martin-Ortega, K. Ramin, J. Spurgeon, J. Atkins, D. Steuerman, M. Jones, J. Maleganos, H. Ding, C. Ochieng, K. Naicker, C. Chikozho, J. Finisdore, and P. Sukhdev. *What Natural Capital Disclosure for Integrated Reporting? Designing and Modelling an Integrated Financial – Natural Capital Accounting and Reporting Framework.* Synergiz–ACTS, Working Paper 2014-01, 2014.

IIRC. *International Integrated Reporting Framework.* London: IIRC, 2013. Accessed July 23 http://integratedreporting.org/wp-content/uploads/2013/12/13-12-08-THE-INTERNATIONAL-IR-FRAMEWORK-2-1.pdf.

Johannsmeier, M.F. *Beeplants of the South-Western Cape: Nectar and Pollen Sources of Honeybees,* 2nd edn. Pretoria: Agricultural Research Council – Plant Protection Research Institute, 2005.

Losey, J.E. and M. Vaughan. "The Economic Value of Ecological Services Provided by Insects". *BioScience* 56 (2006): 311–23.

Klein, A-M., B.E. Vaissière, J.H. Cane, I. Steffan-Dewenter, S.A. Cunningham, C. Kremen and T. Tscharntke. "Importance of Pollinators in Changing Landscapes for World Crops". *Proceedings of the Royal Society B* 274 (2007): 303–13.

Pfiffner, L. and A. Müller. *Wild Bees and Pollination: Fact Sheet.* Research Institute of Organic Agriculture. 2014.

Savage, D. and C. Jasch, *International Guidance Document. Environmental Management Accounting*. New York: IFAC – International Federation of Accountants, 2005.
UNDSD, *Environmental Management Accounting Procedures and Principles.* New York: United Nations, 2001.
vanEngelsdorp, D. and M.D. Meixner, "A Historic Review of Managed Honey Bee Populations in Europe and the United States and the Factors that May Affect Them". *Journal of Invertebrate Pathology* 103 (2010): 580–95.

11

Bee accounting and accountability in the UK

Jill Atkins
University of Sheffield, UK

Elisabetta Barone
Brunel University, UK

Warren Maroun
University of the Witwatersrand, South Africa

Barry Atkins
University of South Wales, UK

This chapter explores the ways in which companies are "accounting for" their activities relating to bee populations. In other words, we explore corporate accountability in relation to the ongoing decline of bee populations and the impact on honey production and agricultural produce. Specifically, we analyse corporate voluntary disclosures relating to bees, including information contained in companies' annual reports, sustainability/corporate social responsibility reports and on corporate websites for a sample of companies listed on the London Stock Exchange. We also discuss corporate disclosures of bee-related information from an integrated general systems theory perspective in order to appreciate the interconnectedness of business activity, species, nature, the ecosystem and ultimately life on Planet Earth.

Understanding social accounting, accounting for biodiversity and integrated reporting

As we discussed in Chapter 1, an issue such as bee decline is one which affects all aspects of life on Earth, if we adopt an integrated or systems approach to analysing the problem. Accounting in all its forms and especially social and environmental accounting may be approached from a systems perspective, which allows us to appreciate the interrelationships between accounting, society, the environment and the planet. In the academic accounting literature, it has been acknowledged that accounting

> ...interacts with systems which we might call "social", "political" and "ethical" as well as being directly related to interactions within and between organisational systems and between those organisational systems and individuals, groups, communities, society, nations and the non-human elements of the planetary natural environment.[1]

The term "social accounting" refers to a treatment of accounting as an important mechanism within society which can be used for purposes of control and management but which also can be "emancipatory". In other words, social accounting has the potential to engender change within society and also within the natural world.[2] Given the ways in which social accounting brings together the impacts of business on human society (employee issues, community impacts) and on the natural environment (air, water, land, biodiversity), it seems a systems approach is a most appropriate way of viewing this rapidly growing field of accounting and accountability. "Accounting" for bees from an anthropocentric as well as from a deep ecology perspective is the subject of many chapters in this book.

If we take a holistic, systems-based approach to bee decline, we can see how recent accounting developments, especially that of integrated reporting, may be the way forward in saving bees, planet and the human race. Given the evolution of social accounting, integrated reporting, business risk models, holistic governance and stakeholder accountability, it is highly likely that all of these financial, accounting, accountability and business systems will also be affected substantially by bee decline.

1 R. Gray *et al.*, *Accounting & Accountability: Changes and Challenges in Corporate Social and Environmental Reporting* (Essex, UK: Pearson Education Limited, 1996), 14.

2 The concept of "emancipatory accounting" is explored in S. Gallhofer and J. Haslam, *Accounting and Emancipation: Some Critical Interventions* (London: Routledge, 2003) and S. Gallhofer *et al.*, "Accounting as Differentiated Universal for Emancipatory Praxis", *Accounting, Auditing & Accountability Journal* 28(5) (2015): 846–74.

Chapter 1 introduced the concept of general systems theory, as well as other perspectives connected to an integrated approach to nature, the planet and ultimately to bee decline. From an academic accounting perspective, accounting, which is a totally human concept and activity,[3] may be viewed as a means of discharging accountability. Any person or organization may attempt to demonstrate their accountability for "something" through disclosing information on their impact on, and activities relating to, that "something". For example, organizations may "account for" (provide publicly available information on) the ways in which their activities impact on marine life. A nuclear power plant can have deleterious impacts on marine life in the vicinity of the plant, given the heavy use of water and temperature effects on fish and other aquatic species. The company operating the plant may choose to report information to stakeholders on the impacts and also on what they are doing to lessen negative impacts. Currently, companies and other organizations are making significant strides in disclosing information about their impacts on biodiversity and on species threatened with extinction, in their integrated reports, sustainability reports and on their company websites.

"Accounting for biodiversity" refers to the public disclosure of information by companies on their impacts on biodiversity and on the ways in which they may be attempting to lessen or prevent these impacts. "Extinction accounting"[4] refers to a relatively new concept, focusing on the ways in which companies report information on their impact on species which are categorized as threatened with extinction, where these species are found on their land. The areas of "accounting for biodiversity" and "extinction accounting"[5] are perfect candidates for a systems approach, as they seek to address the activities and projects in which multinational corporations are involved relating to species conservation, decline and imminent extinction. Any attempt to "account" for biodiversity, either through the application of traditional accounting methods or through more innovative techniques, needs to consider biodiversity within the broadest of contexts possible. Further, ecology, ecosystem services and the natural world all represent systems of components (living and inanimate) which are interrelated, interdependent and interlocking.

3 Thanks to Rob Gray for raising this point!

4 J.F. Atkins, W. Maroun, E. Barone and B.C. Atkins, "From the Big Five to the Big Four? Exploring Extinction Accounting for the Rhinoceros?" working paper presented at Southampton Business School, April 2016.

5 A recent book edited by Michael Jones, *Accounting for Biodiversity* (Abingdon, UK: Routledge, 2014), considers biodiversity from a whole range of different perspectives, as does the editorial to a journal special issue, "Problematising Accounting for Biodiversity". This article adopts implicitly a systems approach, as it considers the problems associated with accounting for biodiversity from the perspectives of philosophy, ecosystems, science and climate change. See M.J. Jones and J.F. Solomon, "Problematising Accounting for Biodiversity", *Accounting, Auditing & Accountability Journal* 26(5) (2013). There is also a forthcoming special forum of the *Accounting, Auditing & Accountability Journal* on the topic of Extinction Accounting & Accountability.

The most recent and extremely significant development in accounting, in terms of systems thinking and a holistic approach to governance, accounting and accountability, is the emergence of integrated reporting. Focusing on a particular species, such as the bee, and the broad impact of its decline, seems a natural extension of discussions about integrated reporting. Indeed, we consider that reporting on and disclosing information about a specific species lies at the very heart of integrated reporting and its original intentions.

Bees and British bee decline

Bees are a big issue in the UK at present, with lobby groups calling for companies, organizations and government to address the decline in bee populations. For example, the Soil Association is running proactive initiatives to prevent further decline in bee populations and hopefully to lead to increases in bee populations within the UK. Their initiative "Keeping Britain Buzzing" includes the following aims: stopping the temporary ban across the European Union on three neonicotinoid pesticides from being overturned as well as campaigning for a permanent ban. The Soil Association is also working with farmers to develop ways of reducing pesticide use and promote organic farming methods. Another example is the British fundraising campaign, "Friends of the Honey Bee", which is being run by the British Beekeepers Association. This campaign includes: funding research into combating the varroa mite; and encouraging the planting of more flowers which provide pollen and nectar for bees. A character named Beattie Bee is their mascot.

The analysis

> Bees—Their Survival is Ours too....[6]

For our sample, we selected a large number of companies which reside in bee-relevant industries, some of which were identified by Stathers but also others we felt were important.[7] Specifically, the industries we selected were: food producers, food

6 Danone website, http://downtoearth.danone.com/2012/06/13/bees-their-survival-is-ours-too/, accessed 10 March 2016.

7 Rick Stathers, *The Bee and the Stockmarket: An Overview of Pollinator Decline and its Economic and Corporate Significance* (Research Paper, London: Schroders, January 2014).

and drug retailers, beverages, luxury goods, general retailers and agrochemicals. We then downloaded the most recent annual reports and sustainability reports for the companies in these sectors. We also searched the companies' websites for any bee-related information. The data were collected for the years 2012, 2013 and 2014. See Table 11.1 for a list of the companies selected for analysis. In order to identify bee-related disclosures, we searched the documents and online information for key "buzzwords" including (but not limited to): bee(s), honey, pollination. The majority of the disclosures were found on corporate websites.

Table 11.1 **Companies analysed**

Name	Sector
AB Foods	Food producers
Ahold	Food and drug retailers
Booker	Food and drug retailers
Britvic	Beverages
Carrefour	Food and drug retailers
Christian Dior	Luxury goods
Coca-Cola	Beverages
Colruyt	Food and drug retailers
Cranswick	Food producers
Dairy Crest	Food producers
Danone	Food producers
Devro	Food producers
Dr Pepper	Beverages
Greencore group	Food producers
Hermes	Luxury goods
Hershey	Food producers
Kraft	Food producers
L'Oreal	Luxury goods
LVMH	Luxury goods
Marks and Spencer	General retailers
Metro Group	Food and drug retailers
WM Morrison	Food and drug retailers
Nestlé	Food producers
Ocado	Food and drug retailers
Pepsico	Beverages
Sainsbury's	Food and drug retailers
Snapple Group	Beverages

Name	Sector
Syngenta	Agrochemicals
Tate & Lyle	Food producers
Tesco	Food and drug retailers
UDG Healthcare	Food and drug retailers
Unilever	Food producers

Our research involved an interpretive content analysis which allowed us to uncover a series of themes in the bee-related disclosures. We found that websites disclosed bee-related information. There was an absence of bee-related information in the formal corporate reports. The disclosure themes we extracted from the data included: an educative theme, which involved training employees and other stakeholders in bees and bee-related issues; a theme of impression management and reputational management; a historical and cultural theme wrapped around the ancient relationship between bees and humans; a theme around initiatives aimed at increasing bee populations; and a theme of debate and discussion relating to chemicals and bee decline. We also found the bee image to be used to represent various ecological initiatives. We discuss these themes in more detail below.

Corporate bee initiatives and partnerships

Our analysis showed that the sample companies were involved in a wide array of bee-focused initiatives and were engaged in partnerships with other organizations with an aim to enhance bee populations and slow bee decline.

Companies particularly in the food retail sector appear to be quite heavily involved in creating partnerships and even calls for action on bee decline. For example, in 2013, the Marks and Spencer website disclosed their involvement with other companies such as the Co-op and B&Q, to call on the British Government to develop a Bee Action Plan. They stress that thousands of members of the British public are calling for action on this issue.

Nestlé disclose on their website how they have developed a partnership with the Northumberland Wildlife Trust to develop a meadow near their Fawdon Factory with the aim of improving biodiversity. They also provide information about their partnership with the Derbyshire Wildlife Trust at the Buxton Nestlé Waters site, which has resulted in the creation of another wild flower meadow. Indeed, the director of "living landscapes" at Derbyshire Wildlife Trust comments in an article posted on the corporate website that,

> Wildflower meadows, once a common sight across the UK, have declined by 97% since the 1930s, as a result of a rapidly expanding population and intensifying agriculture. This is a huge problem for the creatures which depend on them, such as butterflies and bumblebees, which themselves are of huge value to people, pollinating our food crops.[8]

This is a clear illustration of what we have come to call "bee accounting".

Unilever also disclosed detailed information on their corporate website about bee decline and about the various initiatives they have implemented to enhance bee populations. The website states in an article entitled, "Saving Bees",

> It's a stark fact: the world needs bees for food production but the decline in honey-bee colonies is rapidly becoming a global problem... There's no time to waste. At Unilever, we are taking urgent action with our suppliers to save the bees.[9]

The initiatives they discuss include the installation of bee hotels.

Another example we found was Carrefour's explanation on its website that it is encouraging all its producers to adopt integrated pest control, which it defines as an ecological approach including the use of organic pest control methods rather than using pesticides. Furthermore, on its website, Carrefour discusses the 552 active Quality Lines which it has developed as part of its agro-ecological approach. The company defines a Carrefour Quality Line as a partnership between the retailer and its agricultural producers (suppliers) or livestock farmers which is underpinned by a reciprocal commitment. This commitment aims to produce high-quality products which use agro-ecological farming methods on the basis that bees are nature's allies.

Another interesting partnership has been developed between Christian Dior and several initiatives including "Terre d'Abeilles" and CNRS, to address the problem of bee decline. Christian Dior's 2013 annual report states that,

> For several years, Group companies have supported programs to save bees, the natural defenders of our planet's ecosystems. Chaumet has been working with Terre d'Abeilles, a bee protection initiative, since 2002, and Guerlain has signed an environmental funding agreement with a similar organization, the Conservatoire de l'Abeille noire d'Ouessant. Both these brands have longstanding links to bees. In 2011, Louis Vuitton also signed a three-year sponsorship agreement with the CNRS (France's National Centre for Scientific Research) for a project entitled "City bees—Country

8 "Biodiversity Programmes across Manufacturing Sites", Nestlé, accessed 14 April 2016, http://www.nestle.co.uk/csv2013/environmentalimpact/protectingnaturalcapital

9 "Unilever Suppliers: A Closer Look at Biodiversity", Unilever, accessed 14 April 2016, https://www.unilever.com/Images/Unilever-Suppliers_a-closer-look-at-biodiversity_2015_tcm244-423993_en.pdf

bees", the goal of which is to understand why bees fair [*sic*] better today in urban areas than rural ones. A study released in 2012 addressed biodiversity at Louis Vuitton's Cergy 1 and Cergy 3 sites (flora, birds, butterflies).[10]

Another bee-related initiative was identified from the Hermes website, which states that beehives have been installed near water points at their "Maroquinerie Nontronnaise" site. The company also states that the bees' "production" (we assume, honey) is distributed to all personnel working on the site. Further, the website suggests that by incorporating bees and hives into their working environment they are raising employees' awareness of the need to preserve biodiversity and especially the need to protect bees.

The fine fragrance industry certainly seems to be heavily involved in bee-related initiatives. On the LVMH website, there are disclosures relating to a sustainability partnership between Guerlain and the Brittany Black Bee Conservatory (ACANB). The ACANB project aims to protect a species of black bees whose habitat is Ouessant Island. The website also explains that this initiative assists the company in ensuring that natural ingredients used in Guerlain beauty products are sourced in a sustainable manner. It is also interesting to note from the website that the company was given an award for sustainable development-oriented corporate philanthropy by the French Environment Ministry in 2013, for this initiative. This shows the positive reputational impact of biodiversity initiatives on companies.

The educative role of bee disclosures

Disclosures by several companies listed on the London Stock Exchange focused on what we have termed educational initiatives, where the company adopts an educative role towards their readership, demonstrating their knowledge and at the same time apparently seeking to raise the readers' awareness of bee-related issues.

For example, Sainsbury's website provides detailed instructions on how to make a "Bee Hotel".[11] Also on Sainsbury's website there are references to academic research into the decline of pollinators.[12] There may be an element of impression management here, as Sainsbury's demonstrates it is knowledgeable and scientific in its approach to biodiversity conservation and enhancement and this disclosure shows the company in a positive light, emphasizing its commitment to bees. There

10 Christian Dior, *Christian Dior Annual Report as of June 30, 2013* (Paris: Christian Dior, 2013), 73.

11 See www.sainsburys.co.uk/givebeesahome

12 Specifically, L.G. Carvalheiro, "Species Richness Declines and Biotic Homogenisation Have Slowed Down for NW European Pollinators and Plants", *Ecology Letters* 16 (2013): 870–8.

is extensive disclosure on the Sainsbury's website which is informative. The company seems to be taking on the role of an educator in bee science. For example, in an article on the website entitled, "If you build it, they will come", posted on 25 September 2013, Paul Crewe provides detailed information on bees such as the fact that 90% of food crops and 80% of Britain's wild plants are dependent on insect pollination.[13] Later, in the same article, there is a lengthy explanation of the role of solitary bees in pollination which differentiates between solitary bees, which do not live in hives, and the more social honey bees. The website explains that solitary bees do not make honey; instead it is their role as pollinators which interests companies such as Sainsbury's.[14]

An interesting example of what we have termed educative, or educational, reporting may be found on Coca-Cola's website. In an article placed on the website, "The Buzz on Innovation", written in January 2015 and authored by Drew Boyd, the introduction demonstrates a clear desire to "educate" readers, as follows,

> As pollinators, bees are responsible for every third bite of food we eat. Given their importance, a sharp decline in the global bee population could be cause for alarm. But how do scientists track bee populations? And what does this topic have to do with innovation?[15]

The article then continues by discussing an initiative aimed at assisting bee populations called The Great Sunflower Project. The project focuses on amassing bee pollination data in North America and uses volunteers to assist in data collection, with an aim to enhance bee populations.

Another example of educational disclosure may be found on the websites of chemical companies which specialize in pesticide and insecticide production. One such website provides scientific details on the need for continuing the use of neonicotinoids including evidence that these pesticides do not harm bees. The educative role of the web disclosures is evident as the company discusses the need for people to understand the many threats faced by bees and what can be done to help them.

Web disclosures by agrochemical companies include graphs describing the causes of colony mortality as reported by the beekeepers and by the scientists. These graphs and the accompanying scientific discussions are both informative and educational. The websites include information on the different species of bees and provide photographic images and scientific descriptions.

Analysing these disclosures interpretively and critically does raise questions as to whom these disclosures are intended for. Given the basic and informative nature

13 Paul Crewe, "If You Build it, They Will Come…", Sainsbury's, posted 25 September 2013, http://www.j-sainsbury.co.uk/blog/2013/09/if-you-build-it-they-will-come%E2%80%A6/

14 "Give Bees a Home", Sainsbury's, accessed 14 April 2016, https://livewellforless.sainsburys.co.uk/give-bees-a-home/

15 Drew Boyd, "The 'Buzz' on Innovation", Coca-Cola, posted 9 January 2015, http://www.coca-colacompany.com/stories/the-buzz-on-innovation

of some of the disclosures, who do the companies believe will be interested in reading the bee-related material on their website? This is an accounting issue which raised more general and far-ranging considerations relating to the usage and readership of internet reporting, as many of the bee-related disclosures are found on corporate websites.

Another aspect of educational bee-related disclosures is evidence of various training-related disclosures. Some of the bee-related disclosures we found were focused on corporate initiatives aimed at training various groups of stakeholders in bee conservation and even in some cases beekeeping. Some companies appear to have active programmes for training employees to build and keep beehives both at work and at home, for example, Sainsbury's.

Another illustration of what we interpret to be an educative role of bee accounting may be found on the Danone corporate website. The Danone website tells the full story of the reasons why pear trees are now hand-pollinated in Sichuan in China. This case study has been referred to several times within the book (see, for example, the discussion in Chapter 1). The rather frightening story related on the website shows how a natural function such as pollination by honey bees can be destroyed by mono agriculture, by interfering with nature and by using pesticides. This begs the question of why Danone are seeking to "educate" their readers about bees in this way.

The disclosure of bee-related information is in itself educative, as we can see from the above analysis. However, companies are also running educational initiatives, especially where children are concerned. A striking example of this is the disclosure on Nestlé's website of information about their schools-based educational initiative. Specifically, Inder Poonaji, the Head of Environmental Sustainability for the company, was quoted on the website saying,

> It is important to us to enhance the work we have already started by developing schemes which engage communities and schools local to our sites, enabling children to get closer to nature. We're keen for them to understand its true value through experiencing wildflower meadows and the role they play in pollination.[16]

This type of "accounting" is interesting as it raises a number of questions. Again, who is the intended readership? Why are companies involved in these types of educational activities? Is it purely reputation-driven or is this evidence of deeply held views and feelings relating to our need to protect the planet and our natural environment? The use of the accounting function, through corporate web disclosures, for this type of information could be viewed as emancipatory—perhaps by disclosing this information the company's employees believe that they can change people's attitudes and behaviour.

16 "Biodiversity Programmes across Manufacturing Sites"

Disclosures about the historical and cultural relationship between bees and humans

Chapter 2 of this book explored the historical relationship between bees and humans, tracking this relationship back to the earliest times of human civilization. From earliest times of human activity on Earth there are pictures, recordings and later, writings, about the great value people place on bees. In our analysis of corporate disclosures we found evidence to support a deep historical and cultural interest in bees, which was a surprise at the time. For example, in the musings on Sainsbury's website, the company states,

> Just thinking about it, the bees' pollination business is a lot like Sainsbury's retail business in that our products need to be at the right quality to attract customer, just like the trees, plants and flowers need to be at the right quality to attract the bee.[17]

This, we feel, is a fascinating piece of disclosure from a theoretical and philosophical perspective. As discussed in Chapter 2, for centuries, even millennia, people have drawn comparisons and discussed similarities between human activities and bees' activities. This corporate disclosure reveals a company attempting to demonstrate that their business reflects the way in which bees work, so as to show themselves in a positive light. Further, Sainsbury's website states,

> At Sainsbury's we reward our customers with Nectar points; the trees, plants and flowers also reward the bees with nectar too. We have a wide and varied customer base that choose to shop with us and in our gardens we also have a whole variety of bees that choose to pollinate our trees, plants and flowers, with the most famous being the honey bee and bumble bee.[18]

The whole concept of "Nectar points" demonstrates a close linkage, perceived by people and expressed through a business points scheme, between humans and bees.

The historical connection and desire to be somehow similar to bees draws on a deep historical tendency and demonstrates a deep human affection for bees and their devotion to labour. We saw in Chapter 2 how economics and human labour have over time been compared with, and contrasted to bees' tireless activities in the hive. These corporate disclosures seem to demonstrate a deep-seated desire to protect bees which, we argue, is a psychological driver of both bee-related disclosures and bee-oriented corporate initiatives, partnership and conservation efforts. It led us to think that the current trends to assuage bee decline may be fuelled not only by a business case (the need to preserve pollinators for financial reasons) but

17 Paul Crewe, "Bee Happy Now That Spring Has Arrived", Sainsbury's, posted 11 April 2013, http://www.j-sainsbury.co.uk/blog/2013/04/bee-happy-now-that-spring-has-arrived/

18 *Ibid.*

also by a psychological desire to preserve a species which has been held dear by the human race since the dawn of time.

The use of a bee in imagery also reveals the way in which people (and companies through their employees) perceive bees. For example, on its corporate website, Carrefour explains that it used the image of a bee as a symbol to represent the company's campaigns aimed at tackling waste throughout its business. These campaigns include initiatives aimed at encouraging all of the company's employees to reduce waste in their day-to-day activities, including saving paper, water and energy as well as recycling. Carrefour comments that the bee as a symbol represents a balanced ecosystem and community. The bee represents so many different aspects of human activity as we saw in Chapter 2 and companies are calling on the bee as a means of demonstrating their connections with and stewardship over nature.

Summary and discussion

As we can see from the above analysis, bee accounting is quite significant among companies in the food, cosmetics and agrochemical industries. When we began this project we did not expect to find many, if any, example of bee disclosures. We were surprised! Also, over the short time span of our analysis, the quantity of bee-related disclosures has risen substantially. Notably, most of the bee-related disclosures we identified were on websites rather than in the annual or sustainability reports. Our analysis highlights significant differences between the use of the web in contrast to the use of physical printed reports by producers/companies. Indeed, we feel our analysis provides insights into the very different function of the web and published reports as reporting vehicles. Whereas published reports tend to focus on the shareholders' needs and on discharging accountability for material issues through disclosures, we feel that the web is creating a space for a very different form of reporting by companies. We provide evidence to suggest that web disclosures are educative, informative, sometimes defensive, even lobbying in nature, and often emancipatory. We suggest that these features and roles of web disclosures are not limited to bee-related disclosures but may be characteristics pertaining to other areas of sustainability disclosure. Websites appear to be used as debating spaces in some cases. In terms of accounting research and theory, these disclosures and the disclosure "themes" we have identified, raise the question of whom the disclosure is aimed at? What is the intended readership? Research is needed to assess which stakeholders companies really believe these types of bee disclosures, and bee accounting more generally, are aimed at. Similar questions can be asked about all biodiversity disclosures by companies.

We also thought it would be interesting to explore the apparently emergent relationship between bees and companies in the form of drawing analogies between "beekeepers" and "book-keepers" (accountants, basically). As we know, beekeepers care for and keep bees in hives. Book-keepers "care for" the books, accounts and assume responsibility over the figures. Both roles involve taking responsibility. In the case of a book-keeper, this responsibility involves caring for the accounts and ensuring they are up to date on a daily basis whereas the beekeeper tends to the bees under his or her care. Both roles involve stewardship and involve discharging accountability, a duty of care and are thus not dissimilar. The merging of the roles in accounting for bees and corporate activities relating to protecting and growing bee populations raises some interesting thoughts. Whether it is money or bees, people seem to have an inherent need to nurture, audit and take responsibility for their assets. Indeed, historically bees and their honey have been used as a means of payment, a unit of exchange, indicating that beekeepers have been perceived as a form of accountant, caring for their assets, renting them and selling them. Now, accountants are becoming beekeepers in an effort to account for and discharge accountability for their relationship with and impact on bees: the bee has become an identifiable corporate stakeholder. For the first time, perhaps bee accounting is suggestive of a similarity between the role and responsibilities of accountants and beekeepers. Social accounting really does mean that, taking an integrated approach, bees are just as important to "accounting" as financial or calculative figures. Our analysis in this chapter also helps to underline an uncomfortable truth that unless there is dramatic change in corporate activity and in the way bees are treated, there will be no bees. There can be no bee accounting or bee accountability if there are no bees. Bee accounting may represent a crucial emancipatory mechanism which can have the potential of educating stakeholders, demonstrating corporate accountability for bees, and ultimately changing corporate as well as society's attitudes towards bees and their importance to society and the planet's future.

Bibliography

Albright, M.B. "Could Robot Bees Help Save Our Crops?" *National Geographic*, 21 August 2014.

Atkins, J. and I. Thomson. "Accounting for Nature in 19th Century Britain: William Morris and the Defence of the Fairness of the Earth". In *Accounting for Biodiversity*, edited by Jones. Abingdon, UK: Routledge, 2014.

Atkins, J., C. Gräbsch and M.J. Jones. "Biodiversity Reporting: Exploring its Anthropocentric Nature". In *Accounting for Biodiversity*, edited by Jones. Abingdon, UK: Routledge, 2014.

Atkins, J.F., W. Maroun, E. Barone and B.C. Atkins, "From the Big Five to the Big Four? Exploring Extinction Accounting for the Rhinoceros?" working paper presented at Southampton Business School, April 2016.

Boyd, D. "The 'Buzz' on Innovation". Coca-Cola. Posted 9 January 2015. http://www.coca-colacompany.com/stories/the-buzz-on-innovation

Carvalheiro, L.G. "Species Richness Declines and Biotic Homogenisation Have Slowed Down for NW European Pollinators and Plants". *Ecology Letters* 16 (2013): 870–8.

Christian Dior. *Christian Dior Annual Report as of June 30, 2013*. Paris: Christian Dior, 2013, 73.

Constantino, M. *Bees and Beekeeping.* Worcestershire, UK: King Books, 2011.
Crewe, P. "Bee Happy Now That Spring Has Arrived". Sainsbury's. Posted 11 April 2013, http://www.j-sainsbury.co.uk/blog/2013/04/bee-happy-now-that-spring-has-arrived/
Crewe, P. "If You Build it, They Will Come…", Sainsbury's. Posted 25 September 2013. http://www.j-sainsbury.co.uk/blog/2013/09/if-you-build-it-they-will-come%E2%80%A6/
Gallhofer, S. and Haslam, J. *Accounting and Emancipation: Some Critical Interventions.* London, UK: Routledge, 2003.
Gallhofer, S., Haslam, J. and Yonekura, A. "Accounting as Differentiated Universal for Emancipatory Praxis". *Accounting, Auditing & Accountability Journal* 28(5) (2015): 846–74.
Gray, R., Owen, D., Adams, C. *Accounting & Accountability: Changes and Challenges in Corporate Social and Environmental Reporting.* Essex, UK: Pearson Education Limited, 1996.
Jones, M. (ed.). *Accounting for Biodiversity.* Abingdon, UK: Routledge, 2014.
Jones, M.J. and J.F. Solomon. "Problematising Accounting for Biodiversity", *Accounting, Auditing & Accountability Journal* 26(5) (2013): 668-687.
May, A.F. *Beekeeping,* Cape Town, South Africa: Haum, 1969.
Nestlé. "Biodiversity Programmes across Manufacturing Sites". Accessed 14 April 2016. http://www.nestle.co.uk/csv2013/environmentalimpact/protectingnaturalcapital
Palmer-Jones, T and I.W. Foster. "Agricultural Chemicals and the Beekeeping Industry". *New Zealand Journal of Agriculture,* October 1958.
Sainsbury's. "Give Bees a Home". Accessed 14 April 2016. https://livewellforless.sainsburys.co.uk/give-bees-a-home/
Stathers, R. *The Bee and the Stockmarket: An Overview of Pollinator Decline and its Economic and Corporate Significance.* Research Paper, London: Schroders, January 2014.
Unilever. "Unilever Suppliers: A Closer Look at Biodiversity". Accessed 14 April 2016. https://www.unilever.com/Images/Unilever-Suppliers_a-closer-look-at-biodiversity_2015_tcm 244-423993_en.pdf
Wilson, Bee *The Hive: The Story of the Honeybee and Us.* London, UK: John Murray, 2004.

12

Accounting for bees

Evidence from disclosures by US listed companies

Andrea M. Romi and Scott D. Longing
Texas Tech University, USA

The United States Federal Government: the driver of change

The whole of this book is focused on the problems arising from the global decline in bee populations. This chapter focuses on the American context and on the involvement of US corporations, as well as other organizations, in reducing the decline in bee populations. Growing concern about the consequences of this decline has recently garnered attention from an unexpected source, the United States Federal Government. More specifically, in June 2014, the President of the United States developed the Pollinator Health Task Force, co-chaired by the Secretary of Agriculture (US Department of Agriculture, USDA) and the Administrator of the Environmental Protection Agency (EPA). The purpose of this task force was to respond to the losses in the country's pollinators based on the government's belief that these pollinators play a critical role in maintaining diverse ecosystems and in supporting agricultural production.[1]

1 Pollinator Health Task Force, *National Strategy to Promote the Health of Honey Bees and Other Pollinators* (May 19, Washington, DC: The White House, 2015).

In 2015, the task force released its first document, the National Strategy to Promote the Health of Honey Bees and Other Pollinators (Strategy), to communicate its strategy for addressing the pollinator decline problem. The Federal Government, as the largest land manager in the nation, implemented a broad pollinator-friendly landscaping approach across Federal facilities. The government worked along with the EPA to determine the impact of chemical exposure and the USDA to expand pollinator habitats, as well as an abundance of other Federal departments and agencies. These actions, in combination with public–private partnerships, represent the government's initial attempts to achieve three overarching goals:

1. Reduce honey bee colony losses during winter (overwintering mortality) to no more than 15% within 10 years. This goal is informed by the previously released Bee Informed Partnership surveys and the newly established quarterly and annual surveys by the USDA National Agricultural Statistics Service (NASS). Based on the robust data anticipated from the national, statistically based NASS surveys of beekeepers, the Task Force will develop baseline data and additional goal metrics for winter, summer and total annual colony loss.
2. Increase the eastern population of the monarch butterfly to 225 million butterflies occupying an area of approximately 6 hectares in the overwintering grounds in Mexico, through domestic/international actions and public–private partnerships by 2020.
3. Restore or enhance 2.8 million hectares of land for pollinators over the next 5 years through federal actions and public–private partnerships.[2]

In order to achieve these goals, the Task Force developed several action plans.[3] First, the Pollinator Research Action Plan focuses efforts on producing the scientific information needed to better understand pollinator decline. Second, the Task Force developed Best Management Practice guidance to assist in designing and implementing natural landscapes to combat pollinator decline. Finally, the National Seed Strategy for Rehabilitation and Restoration is developing a seed bank of plants that have the greatest potential to support pollinator restoration activities.

The Federal Government's Task Force development and subsequent formal strategy would have no ability to effect change without the necessary resources. In 2015, the President's Budget request to Congress for the 2016 fiscal year included over $82 million in funding for the various federal departments targeting pollinator decline. This is a substantial increase from the $34 million for these same departments in 2015.[4]

2 *Ibid.*, i
3 Pollinator Health Task Force, *National Strategy to Promote the Health of Honey Bees and Other Pollinators.*
4 *Ibid.*

In addition to the President's action plan, the EPA is another governing body heavily involved in combating pollinator decline. With respect to the EPA, the following are actions specifically being taken to protect pollinators:

- Proposing a plan to prohibit the use of all highly toxic pesticides when crops are in bloom and bees are under contract for pollination services.
- Prohibiting the use of certain neonicotinoid pesticides when bees are present.
- Re-evaluating the neonicotinoid family of pesticides, as well as other pesticides.
- Temporarily halting the approval of new outdoor neonicotinoid pesticide uses until new bee data is submitted and pollinator risk assessments are complete.
- Reviewing the new varroa mite control products.
- Developing new bee exposure and effect testing priorities for the registration of new pesticides, new pesticide uses and registration review of existing pesticides.
- Issuing data requirements and risk assessment approaches for pollinators.
- Establishing guidance on best practices for regional, state and tribal inspections of pesticide-related bee deaths.
- Developing a new risk management approach for considering the impacts of herbicides on monarch butterfly habitats and protecting milkweed from pesticide exposure.
- Issuing a benefits analysis of neonicotinoid seed treatments for insect control in US soybean production.
- Providing farmers and beekeepers with EPA's residue toxicity time data as a means of gauging the lengths of time that specific pesticide products may remain toxic to bees and other pollinators.
- Working with pesticide manufacturers to develop new seed-planting technologies that will reduce dust that may be toxic to pollinators during the planting of pesticide-treated seed.
- Incorporating pollinator protection at EPA facilities, on epa.gov, and in other EPA programmes.
- Encouraging pollinator-friendly habitat considerations in land clean-up programmes.[5]

5 Environmental Protection Agency, "EPA Actions to Protect Pollinators", accessed 10 March 2016, http://www2.epa.gov/pollinator-protection/epa-actions-protect-pollinators.

Two additional US federal agencies involved in promoting the health of pollinators are the US Department of Agriculture (USDA) and the US Fish and Wildlife Service (USFWS). The Natural Resource Conservation Service (NRCS) is the component of the USDA charged with promoting pollinator health through guidance and support to farmers and ranchers implementing conservation practices. In 2014 the NRCS provided close to $3 million and in 2015 provided $4 million for technical and financial assistance to farmers interested in safe and diverse food resources for honey bees. The NRCS also maintains numerous Plant Materials Centers that focus on selecting plants used by pollinators and providing recommendations on plants that will enhance pollinator populations. The USFWS is one of 26 conservation, education and research partners in the US that make up the Monarch Joint Venture aimed at monarch butterfly conservation. In 2015, the USFWS partnered with the National Wildlife Federation and the National Fish and Wildlife Foundation in pledging $2 million for on-the-ground conservation projects for monarch conservation.

Public–private partnerships

Despite a significant increase in resources, the US Federal Government recognizes their limitation in tackling pollinator decline. For this reason, the President identified the need for public–private partnerships. In the recently communicated detailed strategy, the Task Force recommends the coordination of a multifaceted portfolio of public education and outreach strategies intended to attract multiple audiences (individuals, schools, libraries, museums and other educational venues, demographically diverse audiences, organic certifiers, etc.) and particularly targeting corporations. Governmental as well as non-governmental organizations will deliberately focus public education tactics towards corporations, in order to promote change.

The materials developed to promote change involve four core principles to guide the scope of intended actions:

1. Pollinator conservation is a shared national responsibility.
2. The demographically diverse US public requires customized strategies of communication, education, and outreach. The key messages should be relevant to each target audience and well understood by multicultural audiences.
3. The actions of a single person can make a difference—every citizen can contribute to pollinator conservation and should have the opportunity to become engaged in ways that are meaningful.

4. Agencies involved in implementing the Presidential Memorandum should seek to educate and empower citizens as partners in pollinator conservation.[6]

In the US, two main entities answering the call to take responsibility for pollinator decline are non-governmental organizations (NGO) and corporations.

Non-governmental organizations

There are several NGOs involved in pollinator conservation activities within the US. In particular, some of these have been publicly pressuring the EPA to ban neonicotinoid pesticides for some time, as these pesticides have been shown to have lethal and sub-lethal effects on honey bees and can be contacted by bees through multiple routes of exposure in agricultural fields.[7] In addition, many NGOs also partake in conservation actions and educating the public about the ways to promote pollinator health. The list of NGOs includes, but is certainly not limited to: the Natural Resources Defense Council, Earthjustice, the National Wildlife Federation, the Xerces Society, and The Pollinator Partnership.

The Natural Resources Defense Council (NRDC) was founded in 1970 to protect the nation's air, land and water from the forces of pollution and corporate greed. They are considered the nation's most effective environmental action group, combining the grass-roots power of more than 2 million members and online activists with the courtroom clout and expertise of nearly 500 lawyers, scientists and other professionals.[8] The NRDC has been instrumental in pressuring the EPA to ban neonicotinoid pesticides. In addition to their substantial lobbying efforts, the NRDC also provides a wealth of public information related to colony collapse disorder, information pertaining to the reasons bees are important and what plants/foods pollination affects, in addition to individual tips such as how to attract bees to a home garden.

Joining the NRDC in their fight to impact policy is Earthjustice. Earthjustice uses the power of law and the strength of partnership to protect people's health; to preserve magnificent places and wildlife; to advance clean energy; and to combat

6 Pollinator Health Task Force, *National Strategy to Promote the Health of Honey Bees and Other Pollinators.*

7 C.H. Krupke *et al.*, "Multiple Routes of Pesticide Exposure for Honey Bees Living Near Agricultural Fields", *PLoS ONE* 7(1) (2012): e29268, doi:10.1371/journal.pone.0029268.

8 Natural Resources Defense Council, "The NRDC Story", accessed 10 March 2016, http://www.nrdc.org/about/nrdc-story.asp.

climate change.[9] In addition to lobbying the EPA, Earthjustice also spends time representing the beekeeping industry, updates the public concerning all of their current court proceedings in relation to the protection of pollinators, and provides public access to research related to pollinator decline and colony collapse disorder. They also attempt to involve the public by providing access to several petitions targeting the eradication of neonicotinoids.

While both the NRDC and Earthjustice are focused on the legal battles associated with pollinator decline, there are other NGOs targeting a more diverse audience. First, The National Wildlife Federation (NWF) is a nationwide federation of state and territorial affiliate organizations with nearly 6 million members and supporters across the country, representing a voice for wildlife, dedicated to protecting wildlife and habitat and inspiring the future generation of conservationists.[10] They have joined the battle to conserve the declining North America pollinator populations in many ways, one of which is their Million Pollinator Garden Challenge. In support of the President's recent call, the NWF joined dozens of conservation and gardening organizations as well as seed groups to form the National Pollinator Garden Network and launch this new nationwide campaign, designed to accelerate growing efforts across America to reverse the alarming decline of pollinators, including honey bees and native bees.[11] The NWF will work with the Network to rally hundreds of thousands of gardeners, horticultural professionals, schools and volunteers to help reach a million pollinator gardens in the near future. Specifically, the NWF outlined in their 2015 annual meeting, that they support increasing and, in the long term, sustaining pollinator populations such as bees, butterflies and other species through a variety of means including scientific assessment, policy and practice reforms, monitoring, public–private partnerships, public education, grass-roots activism, volunteer programme development, species-specific campaigns and home, school and community initiated habitats.

Two NGOs specifically developing programmes to partner with businesses in the US are the Xerces Society and the Pollinator Partnership. For over 40 years, the Xerces Society has worked to protect wildlife through the conservation of invertebrates

9 Earthjustice, "Our Story", accessed 10 March 2016, http://earthjustice.org/about.

10 National Wildlife Federation (NWF), "Our Work: Protecting Wildlife, Inspiring Future Generations", accessed 10 March 2016, http://www.nwf.org/What-We-Do.aspx.

11 David Mizejewski, "Take the Million Pollinator Garden Challenge", accessed 3 June 2015, http://blog.nwf.org/2015/06/million-pollinator-garden-challenge/?s_subsrc=Web_Biggie__Home_MPGC.

and their habitat.[12] Xerces delivers this action by harnessing information produced from scientific studies and citizen science programmes to implement conservation. Xerces manages the largest pollinator conservation programme in the world, referred to as the Xerces Society Pollinator Conservation Program (PCP). The PCP has restoration projects throughout North America, Europe and Asia and has worked diligently to create significant acreage dedicated to pollinator habitat, conducted hundreds of native pollinator workshops with training and education, written many publications on pollinator conservation, created a bumblebee watch programme to find new populations of at-risk bees, and developed reduced-risk pest control strategies. One important contribution has been their outreach with business. The Xerces Society works with mission-driven companies to source pollinator-safe ingredients for their products and assists these organizations in developing pollinator-friendly supply chains. This Xerces–business coalition includes more than 40 businesses to date, working towards making pollinator conservation a mainstream business practice.[13]

The second NGO dedicated to working with US businesses to impact pollinator decline is the Pollinator Partnership (PP), the largest non-profit organization in the world dedicated exclusively to the protection and promotion of pollinators and their ecosystems.[14] The Pollinator Partnership's mission is to promote the health of pollinators through conservation, education and research. Among many initiatives, PP manages Business for Bees: American Business Collaboration for Pollinator Conservation Action. Business for Bees is a joint effort, applying business management tactics to cooperative conservation efforts. Specifically, the Business for Bees concept includes: creating a peer-to-peer network of businesses that commit to taking action to foster the recovery and sustainment of pollinators and their habitat; enlisting new voices in a coordinated vision of pollinators as a vital part of the American landscape; providing feedback and connection to the White House and other government and NGO initiatives; and engaging employees, customers, communities and industries in these efforts.[15] The overall

12 Xerces Society for Invertebrate Conservation, "Protecting the Life that Sustains Us", accessed 10 March 2016, http://www.xerces.org/.

13 Xerces Society for Invertebrate Conservation, *The Xerces Society for Invertebrate Conservation: Pollinator Conservation Program*, accessed 10 March 2016, http://www.xerces.org/wp-content/uploads/2015/03/PollinatorProgramFactsheet_Mar2015_web.pdf.

14 Pollinator Partnership, "Our Mission", accessed 10 March 2016, http://www.pollinator.org/.

15 Business for Bees, "Business for Bees: American Business Collaboration for Pollinator Conservation Action", accessed 10 March 2016, http://www.pollinator.org/bizforbees.htm.

strategy of this effort is to utilize a business approach to problem solving, thus reaching constituencies needed to be successful in the pollinator conservation effort.

The US corporate landscape in relation to pollinator conservation

Pollinator decline is not solely a governmental or individual problem but one which urgently requires a multi-organizational solution. Corporations have both a social and economic responsibility to assist in developing this solution. First, from a social perspective, corporations not only have a direct impact on pollinator decline through use of insecticides (agricultural companies), but also have an indirect impact by means of climate change. Through the increasing greenhouse gas emissions associated with corporate activity, the greening, flowering and ageing cycles of plants are altered, impacting pollinators and multiplying the impacts of habitat loss.[16]

Second, from a purely economic perspective, corporations will directly or indirectly witness pollinator decline impacts on their bottom line. Pollinator decline is projected to increase over time, with a current estimated economic impact of approximately $186 billion.[17] A direct consequence of this problem includes an increase in raw material prices, significantly affecting cash flows of US corporations.[18] Other firms will face indirect consequences including, but not limited to: impacts to health services from the decline in available fruits, vegetables, and nuts; impacts to the transportation industry dependent on the delivery of agricultural products; and impacts to the recreation industry due to the decline in human food consumption. It is these types of risks from pollinator decline that should be included in firm-specific disclosures under current US accounting and reporting standards. As we can see from earlier chapters in this book, pollinator, and especially bee, decline is a material financial risk for both companies and their major shareholders. Any financially material and significant issue should be discovered

16 R. Stathers, *The Bee and the Stockmarket: An Overview of Pollinator Decline and its Economic and Corporate Significance* (London: Schroder Investment Management Limited, 2014).

17 N. Gallai *et al.*, "Economic Valuation of the Vulnerability of World Agriculture Confronted with Pollinator Decline", *Ecological Economics* 68 (2009): 810–21.

18 Stathers, *The Bee and the Stockmarket.*

by a firm's internal control and governance system and should therefore be discussed in public corporate filings.

In the US, the Securities and Exchange Commission (SEC), along with the accounting oversight bodies, provide firms with guidance on the appropriate items to disclose to the public. Reporting practices within the US are guided by the conceptual framework. In the face of reporting discretion, where management is not certain what information should be reported, the conceptual framework offers direction. Specifically, when considering disclosure, there is a trade-off between the cost of providing information and the usefulness of such information for decision-making.[19] Costs of providing information include the actual cost of services to gather and print the information, but may also include proprietary costs. As for the usefulness of information, among other things, the conceptual framework requires that information be relevant. Relevance is defined as information exhibiting predictive value, confirmatory value and materiality. In other words, in the face of a decision to include information in public reports, firms must determine whether the information can be predicted and measured, can be verified and whether it is material to the firm. Based on estimates of economic impact and both direct and indirect corporate consequences, pollinator decline is likely to be a relevant issue for all US firms.

After a firm determines that pollinator decline poses a risk and should be disclosed, it must decide in what form to report such information. The SEC specifically requests firms to provide information about the operations of the business as well as addressing firm risks in the annual report (10K). The annual report consists of four distinct parts, two of which include requirements for risk discussions. In Part I, Item 1A "Risk Factors" includes information about the most significant risks that apply to the company. This section focuses on the risks themselves, not how the company addresses those risks, and can be discussed in relation to the entire industry. A second area in the 10-K where risk is addressed is in Part II, Item 7 "Management's Discussion and Analysis of Financial Condition and Results of Operations" (MD&A). In this section, management discusses their views of key business risks and what is being done to address these risks. Given the severity of pollinator decline, all public US firms should have some disclosure in one of these two areas of the annual report.

19 Financial Accounting Standards Board, *Statements of Financial Accounting Concepts No. 8* (Norwalk, CT: The Financial Accounting Foundation, 2010); International Accounting Standards Board (IASB), *The Conceptual Framework for Financial Reporting 2010* (IASB, 2010).

In addition to disclosures in a firm's annual report, disclosures concerning pollinator decline will likely be found in each firm's stand-alone sustainability report. These reports, also referred to as corporate responsibility reports, citizenship reports, accountability reports, etc., provide stakeholders with a glimpse into each firm's environmental and social activities. Within the US, stand-alone sustainability reports are voluntary, are often not assured, vary in length and informational inclusion, and have been used in many instances as an image-creating marketing tool (i.e. greenwashing). If a firm is involved in any activities attempting to tackle the pollinator decline problem, it would be considered a positive attribute of a firm's sustainability behaviour and would likely be disclosed. Additionally, given the call by the US Government for businesses to focus on pollinator decline and the involvement by NGOs in cooperating with US firms towards finding solutions to pollinator decline, we anticipate US corporate involvement and disclosure to be strong.

In order to determine whether firms are, in fact, disclosing information required by the SEC, we search US public company 10-Ks and stand-alone sustainability reports for disclosures related to pollinator decline. We specifically search for instances of bee disclosures, any conservation efforts they are involved in, the dollar value associated with any conservation efforts, the net present value of the firm-specific risks, etc. We also determine what factors lead some firms to provide information and the extent of the information provided, including the industries that may be more prone to informative disclosures. This is certainly a broad overview of the current corporate pollinator decline disclosure landscape in the US.

US firm disclosures

Since the size of a firm often predicts increased voluntary disclosures, we chose the S&P 100 firms as our sample of interest. The S&P 100 is a stock market index of US stocks, representing approximately 45% of the market capitalization of the US equity markets. These firms are some of the largest and most established firms in the US. We focus on public firms because they have comparable forms of disclosure which makes analysis among firms easier. It is likely that private firms also disclose similar, if not more, information related to bees, but finding that information is more difficult. We also focus on the most current year's disclosures as pollinator decline disclosures are likely to improve over time. After searching each firm's 2014 annual report, we were disappointed and somewhat surprised to find absolutely no disclosure of any type related to firm-specific risks linked to pollinator decline. Even with the significant economic impact to the US estimated at $186 billion, the direct consequence of increases in raw material prices predicted to significantly affect future cash flows of US corporations, and the indirect consequences resulting from impacts to health, transportation, and recreation services, there was absolutely no discussion in any annual report of such risks.

Given the non-existent representation in annual reports, we turn to the S&P 100's stand-alone sustainability reports for the current corporate discussion. While the

US does appear to be entering this arena, representation remains weak. Figure 12.1 shows the breakdown of disclosures per sustainability report.

Figure 12.1 **Instances of disclosure per sustainability report**

Source: Authors' own

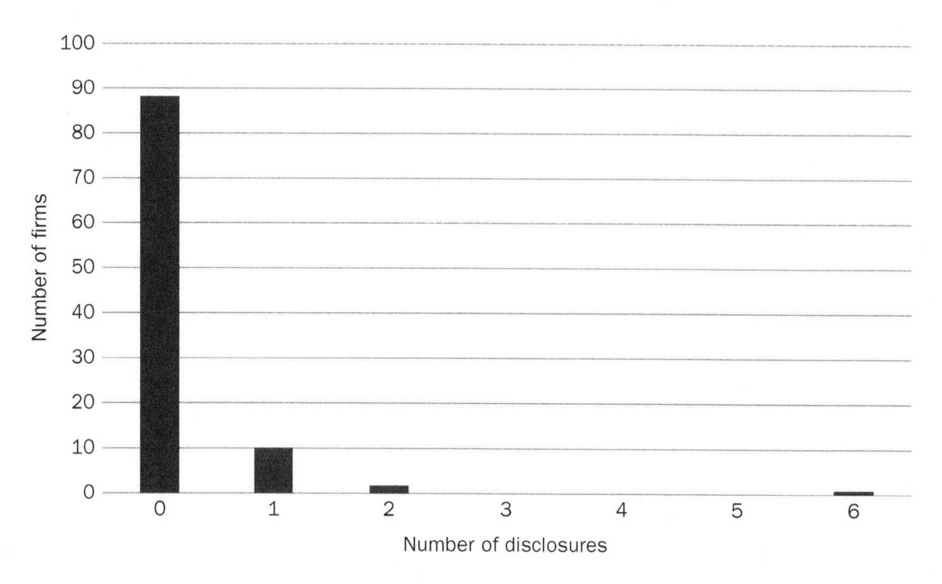

As indicated, an overwhelming majority of the largest public firms in the US do not disclose anything about bees in their sustainability reports (88 of the 100 firms). Overall, there were a total of 12 firms disclosing information about bees or the problems associated with pollinator decline. Of those 12, 10 firms only mentioned bees once, with the average sustainability report in our sample at approximately 111 pages. One firm discusses bees on two separate occasions within their sustainability report and one firm featured extensive discussions in six different areas within their report.

Figure 12.2 provides a breakdown of what is included in each of these firm-specific disclosures. Some firms, while only discussing bees on one occasion, may have discussed more than one activity.

Figure 12.2 **Specifics of disclosures within the sustainability report**

Source: Authors' own

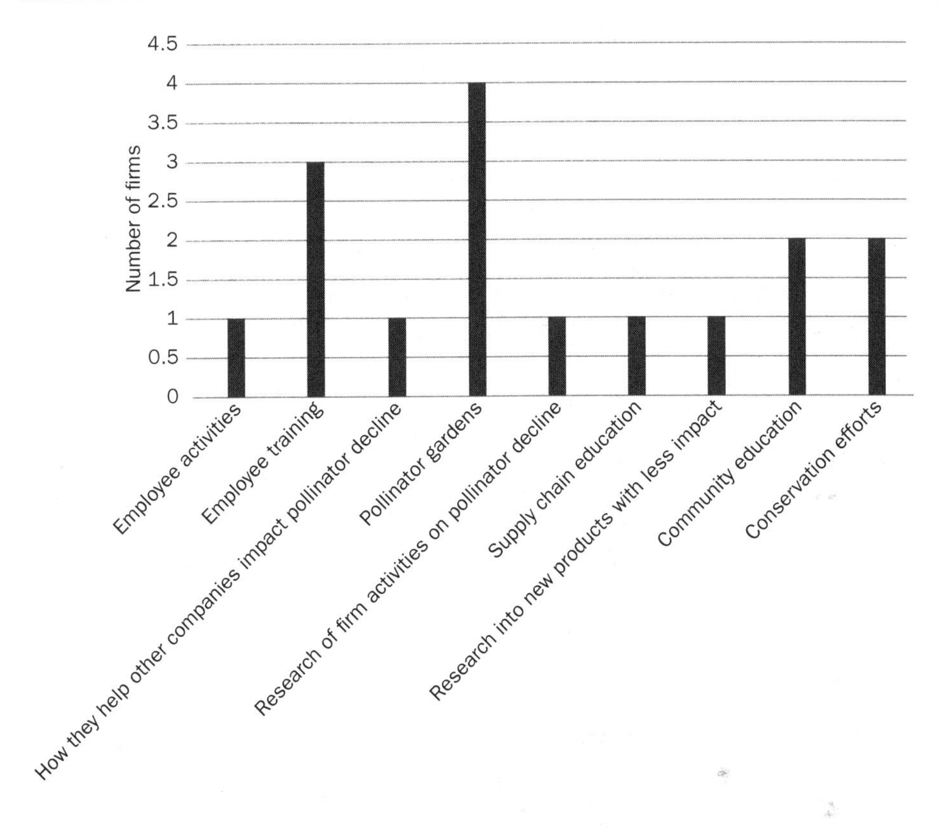

We analysed all discussions of bee-related activities within stand-alone sustainability reports and determined there were nine distinct activities communicated. In the case of disclosures related to employee activities, how the company assists other companies to impact pollinator decline, research specifically focused on how the firm's activities are impacting pollinator decline, providing supply chain education, and activities related to researching new product with the potential for reduced impact on bees, we found that only one firm discussed each of these issues.

Hewlett Packard's *2014 Living Progress Report* provides discussions about an employee Global Volunteer Survey it conducted to determine the type of volunteer activities HP employees participate in. Survey results indicated that some employees volunteered by building pollinator habitats in Toronto, Canada.[20] This disclosure was not a company activity *per se*, but did include the volunteer work of employees within the company. FedEx was one US firm with a unique disclosure

20 Hewlett Packard, *HP 2014 Living Progress Report* (Palo Alto, CA: Hewlett Packard, 2014), 1–136.

relating to bees. Instead of specifically discussing something they had done related to combating pollinator decline, they provided a discussion, under the heading "Mann Lake Fights Against Colony Collapse Disorder", of a company in Minnesota that is the world's largest supplier of beekeeping equipment. FedEx then discussed colony collapse disorder and the potential long-term consequences of the problem and how this company in Minnesota is working to combat the problem by providing beekeeping equipment to both hobbyists and commercial beekeepers. FedEx then states,

> The location of its Minnesota headquarters, far from urban areas of any size, doesn't impede delivery and distribution—the company relies on a combination of FedEx Express, FedEx Ground and FedEx Freight to get its products to beekeepers and distributors.[21]

This is certainly an instance where FedEx is disclosing their indirect involvement in the problem.

American Electric Power (AEP) was the only firm to discuss participating in research on how their own activities may be impacting pollinator decline. Specifically, in their *2014 Corporate Accountability Report*, AEP states, "Recently, AEP partnered with the Electric Power Research Institute to evaluate the ecosystem services provided by the site and the possible impacts that shale gas fracking could have on these resources". AEP specifically includes pollination as one of the resources supplied by the ecosystem, benefiting mankind, that they are interested in studying. They go on to assure the reader that, "Results of the study to date have indicated no long lasting impacts".[22]

Within the US, Walmart has long been a driving force behind public discussions concerning supply chain sustainability issues. It appears the issues of bees have also caught its attention. In the *2015 Global Responsibility Report*, Walmart discusses its bee activities under the heading "Sustainable Food Chain". Specifically, it states,

> Walmart is known for lowering the cost of food for our customers. We have been working hard to do it in a way that also lowers the true cost to society—meaning improving yields while reducing GHG emissions and preserving natural capital (oceans, forests, water, air quality, pollinator health), and enhancing farmer livelihoods.[23]

21 FedEx, *2014 Global Citizenship Report: Moving Possibilities* (Memphis, TN: FedEx, 2014), 22.

22 American Electric Power, *2014 Corporate Accountability Report* (Columbus, OH: American Electric Power, 2014), 27.

23 Walmart, *2015 Global Responsibility Report* (Bentonville, AR: Walmart Stores, Inc., 2015), 76.

They go on to discuss the ways in which they are achieving this goal. We deemed this activity, as disclosed, to be a type of supply chain education, both by educating themselves on how to achieve this goal, as well as educating those in their supply chain.

Monsanto was actually one of the most forthcoming firms in the US with regard to pollinator disclosures. They disclosed six separate discussions, one of which centred on research into new products which would likely have less impact on pollinator decline. In their *2014 Sustainability Report*, Monsanto states, "Monsanto's agricultural biologicals research focuses on creating products derived from nature to support plant health and protection from pests". They further disclose, "Through BioDirect technology, researchers are using their knowledge of plant and pest genetics to develop new topical solutions for controlling weeds, insects and viruses as well as protecting honey bee health".[24]

In addition to each of the aforementioned activities, community education and conservation efforts in general were also disclosed, both on two separate occasions. An example of community education can be found in Monsanto's discussion of their involvement in a coalition of researchers, advisers and other stakeholders in an effort towards "working with beekeepers to better understand how to keep healthier bees".[25] General conservation effort disclosures were also found in Monsanto's Sustainability Report. This disclosure was the only one providing a dollar amount of investment. Specifically, Monsanto states,

> Almond crops are a crucial part of the California agriculture industry that requires 800 commercial beekeepers and more than 1.8 million honey bee colonies—transported from around the country—for pollination each year. By placing a variety of early-blooming flowers in areas adjacent to almond groves and other farmland, bees have access to a nutritious food source during peak pollination times. Our three-year partnership with PAm, a leading organization dedicated to improving the vitality of honey bee colonies, includes a $250,000 investment in educating California almond growers and landowners on the value of honey bee forage, as well as planting forage itself. With 2014 marking the second year of the partnership, together we have planted more than 3,000 acres of forage.[26]

Employee training was also a popular disclosure among our 12 firms. Within our sample, three firms discussed providing employees with training on the pollinator decline problem and/or how they could assist in combating the problem. Bristol-Myers Squibb discussed employee training twice within their *2014 Foundation and Corporate Social Responsibility Report*. First,

24 Monsanto, *From the Inside Out: Monsanto 2014 Sustainability Report* (St Louis, MO: Monsanto, 2014), 24.
25 Monsanto, *From the Inside Out: Monsanto 2014 Sustainability Report*, 67.
26 Monsanto, *From the Inside Out: Monsanto 2014 Sustainability Report*, 67.

> Each year during Earth Week in April, Bristol-Myers Squibb employees around the world demonstrate their commitment to environmental stewardship and a sustainable future by participating in activities ranging from planting trees to attending workshops about the benefits of keeping honeybees and the conservation of energy and water.[27]

They went on to disclose,

> To raise awareness about biodiversity and the essential role that bees play in the ecosystem, the company introduced employees and their children to its apiary. Workshops and discussions were led by professional beekeepers who introduced the hive and the role of the bees and their queen. Thirty-five children participated in the programme, which also included a presentation of the movie "More than Honey," an award-winning documentary that takes an in-depth look at honeybee colonies around the world.[28]

Among all of the pollinator disclosures, the most frequently disclosed activity involved developing pollinator gardens. One example of this type of disclosure is General Motors (GM). In their *2014 Sustainability Report*, GM discusses an activity under the heading "Leadership in natural habitat", "A 27,000-square-foot garden was designed to benefit pollinators, such as bees and butterflies, that are critical to the reproduction of 90 percent of flowering plants and one-third of human food crops worldwide".[29]

One of the firm-specific characteristics, beyond that of size, likely to impact the decision to participate in and disclose bee-related activities, is the industry in which the firm operates. This is due to the fact that different industries are likely to be affected more by pollinator decline than others. Similarly, the activities of firms in different industries are likely to impact pollinator decline in different ways. Figure 12.3 provides a breakdown of the industry membership within our disclosing sample.

27 Bristol-Myers Squibb, *Foundation and Corporate Social Responsibility: Innovation and Inspiration—2014 Report* (New York: Bristol-Myers Squibb, 2014), 30.

28 *Ibid.*

29 General Motors, *A Driving Force: 2014 Sustainability Report* (Detroit, MI: General Motors, 2014), 65.

Figure 12.3 **Industry representation**
Source: Authors' own

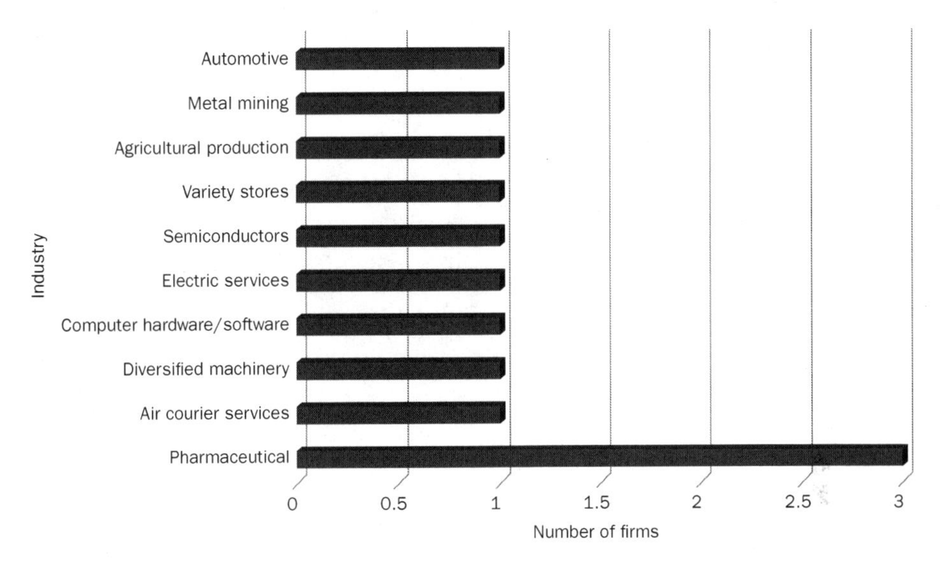

As indicated in Figure 12.3, a majority of industries in our sample are only represented once (automotive, metal mining, agricultural production, variety stores, semiconductors, electric services, computer hardware/software, diversified machinery, and air courier services). The pharmaceutical industry was the most represented industry within our sample of large, US public firms, with a total of three firms disclosing bee-related information.

While the US bee disclosures appear to be minimal, some disclosure is a good sign. US public firms often lag other countries in reporting environmental and social activities and wait for other countries to provide direction. Another possible catalyst for US corporate involvement is the actual cooperation or partnering of the Federal Government with NGOs to encourage firm participation in bee conservation activities. Often referred to as cross-sector social partnerships (CSSP), this informal tri-sector partnership will likely educate firms on the risks to their operations from pollinator decline, encouraging firm-specific initiatives related to bees. Romi *et al.* examine pollinator disclosures (including bees and butterflies) in US S&P 500 firms from 2013 to 2015, and find that the formation of a CSSP between the US Government, NGOs and public firms to combat pollinator decline significantly increased pollinator disclosures, with a disclosure rate increasing from 6 to 14% during that timeframe.[30]

30 A.R. Romi *et al.*, *Fighting Against Pollinator Extinction: An Examination of the Corporate Disclosure Response to the U.S. Pollinator Decline Policy* (Working Paper, Texas Tech University, 2016).

Finally, as the current trend in responsible investment continues to rise, the impacts of biodiversity, in this case pollinator decline, are increasingly being incorporated into investment and financing decisions.[31] As the expected costs of issues such as pollinator decline continue to rise, corporate cash flows are likely to be affected.[32] Even the Big four accounting firms have begun to take notice. Each firm has published documents informing corporations of the potential impact of biodiversity on business, and that firm-specific information related to these issues is increasingly being requested by stakeholders.[33] The combination of international activity, CSSP involvement, responsible investment and financing decisions, and the isomorphic pressure of disclosures among peer firms, will likely continue to increase US firm-specific bee disclosures in the future, Additionally, as more stakeholders become aware of the pollinator decline issue and its potential impact on firms, they too will demand greater action and accountability/transparency.

Bibliography

American Electric Power. *2014 Corporate Accountability Report.* Columbus, OH: American Electric Power, 2014, 1–259.

Bristol-Myers Squibb. *Foundation and Corporate Social Responsibility: Innovation and Inspiration—2014 Report.* New York: Bristol-Myers Squibb, 2014, 1–37.

Business for Bees. "Business for Bees: American Business Collaboration for Pollinator Conservation Action". Accessed 10 March 2016. http://www.pollinator.org/bizforbees.htm.

Deloitte. *Business and the Biodiversity Challenge: A Study of Actions among the Fortune Global 50 Companies.* Deloitte Global Services Limited, 2012.

31 Millennium Ecosystem Assessment, *Ecosystems and Human Well-being: Opportunities and Challenges for Business and Industry* (Washington, DC: World Resources Institute, 2005), 1–34; PWC, *Result Smarter Growth: 13th Annual Global CEO Survey, Setting a Smarter Course for Growth* (UK: PricewaterhouseCoopers International Limited, 2010); I. Mulder and P. Clements-Hunt, *CEO Briefing: Demystifying Materiality—Hardwiring Biodiversity and Ecosystem Services into Finance* (Geneva, Switzerland: UNEP Finance Initiative, 2010).

32 Millennium Ecosystem Assessment, *Ecosystems and Human Well-being.*

33 KPMG Sustainability, *Biodiversity and Ecosystem Services: Risk and Opportunity Analysis within the Pharmaceutical Sector* (The Netherlands: KPMG Advisory N.V., 2011); KPMG, *Sustainable Insight: The Nature of Ecosystem Service Risks for Business* (The Netherlands: KPMG Advisory N.V., 2011); Deloitte, *Business and the Biodiversity Challenge: A Study of Actions among the Fortune Global 50 Companies* (Deloitte Global Services Limited, 2012); EY, *Six Growing Trends in Corporate Sustainability: An EY Survey in Cooperation with GreenBiz Group* (London, UK: EYGM Limited, 2013); PWC, "Corporate Biodiversity Strategy: Managing Biodiversity Risk for Business", accessed September 22, 2015, http://www.pwc.co.uk/services/sustainability-climate-change/insights/how-will-your-business-manage-biodiversity-risk.html.

Earthjustice. "Our Story". Accessed 10 March 2016. http://earthjustice.org/about.
Environmental Protection Agency. "EPA Actions to Protect Pollinators". Accessed 10 March 2016. http://www2.epa.gov/pollinator-protection/epa-actions-protect-pollinators.
EY. *Six Growing Trends in Corporate Sustainability: An EY Survey in Cooperation with GreenBiz Group*. London, UK: EYGM Limited, 2013.
FedEx. *2014 Global Citizenship Report: Moving Possibilities*. Memphis, TN: FedEx, 2014, 1–86.
Financial Accounting Standards Board. *Statements of Financial Accounting Concepts No. 8*. Norwalk, CT: The Financial Accounting Foundation, 2010.
Gallai, N., J-M. Salles, J. Settele and B. Vaissière. "Economic Valuation of the Vulnerability of World Agriculture Confronted with Pollinator Decline". *Ecological Economics* 68 (2009): 810–21.
General Motors. *A Driving Force: 2014 Sustainability Report*. Detroit, MI: General Motors, 2014, 1–117.
Hewlett Packard. *HP 2014 Living Progress Report*. Palo Alto, CA: Hewlett Packard, 2014, 1–136.
International Accounting Standards Board (IASB). *The Conceptual Framework for Financial Reporting 2010*. IASB, 2010.
KPMG. *Sustainable Insight: The Nature of Ecosystem Service Risks for Business*. The Netherlands: KPMG Advisory N.V., 2011.
KPMG Sustainability. *Biodiversity and Ecosystem Services: Risk and Opportunity Analysis within the Pharmaceutical Sector*. The Netherlands: KPMG Advisory N.V., 2011.
Krupke, C.H., G.J. Hunt, B.D. Eitzer, G. Andino, and K. Given. "Multiple Routes of Pesticide Exposure for Honey Bees Living Near Agricultural Fields". *PLoS ONE* 7(1) (2012): e29268, doi:10.1371/journal.pone.0029268.
MA (Millennium Ecosystem Assessment). *Ecosystems and Human Well-being: Opportunities and Challenges for Business and Industry*. Washington, DC: World Resources Institute, 2005, 1–34.
Mizejewski, David. "Take the Million Pollinator Garden Challenge". Accessed 3 June 2015. http://blog.nwf.org/2015/06/million-pollinator-garden-challenge/?s_subsrc=Web_Biggie__Home_MPGC.
Monsanto. *From the Inside Out: Monsanto 2014 Sustainability Report*. St Louis, MO: Monsanto, 2014, 1–169.
Mulder, I. and P. Clements-Hunt. *CEO Briefing: Demystifying Materiality—Hardwiring Biodiversity and Ecosystem Services into Finance*. Geneva, Switzerland: UNEP Finance Initiative, 2010.
Natural Resources Defense Council. "The NRDC Story". Accessed 10 March 2016. http://www.nrdc.org/about/nrdc-story.asp.
National Wildlife Federation (NWF). "Our Work: Protecting Wildlife, Inspiring Future Generations". Accessed 10 March 2016. http://www.nwf.org/What-We-Do.aspx.
Pollinator Health Task Force. *National Strategy to Promote the Health of Honey Bees and Other Pollinators*. May 19, 2015. Washington, DC: The White House, 2015.
Pollinator Partnership. "Our Mission" Accessed 10 March 2016. http://www.pollinator.org/.
PWC. *Result Smarter Growth: 13th Annual Global CEO Survey, Setting a Smarter Course for Growth*. UK: PricewaterhouseCoopers International Limited, 2010.
PWC. "Corporate Biodiversity Strategy: Managing Biodiversity Risk for Business". Accessed September 22, 2015 http://www.pwc.co.uk/services/sustainability-climate-change/insights/how-will-your-business-manage-biodiversity-risk.html.
Romi, A.R, G. Michelon and S.D. Longing. *Fighting Against Pollinator Extinction: An Examination of the Corporate Disclosure Response to the U.S. Pollinator Decline Policy*. Working Paper, Texas Tech University, 2016.
Stathers, R. *The Bee and the Stockmarket: An Overview of Pollinator Decline and its Economic and Corporate Significance*. London: Schroder Investment Management Limited, 2014.

U.S. Securities and Exchange Commission (SEC). "How to Read a 10-K". Accessed 10 March 2016. http://www.sec.gov/answers/reada10k.htm.

Walmart. *2015 Global Responsibility Report.* Bentonville, AR: Walmart Stores, Inc., 2015, 1–138.

Xerces Society for Invertebrate Conservation. "Protecting the Life that Sustains Us". Accessed 10 March 2016. http://www.xerces.org/.

Xerces Society for Invertebrate Conservation. *The Xerces Society for Invertebrate Conservation: Pollinator Conservation Program.* Accessed 10 March 2016. http://www.xerces.org/wp-content/uploads/2015/03/PollinatorProgramFactsheet_Mar2015_web.pdf.

13

No bees in their bonnet

On the absence of bee-reporting by South African listed companies

Warren Maroun
University of the Witwatersrand, South Africa

South Africa is the third most biologically diverse country in the world with approximately 1 million species on record, many of which are unique to the region, and nine terrestrial biomes.[1] As a result, the conservation of the country's biodiversity is considered paramount from a deep ecological perspective.[2] From an anthropocentric point of view, management of the country's natural resources is also an imperative, especially when considering that ecosystem services ensure the livelihood of

1 R. Wynberg, "A Decade of Biodiversity Conservation and Use in South Africa: Tracking Progress from the Rio Earth Summit to the Johannesburg World Summit on Sustainable Development: Review Article", *South African Journal of Science* 98 (5 & 6) (2002): 233–43; J.K. Turpie, "The Existence Value of Biodiversity in South Africa: How Interest, Experience, Knowledge, Income and Perceived Level of Threat Influence Local Willingness to Pay", *Ecological Economics* 46 (2) (2003): 199–216.

2 Department of Environmental Affairs and Tourism, *Convention on Biological Diversity: South African National Report to the Fourth Conference of the Parties Department* (South Africa: Department of Environmental Affairs and Tourism, 1998); Wynberg, "A Decade of Biodiversity Conservation and Use in South Africa".

millions of South Africans,[3] adding an estimated ZAR73 billion per annum (7%) to the country's GDP.[4] Key among these ecosystem services are the pollination activities carried out by a variety of species on which South African agriculture is heavily dependent.

Crop pollination by insects, predominantly bees, is estimated to increase the production yield and quality of 35% of crops worldwide and has made a significant contribution to American and European agriculture. Research examining the importance of pollination ecosystem services for African crop production is limited compared with the number of studies carried out in the USA and Europe.[5] Nevertheless, what research has been completed points to the significant role of bees (and other pollinating insects) in South African commercial farming; the importance of determining trends in the population of these animals; developing sustainable solutions for growing essential crops; and preserving the habitats and foraging resources on which pollinators depend.[6] With pollination ecosystem services linked directly with the variety and yield of crops, food security and financial viability of agriculture-dependent industry,[7] there is a strong anthropocentric case for preserving pollination ecosystem services, highlighting the need to examine how South African companies in the food and agriculture industry are responding to the threats posed to pollinators by, *inter alia*, habitat loss, climate change and the emergence of new diseases.[8]

3 Anthropocentrism frames the value of biodiversity in terms of the utility it provides to human beings rather than its inherent biological value. For details see World Conservation Monitoring Centre, *Global Biodiversity: Status of the Earth's Living Resources*, 1992, accessed 1 March 2015, http://www.biodiversitylibrary.org/item/97636#page/5/mode/1up; and G. Samkin *et al.*, "Developing a Reporting and Evaluation Framework for Biodiversity", *Accounting, Auditing & Accountability Journal* 27 (3) (2014): 527–62.

4 Department of Environmental Affairs and Tourism, *South Africa's Fourth National Report to the Convention on Biological Diversity* (South Africa: Department of Environmental Affairs and Tourism, 2009).

5 A. Melin *et al.*, "Pollination Ecosystem Services in South African Agricultural Systems", *South African Journal of Science* 110 (11/12) (2014): 25–33.

6 S.K. Gess and F.W. Gess, Potential Pollinators of the Cape Group of Crotalarieae (sensu Polhill) (Fabales: Papilionaceae), with Implications for Seed Production in Cultivated Rooibos Tea (Entomological Society of Southern Africa, 1994); M.H. Allsopp *et al.*, "Valuing Insect Pollination Services with Cost of Replacement", PLoS ONE 3 (9) (2008): e3128; Melin *et al.*, "Pollination Ecosystem Services in South African Agricultural Systems".

7 C.J. Crous and F. Roets, "Realising the Value of Continuous Monitoring Programmes for Biodiversity Conservation", *South African Journal of Science* 110 (11/12) (2014): 7–11.

8 Cf. M.J. Jones, "Accounting for the Environment: Towards a Theoretical Perspective for Environmental Accounting and Reporting", *Accounting Forum* 34 (2) (2010): 123–38; Samkin *et al.*, "Developing a Reporting and Evaluation Framework for Biodiversity".

The focus on this jurisdiction is also informed by the fact that what little research has been done on biodiversity reporting has largely overlooked South Africa and other leading economies on the continent.[9] This is despite the significant environmental resources found in Africa and the significant strides being made by South Africa when it comes to non-financial reporting.[10] In particular, the country has well-established principles of corporate governance which promote a stakeholder-centric approach to corporate reporting. Since 2002, there has been a steady increase in the emphasis placed on environmental, social and governance disclosures (ESG).[11] From 2010, the country's stock exchange became the first in the world to introduce a *de facto* requirement for listed companies to prepare integrated reports which include financial and non-financial performance indicators and highlight their interconnections in the context of generating sustainable returns.[12] In this way, although South Africa is often described as an emerging African market, it offers a sophisticated reporting environment which is sufficiently mature for investors and other stakeholders reasonably to expect biodiversity-related concerns to feature in corporate communications.

In this context, this chapter offers an initial review of a particular aspect of corporate accounting and accountability for species and biodiversity by exploring the disclosure practices of some of South Africa's best-known companies. It examines disclosures on pollination ecosystem services, focusing mainly on bees, by organizations in the farming, food and retail sectors. The aim is not to develop a guideline for reporting on the role of pollinators in companies' business models but to examine the disclosures on pollination services by local companies and the steps being taken to assist with conserving bees and other pollinating insects. Before proceeding with an analysis of results, the section below discusses the role of the

9 M.J. Jones and J.F. Solomon, "Problematising Accounting for Biodiversity", *Accounting, Auditing & Accountability Journal* 26 (5) (2013): 668–87.

10 J. Solomon and W. Maroun, *Integrated Reporting: The New Face of Social, Ethical and Environmental Reporting in South Africa?* (London: The Association of Chartered Certified Accountants, 2012); J. Atkins and W. Maroun, *South African Institutional Investors' Perceptions of Integrated Reporting* (London: Association of Chartered Certified Accountants, 2014); J. Atkins and W. Maroun, "Integrated Reporting in South Africa in 2012: Perspectives from South African Institutional Investors", *Meditari Accountancy Research* 23 (2) (2015): 197–221.

11 Institute of Directors in Southern Africa (IOD), *The King Report on Corporate Governance in South Africa—2002 (King-II)* (Johannesburg, South Africa: Lexis Nexus South Africa, 2002); J. Solomon, *Corporate Governance and Accountability* (3rd edition, Chichester, UK: John Wiley and Sons Ltd, 2010).

12 Integrated Reporting Committee of South Africa (IRCSA), "Framework for Integrated Reporting and the Integrated Report", 2011, accessed 5 June 2012, www.sustainabilitysa.org; Solomon and Maroun, *Integrated Reporting*.

honey bee in South African agriculture and touches on some of the local projects to manage and preserve better pollination ecosystem services. This is followed by a brief review of the ways in which South African companies report information on biodiversity and the chosen research method. The chapter concludes with findings, recommendations and areas for future research.

The importance of pollinators for South African agriculture

There are only nine studies dealing with the importance of pollinator assemblages for crop production in South Africa, with most research concentrating on mango, sunflower and rooibos cultivation.[13] What little has been done finds that the diversity of flower-visiting insects is important for improving crop productivity and not only the frequency of visits by a single species.[14] The role of the honey bee in South African agriculture should not, however, be understated, given that it is the most common pollinator in local crop fields.[15] Furthermore, while some crops (such as rooibos, lucerne and mango) can be pollinated by other species, where honey bees are present, they contribute significantly to the pollination of crops, often working with other pollinating insect species to improve yields.[16]

The significant role played by bees in South African agriculture is clearly evident when considering the Western Cape fruit industry which is home to just over half

13 Rooibos is grown in South Africa and is consumed as a herbal tea but is also used in skincare and other household products. The therapeutic properties of the plant are currently under investigation.

14 L.G. Carvalheiro *et al.*, "Natural and Within-Farmland Biodiversity Enhances Crop Productivity", *Ecology Letters* 14 (3) (2011): 251–9; L.G. Carvalheiro *et al.*, "Creating Patches of Native Flowers Facilitates Crop Pollination in Large Agricultural Fields: Mango as a Case Study", *Journal of Applied Ecology* 49 (6) (2012): 1373–83.

15 Melin *et al.*, "Pollination Ecosystem Services in South African Agricultural Systems".

16 Gess and Gess, *Potential Pollinators of the Cape Group of Crotalarieae*; R.H. Watmough, "The potential of Megachile gratiosa Cameron, Xylocopa caffra (Linnaeus) (Hymenoptera: Megachilidae and Anthophoridae) and other solitary bees as pollinators of alfalfa, *Medicago sativa* L. (Fabaceae), in the Oudtshoorn District, South Africa" (Entomological Society of Southern Africa, 1999); Melin *et al.*, "Pollination Ecosystem Services in South African Agricultural Systems".

of the country's 77,800 hectares of deciduous fruit farms.[17] Fruit species such as apples or plums require two or six hives per cultivated hectare, respectively, with the result that some 30,000 hives are required to provide essential pollination services with an estimated value of US$28 to $122.8 million for which only $1.8 million is being paid. This is estimated at only 41% of the internationally recommended units per hectare, possibly as a result of the above average number of wild hives in the region which contribute between US$49.1 and US$310.9 million in value.[18]

Presently, a surplus pollinating capacity of 20% is thought to be available, suggesting that South Africa enjoys an excess number of bees for optimal pollination, but it is unclear if this is sufficient to cover contingencies such as the outbreak of disease or changes in weather. Furthermore, although South Africa has not reported declines in managed bee populations comparable to those seen in the USA, the current consensus is that it would be premature to conclude that this important ecosystem service is not at risk. The last detailed bee census was carried out in 1975. Ongoing funding shortages by the Agricultural Research Council mean that the lack of longitudinal data necessary for assessing changes in the bee population is unlikely to be addressed in the near future.[19] In addition, there is evidence of colony losses from 2009 to 2011, which would be higher than tolerable levels in the USA and Europe, and the recent emergence of an American bee disease (American foulbrood disease) in the Western Cape has the potential to result in significant future losses.[20] Even if it is argued that scientific research is not conclusive and that colony collapse disorder is not currently affecting the sustainability of managed and wild hives in South Africa, at a minimum, the demand for pollinator-dependent crops is on the increase, while a decline in the variety and availability of suitable natural habitats—coupled with the outbreak of disease—is constraining the capacity of existing natural and managed pollination services.[21]

17 The Western Cape is one of nine provinces in South Africa. The provincial capital is Cape Town.

18 Allsopp *et al.*, "Valuing Insect Pollination Services with Cost of Replacement"; Melin *et al.*, "Pollination Ecosystem Services in South African Agricultural Systems".

19 Melin *et al.*, "Pollination Ecosystem Services in South African Agricultural Systems".

20 Wossler, "South Africa: American Disease That's Wreaking Havoc on the Cape's Honeybee Population", *All Africa*, January 2015, accessed 22 September 2015, http://allafrica.com/stories/201508020001.html; C.W. Pirk *et al.*, "A Survey of Managed Honey Bee Colony Losses in the Republic of South Africa—2009 to 2011", *Journal of Apicultural Research* 53 (1) (2014): 35–42.

21 M.A. Aizen and L.D. Harder, "The Global Stock of Domesticated Honey Bees Is Growing Slower Than Agricultural Demand for Pollination", *Current Biology* 19 (11) (2009): 915–8; Melin *et al.*, "Pollination Ecosystem Services in South African Agricultural Systems"; Wossler, "South Africa: American Disease That's Wreaking Havoc on the Cape's Honeybee Population".

In this context, the Department of Environmental Affairs and Tourism and the South African National Biodiversity Institute (SANBI) have paid considerable attention to pollination services provided by honey bees[22] as part of the multinational Global Pollination Project (GPP) being coordinated by the United Nations Environment Programme (UNEP) and the Food and Agricultural Organisation of the United Nations (FAO).[23] In South Africa, the GPP has been linked to the African Pollinator Initiative, founded in 1999, in order to promote participation by African countries on the GPP and improve conservation and study of pollinators on the continent. Added to this are a number of research and conservation projects and initiatives designed to promote sustainable agricultural practices such as the Honeybee Forage Project, the Right Rooibos Initiative and Farming for the Future Project.[24]

The last two schemes are examples of sustainable farming initiatives designed to minimize the impact of agricultural activities on ecosystems and promote sustainable growing practices with an indirect benefit for pollinators.[25] In contrast, the Honeybee Forage Project is a direct response to the need to understand better the interaction between crop yields and pollination services. Recognizing the growing importance of managed hives for commercial farming in South Africa, SANBI—with the Working for Water Programme and Agricultural Research Council—is developing recommendations for improving and managing foraging resources for honey bees. The national project includes research on the long-term trends in bee populations, pollination deficits in target crops, the quality and quantity of foraging resources and the conservation of pollination services.[26]

22 SANBI. *Biodiversity Information Policy Document—Policy Series: Principle and Guidelines* (South Africa: SANBI, 2010).

23 The seven countries participating in the project, which concluded in 2013, were Brazil, Ghana, India, Kenya, Nepal, Pakistan and South Africa.

24 Agricultural Research Council, *The African Pollinator Initiative* (Agricultural Research Council, 2014), accessed 2 January 2015, http://www.arc.agric.za/arc-ppri/Pages/Biosystematics/African-Pollinator-Initiative-(API).aspx.

25 Conservation South Africa, *Green Choice: Serving Nature's Bounty*, 2011, accessed 12 May 2015, http://biodiversityadvisor.sanbi.org/wp-content/uploads/2014/07/GreenChoice_brochure-2011_low-res.pdf; Conservation South Africa and South African National Biodiversity Institute, *Crop Agriculture, Pollination and the Honeybee*, 2014, accessed 14 March 2016, http://biodiversityadvisor.sanbi.org/wp-content/uploads/2012/11/Crop-Agric-Pollination-and-the-honeybee-Brochure.pdf.

26 SANBI, *Crop Agriculture, Pollination and the Honeybee.*

Biodiversity reporting by South African companies

The pivotal role being played by honey bees begs the question: What are companies in the South African food and agriculture sectors reporting to their stakeholders? Guidelines dealing specifically with reporting on bees (or other species) have not yet been developed.[27] The International Integrated Reporting Council (IIRC) has, however, issued a principles-based framework for preparing an integrated report which, *inter alia,* requires companies to explain how they are managing different forms of capital in order to generate value.[28] This includes an assessment of the relevance of "natural capital" which the IIRC describes as "all renewable and non-renewable environmental resources and processes that provide goods or services that support the past, current or future prosperity of an organization" specifically including "biodiversity and eco-system health".[29] The Global Reporting Initiative (GRI) also offers indirect guidance. It defines "biodiversity" and outlines biodiversity reporting indicators including strategies for managing habitat impact, species affected by corporate activity and the nature of operations located close to areas of high biodiversity value.[30] The extent to which South African companies have been actively reporting on key species in their annual or integrated reports is, however, unclear.

Research carried out in Europe and Australia has shown that most companies provide little if any information on the impact their operations are having on local ecosystems.[31] Disclosures are usually generic and, contrary to the recommendations of the IIRC, fail to highlight clearly the interconnection between biodiversity loss and management of ecosystem services (on the one hand) and the operations and financial returns of the respective organizations on the other.[32] Equally worrying is the fact that, while financial risk identification and analysis is

27 Jones and Solomon, "Problematising Accounting for Biodiversity".

28 International Integrated Reporting Council (IIRC), *The International Framework: Integrated Reporting,* 2013, accessed 1 October 2013, http://www.theiirc.org/wp-content/uploads/2013/12/13-12-08-THE-INTERNATIONAL-IR-FRAMEWORK-2-1.pdf.

29 IIRC, *The International Framework: Integrated Reporting.*

30 M. King, "Comments on: Integrated Reporting and the Integrated Report", *International Corporate Governance Conference, Johannesburg, South Africa, 23 October 2012*; Global Reporting Initiative (GRI), *G4 Sustainability Reporting Guidelines,* 2014, accessed 10 February 2015, https://www.globalreporting.org/reporting/g4/Pages/default.aspx.

31 Jones, "Accounting for the Environment"; Jones and Solomon, "Problematising Accounting for Biodiversity"; Samkin *et al.*, "Developing a Reporting and Evaluation Framework for Biodiversity".

32 C. Grabsch *et al.*, *Accounting for Biodiversity in Crisis: A European Perspective* (Working Paper, London: Kings College, 2012); G. Rimmel and K. Jonäll, "Biodiversity Reporting in Sweden: Corporate Disclosure and Preparers' Views", *Accounting, Auditing & Accountability Journal* 26 (5) (2013): 746–78; D. van Liempd and J. Busch, "Biodiversity Reporting in Denmark", *Accounting, Auditing & Accountability Journal* 26 (5) (2013): 833–72.

well-developed,[33] most companies do not define their key risks to include natural capital. They overlook the important benefits offered by protecting threatened ecosystem services [34] despite scientific research highlighting the significant financial contribution made by these systems.[35]

Whether or not these findings are applicable in South Africa is unclear. There has been no research on biodiversity reporting by South African companies. Prior research has reported an increase in the extent to which companies are complying with reporting guidelines provided by the GRI or King-III.[36] Some scholars have also examined changes in the nature and extent of non-financial reporting [37] or the value relevance of this information.[38] Biodiversity disclosure, in general, and the impact of specific ecosystem services or species have, however, been completely neglected.[39]

33 K. Raemaekers *et al.*, "Risk Disclosures by South African Listed Companies Post-King III", *South African Journal of Accounting Research* (2015): 1–20.

34 Jones and Solomon, "Problematising Accounting for Biodiversity"; van Liempd and Busch, "Biodiversity Reporting in Denmark".

35 Allsopp *et al.*, "Valuing Insect Pollination Services with Cost of Replacement"; Melin *et al.*, "Pollination Ecosystem Services in South African Agricultural Systems".

36 For examples, see T. Hindley and P. Buys, "Integrated Reporting Compliance With The Global Reporting Initiative Framework: An Analysis Of The South African Mining Industry", *International Business & Economics Research Journal* 11 (11) (2012): 1249–60; Solomon and W. Maroun, *Integrated Reporting*; C. Carels *et al.*, "Integrated Reporting in the South African Mining Sector", *Corporate Ownership and Control* 11 (1) (2013): 991–1005.

37 see C.J. de Villiers and P. Barnard, "Environmental Reporting in South Africa from 1994 to 1999: A Research Note", *Meditari Accountancy Research* 8 (1) (2000): 15–23; G. Samkin, "Changes in Sustainability Reporting by an African Defence Contractor: A Longitudinal Analysis", *Meditari Accountancy Research* 20 (2) (2012): 134–66; B. Loate *et al.*, "Acid Mine Drainage in South Africa: A Test of Legitimacy Theory", *Journal of Governance and Regulation* 4 (2) (2015): 26–40.

38 M. de Klerk and C. de Villiers, "The Value Relevance of Corporate Responsibility Reporting: South African Evidence", *Meditari Accountancy Research* 20 (1) (2012): 21–38.

39 J. Atkins *et al.*, "Exploring Rhinoceros Conservation and Protection: Corporate Disclosures and Extinction Accounting by Leading South African Companies", *Meditari Accountancy Research*, 2 July 2015; H. Mansoor and W. Maroun, "Biodiversity Reporting in the South African Food and Mining Sectors", in*: 27th International Congress on Social and Environmental Accounting Research, 27 August 2015, Egham, United Kingdom* (United Kingdom: Centre for Social and Environmental Accounting Research).

Analysing bee disclosures by South African corporates: research method and approach

The exploratory nature of the study required a flexible approach to data collection.[40] As a result, the unit of analysis was not specific words found in corporate communications but rather phrases, images and themes which related directly or indirectly to pollination ecosystem services.

A systematic approach was followed to collect data. The integrated or annual reports of the selected companies were read several times to identify any content dealing with biodiversity. For this purpose, the researchers relied on the GRI reporting guidelines and the prior research on biodiversity disclosures to identify biodiversity-related content.[41] This included, for example, information on specific pollinators, colony collapse disorder, honey production, habitat descriptions, environmental incidents, biodiversity action plans and environmental mission statements.[42] Identified text was included on a coding sheet according to source of the information.[43] Due to the limited number of disclosures dealing with pollination ecosystem services, a more refined coding system was not required and the frequency of disclosures was not calculated.

The research included integrated/annual reports from 2011 to 2013, the related sustainability reports and the content of companies' websites. Periods before 2011 were excluded as these predated the introduction of the latest code of corporate governance and release of the discussion papers on integrated reporting. The study dealt only with companies in the food and agriculture sector, with a focus on the following sub-sectors: (1) broad retail, (2) drug retail, (3) food production, (4) food retail and wholesale, (5) farming and fishing. The population comprised 31 companies of which 19 were included in the final analysis, those being companies with an above average market capitalization. This was based on the assumption that larger companies have more resources at their disposal and more sophisticated reporting systems which may increase the likelihood of comprehensive reporting that would deal with specific biodiversity issues.[44]

40 D. Merkl-Davies *et al.*, "Text Analysis Methodologies in Corporate Narrative Reporting Research", in *23rd CSEAR International Colloquium, 2011, St Andrews, United Kingdom*.
41 Samkin *et al.*, "Developing a Reporting and Evaluation Framework for Biodiversity".
42 Grabsch *et al.*, *Accounting for Biodiversity in Crisis: A European Perspective*; van Liempd and Busch, "Biodiversity Reporting in Denmark".
43 This follows a comparable approach to that used by Carels *et al.*, "Integrated Reporting in the South African Mining Sector", Samkin *et al.*, "Developing a Reporting and Evaluation Framework for Biodiversity" and Loate *et al.*, "Acid Mine Drainage in South Africa".
44 Grabsch *et al.*, *Accounting for Biodiversity in Crisis: A European Perspective*; van Liempd and Busch, "Biodiversity Reporting in Denmark".

Bee disclosures in South African corporate reports

Of the 19 companies analysed, only two reported on pollination ecosystem services. Pick n Pay included a short article on their webpage titled: "Where have all the bees gone?"[45] The piece touches on the importance of bees as sources of honey, beeswax and vital pollination services. It also deals with possible causes of the decline in global bee populations including the use of pesticides, electromagnetic radiation from cellular phones and global warming. The article concludes:

> So why should we be worried about all of this? Aside from the very real threat to our food supply, landscapes will begin to change as bee-pollinated flora is replaced by more resilient species adapted to the environment. Biodiversity will surely suffer too, as wildlife finds itself ill-suited to this strange new world. Insects may not be first on the usual conservation agenda, but without due intervention, the interdependence of life, great and small, will be exhibited to catastrophic effect.[46]

Interestingly, however, the risk of biodiversity loss and threat to the food supply do not feature in the integrated report published in the same financial year. The strategic priorities of the group are focused exclusively on financial and manufactured capital such as growing selling space ahead of the market, building customer relationships and developing cost-effective replenishment systems. The implications of a decline in bee populations for supply of core products and their price are not identified as key risks despite sales growth, gross margin and product availability being clearly listed as key performance indicators.[47]

What is also peculiar is the fact that pollination ecosystem services do not feature in the discussion on food security. Fuel and commodity prices, energy usage, water consumption and soil quality are touched on as is the emerging consumer awareness of the importance of sustainable food options. The report identifies an additional six strategic focal points including sustainability of product lines, "providing a resilient supply base" and "working for a clean and healthy environment". It is interesting to note that these are not included as the primary strategic indicators (identified on page 3 of the integrated report) but are included towards the end of the document as "second tier" considerations.[48] In addition, plans for dealing with a decline in bee populations[49] or warnings about the impact which this would have on the profitability of the group (and industry) are completely ignored.

45 Pick 'n Pay, "Where have all the bees gone?" accessed 1 March 2015, http://www.picknpay.co.za/green-news/where-have-all-the-bees-gone.
46 *Ibid.*
47 Pick 'n Pay, *Pick 'n Pay: Integrated Annual Report: 2012*, accessed 1 March 2015, http://www.picknpayinvestor.co.za/financials/annual_reports/2012/downloads/integrated_ar/pnp_integrated_ar.pdf.
48 *Ibid.*
49 Melin *et al.*, "Pollination Ecosystem Services in South African Agricultural Systems".

The second example of disclosures on bees comes from Woolworths. As part of the "Sustainable Agriculture and Management of Biodiversity" section of its *2011 Good Business Journey Report,* the company refers to badger friendly honey.[50] Its webpage also discusses predator-friendly and other sustainable farming practices which could, indirectly, benefit pollinators. As with Pick n Pay, however, the company does not deal with bees in their integrated reports. Strategic areas concentrate mainly on financial capital. The "good business journey" is included in the strategy and risk sections of, for example, the 2012 integrated report, but the emphasis is on social and governance metrics rather than on environmental indicators.[51]

This is not to say that environmental challenges are ignored. The company discusses briefly sustainable farming practices and the formation of business partnerships to manage biodiversity.[52] Consider, for example, the following extracts from the company's 2012 Integrated Report which make reference to "biodiversity":

> Experience has shown us that most sustainability challenges cannot be solved by one company acting alone. Our sustainability initiatives have often been a natural extension of the longstanding partnerships we share with our suppliers and corporate partners and have allowed us to formulate unique responses to biodiversity, transformation and other objectives.[53]
>
> Our Farming for the Future programme is externally audited by Enviroscientific and assists farmers to grow quality produce while protecting the environment. Farming for the Future focuses on soil health, water efficiency and the protection of biodiversity—crucial issues for agriculture in South Africa. We will be extending our Farming for the Future programme to 50% of the Food business by 2015, thereby reducing pesticide, chemical and fertiliser usage further and building soil quality. We will continue to appropriately manage the impacts of farming and fishing on biodiversity.[54]

These statements give some indication of the efforts being made to improve the sustainability of existing farming practices and suggest that the company is aware of the need to manage the country's biodiversity mass. The commentary is, however, generic and—similar to the Pick n Pay integrated reports—does not identify specific threats to pollination ecosystem services, despite the impact which declining bee populations would have on availability and the price of certain products.

50 Woolworths, *Woolworths Holdings Limited: 2011 Good Business Journey Report,* accessed 1 March 2015, http://www.woolworthsholdings.co.za/downloads/2011_good_business_journey_report.pdf.

51 For example, see Woolworths, *Woolworths Holding Limited: 2012 Integrated Report,* accessed 1 March 2015, http://www.woolworthsholdings.co.za/downloads/2012/2012_integrated_report.pdf, pp. 6 and 27.

52 *Ibid.*

53 *Ibid.*, 35.

54 *Ibid.*, 71.

None of the other companies included bee-related disclosures in its integrated or sustainability reports or on its website. The almost complete absence of disclosures dealing with bees and pollination services by South African corporates is consistent with international research suggesting that companies are paying little attention to biodiversity management in their communications with stakeholders.[55] The results are, however, at odds with the fact that the South African capital market has a long history of applying codes of best practice which recommend reporting on significant environmental issues.[56] If pollination services are an essential component of South African agriculture, as explained by a mounting body of scientific research,[57] these should be featuring prominently in integrated reports as part of the assessment of the interaction between natural and financial capital in the business models of companies directly or indirectly dependent on agricultural activities. Several factors may explain poor reporting practices.

Limited understanding of biodiversity reporting and its role in integrated reports

As touched on earlier in the chapter, the absence of a clear biodiversity reporting typology can provide one explanation for poor disclosures on pollination services by South African corporates. A preliminary study finds that local food and mining companies are having difficulties defining "biodiversity" and identifying precisely the types of information which should be communicated to stakeholders

55 Grabsch *et al.*, *Accounting for Biodiversity in Crisis: A European Perspective*; van Liempd and Busch, "Biodiversity Reporting in Denmark"; Samkin *et al.*, "Developing a Reporting and Evaluation Framework for Biodiversity".

56 de Villiers and Barnard, "Environmental Reporting in South Africa from 1994 to 1999: A Research Note"; IOD, *The King Code of Governance for South Africa (2009) and King Report on Governance for South Africa (2009) (King-III)* (Johannesburg, South Africa: Lexis Nexus South Africa, 2009); King, "Comments on: Integrated Reporting and the Integrated Report"; Carels *et al.*, "Integrated Reporting in the South African Mining Sector"; Atkins and Maroun, "Integrated Reporting in South Africa in 2012".

57 Gess and Gess, *Potential Pollinators of the Cape Group of Crotalarieae*; Melin *et al.*, "Pollination Ecosystem Services in South African Agricultural Systems"; Pirk *et al.*, "A Survey of Managed Honey Bee Colony Losses in the Republic of South Africa"; SANBI, *Honeybees in South Africa—What Landowners Can Do to Help*, accessed 8 May 2015, http://www.sanbi.org/sites/default/files/documents/documents/honbeybees-sa-brochure-28-may-eng-proof.pdf.

interested in understanding the impact of biodiversity loss on business operations.[58] Compounding this challenge is the fact that King-III and the Integrated Reporting Framework provide a principles-based approach to reporting and do not give a clear indication of how the contribution of ecosystem services to a firm's natural capital should be measured and interconnected with other forms of capital under an organization's control. There have been some efforts to create a hierarchy which can be used to describe the natural resources for which an entity is held accountable[59] but research on how biodiversity loss can be integrated with traditional measures of a firm's financial position and performance is limited.[60] Biodiversity reporting can, therefore, be interpreted as an emerging area of corporate governance research which, arguably, companies in both South Africa and international jurisdictions are still grappling with.[61] The same could be said of the integrated reporting initiative as a whole.

Although integrated reporting is in its fifth year in South Africa, both the IIRC and the Integrated Reporting Committee of South Africa (IRCSA) acknowledge that it will take time to identify the information needs of stakeholders and find an appropriate balance of financial and non-financial indicators which demonstrate clearly how the chosen business model generates sustainable returns.[62] This has been confirmed by initial research on the views of South African institutional investors and preparers of the first sets of integrated reports. The findings show that, while progress is being made, companies are struggling to refine their reporting systems in order to collect relevant data and find the most suitable manner of presenting this information to stakeholders.[63] It can, therefore, be argued that it is no surprise that non-financial reporting on a very specific biodiversity issue is limited.

58 Mansoor and Maroun, "Biodiversity Reporting in the South African Food and Mining Sectors".

59 M.J. Jones, "Accounting for Biodiversity: A Pilot Study", *The British Accounting Review* 28 (4) (1996): 281–303; M.J. Jones, "Accounting for Biodiversity: Operationalising Environmental Accounting", *Accounting, Auditing & Accountability Journal* 16 (5) (2003): 762–89; J. Siddiqui, "Mainstreaming Biodiversity Accounting: Potential Implications for a Developing Economy", *Accounting, Auditing & Accountability Journal* 26 (5) (2013): 779–805.

60 Jones and Solomon, "Problematising Accounting for Biodiversity".

61 T. Cuckston, "Bringing Tropical Forest Biodiversity Conservation into Financial Accounting Calculation", *Accounting, Auditing & Accountability Journal* 26 (5) (2013): 688–714; Samkin *et al.*, "Developing a Reporting and Evaluation Framework for Biodiversity"; Mansoor and Maroun, "Biodiversity Reporting in the South African Food and Mining Sectors".

62 IRCSA, "Framework for Integrated Reporting and the Integrated Report"; King, "Comments on: Integrated Reporting and the Integrated Report"; Solomon and W. Maroun, *Integrated Reporting.*

63 Atkins and Maroun, *South African Institutional Investors' Perceptions of Integrated Reporting.*

The relevance of academic research

The shortage of academic research may be compounding difficulties. In particular, there is little interdisciplinary work which interprets scientific findings on pollination ecosystem services in a commercial context. At the same time, the existing research on environmental accounting and biodiversity reporting is more focused on the development of abstract theory than offering normative (but practical) recommendations on how leading scientific research can be integrated into the risk analysis and integrated reporting process.[64] As a result much of the environmental accounting research, despite being of excellent quality, is not being accessed by accounting and corporate reporting practitioners, limiting its impact.

In South Africa, this problem is aggravated by the fact that American and European accounting scholars are more active than their local counterparts who are producing far fewer academic papers on pressing environmental accounting issues.[65] Consequently, much of the corporate governance research carried out to date has a strong European or American focus with little on the functioning of mechanisms of accountability from Africa.[66] Even in those cases where South African-specific studies do engage with important governance and environmental issues, this normally avoids employing an interpretive or critical approach capable of engaging new areas of enquiry and offering *useful* recommendations to those in practice.[67]

No news is good news

From a different perspective, the absence of detailed reporting on South African pollinators could be an indication that South African companies operating in the food and retail sector do not face an environmental challenge. There is a considerable body of international research showing that companies respond to

64 This limitation was discussed at length at the Centre for Environmental, Social and Accountability Research Conference (June 2015).

65 A. West, "A Commentary on the Global Position of South African Accounting Research", *Meditari Accountancy Research* 14 (1) (2006): 121–37.

66 N. Brennan and J. Solomon, "Corporate Governance, Accountability and Mechanisms of Accountability: An Overview", *Accounting, Auditing & Accountability Journal* 21 (7) (2008): 885–906.

67 D. Coetsee and N. Stegmann, "A Profile of Accounting Research in South African Accounting Journals", *Meditari Accountancy Research* 20 (2) (2012): 92–112; W. Maroun and C. Jonker, "Critical and Interpretive Accounting, Auditing and Governance Research in South Africa", *Southern African Journal of Accountability and Auditing Research* 16 (2014): 51–62.

environmental issues with additional non-financial reporting.[68] The same is true in a South African setting where, in particular, the mining industry has an established history of including additional information in their integrated reports on emerging environmental problems.[69] If companies are preparing sound integrated reports,[70] the lack of detailed bee-related disclosures may be interpreted positively. Limited reporting could be an indication of the fact that leading food producers do not regard issues such as colony collapse disorder as posing significant risks to the respective organizations' ability to generate sustainable returns. This view may be supported by the fact that a decline in bee populations has been reported but, as discussed earlier in the chapter, the absence of longitudinal data means that results are not definitive. Furthermore, what losses have been reported are, historically, smaller than those being experienced in the USA and Europe[71] while the precise causes of decreases in the local bee population are still being debated and cannot be attributed specifically to the activities of the organizations under review.[72]

It is also possible that, although companies are not reporting in detail on the risks posed by threatened pollination ecosystem services, some action is being taken. In general the prior research shows that the country's codes of corporate governance and integrated reporting initiative have resulted in an increase in the extent of non-financial reporting and a growing appreciation of the importance of ESG information by investors and other stakeholders.[73] At the same time, the companies under review may not deal specifically with pollinators and other biodiversity-related issues but, in general, do indicate an awareness of the need for improved agricultural practices and the sustainable management of natural capital. In this context, an absence of bee-specific reporting may simply be indicative of the difficulties in determining precisely what information to include in an integrated or sustainability report, especially when considering that detailed guidance on materiality in a non-financial reporting setting is not available. It must also be kept in mind that

68 G. O'Donovan, "Environmental Disclosures in the Annual Report", *Accounting, Auditing & Accountability Journal* 15 (3) (2002): 344–71; D.M. Patten, "The Relation between Environmental Performance and Environmental Disclosure: A Research Note", *Accounting, Organizations and Society* 27 (8) (2002): 763–73.

69 See, for example, C. de Villiers and C.J. van Staden, "Can Less Environmental Disclosure Have a Legitimising Effect? Evidence from Africa", *Accounting, Organizations and Society* 31 (8) (2006): 763–81; Carels *et al.*, "Integrated Reporting in the South African Mining Sector"; and Loate *et al.*, "Acid Mine Drainage in South Africa".

70 Atkins and Maroun, "Integrated Reporting in South Africa in 2012".

71 Melin *et al.*, "Pollination Ecosystem Services in South African Agricultural Systems".

72 See Gess and Gess, *Potential Pollinators of the Cape Group of Crotalarieae* and Melin *et al.*, "Pollination Ecosystem Services in South African Agricultural Systems".

73 de Klerk and de Villiers, "The Value Relevance of Corporate Responsibility Reporting"; Carels *et al.*, "Integrated Reporting in the South African Mining Sector"; Atkins and Maroun, *South African Institutional Investors' Perceptions of Integrated Reporting*; K. Raemaekers and W. Maroun, "Trends in Risk Disclosure: Practices of South African Listed Companies", *Accounting Perspectives in Southern Africa* 2 (1) (2014): 8–15; Atkins and Maroun, "Integrated Reporting in South Africa in 2012".

the causes and extent of colony collapse is a complex scientific issue and a commercially orientated organization may legitimately require time to internalize and to develop a suitable operational and strategic response. In other words, limited reporting does not necessarily mean that the local food and agriculture sectors are indifferent to the plight of bees but simply that the formulation of an action plan—and its inclusion in integrated reports—is going to take time. As discussed below, more critical views are, however, also possible.

Impression management and private stakeholder engagement

There is a growing body of research which argues that environmental, social and governance reporting is frequently employed to influence favourably the perception of an organization and win the support of stakeholders.[74] On the one hand, impression management can take the form of unintentional concealment by companies of "negative truths from themselves and, consequently, from shareholders".[75] The intention is not deliberately to mislead stakeholders. Instead, corporate reporting reflects a natural tendency to avoid presenting information which is inconsistent with existing impressions or discussing negative outcomes which could contradict the image of sustainable business.[76]

As explained by leading social science researchers, the continued existence of modern organizations is predicated on the "decoupling" of the image of a technical rational structure from the possibility of decreased internal coordination and control to preserve a logic of confidence and good faith.[77] This means that the image of an organization as socially or environmentally responsible is not always entirely consistent with its internal processes and operations but that these inconsistencies

74 R. Gray, "Back to Basics: What Do We Mean by Environmental (and Social) Accounting and What Is It For? A Reaction to Thornton", *Critical Perspectives on Accounting* 24 (6) (2013): 459–68; J.F. Solomon *et al.*, "Impression Management, Myth Creation and Fabrication in Private Social and Environmental Reporting: Insights from Erving Goffman", *Accounting, Organizations and Society* 38 (3) (2013): 195–213.

75 Solomon *et al.*, "Impression Management, Myth Creation and Fabrication in Private Social and Environmental Reporting".

76 *Ibid.*

77 J.W. Meyer and B. Rowan, "Institutionalized Organizations: Formal Structure as Myth and Ceremony", *American Journal of Sociology* 83 (2) (1977): 340–63; M.C. Suchman, "Managing Legitimacy: Strategic and Institutional Approaches", *The Academy of Management Review* 20 (3) (1995): 571–610.

seldom become apparent to stakeholders, allowing the organization to maintain legitimacy.

In the context of biodiversity reporting, limited bee-related disclosures can, therefore, be interpreted as unwillingness by companies to engage with a potentially catastrophic environmental issue which can undermine the carefully constructed image of a sustainable business model.[78] To paraphrase Solomon *et al.*, with sustainability becoming an increasingly integral part of corporate reporting, a balance must be struck between discussing environmental issues and preserving the taken-for-granted assumption that some of the country's largest organizations have environmental risks, like colony collapse disorder, under control.[79] Investees are co-opted in this impression management exercise, sharing—rather than challenging—the experiences and views of the organization, as a result "creating and disseminating a dual myth of social and environmental accountability".[80]

This is not to say that all impression management is about constructing an image of a socially acceptable organization without the intention of manipulating stakeholders.[81] For example, protecting past accomplishments, presenting circumstances as extraordinary or beyond the organization's control and denying either the relevance of or responsibility for a negative event are common legitimization tactics used deliberately to manage impressions and preserve the image of a credible entity.[82] In this context, organizations go to considerable length to demonstrate compliance with best reporting and operational practices in order to construct an image of a socially and environmentally responsible business and either justify or deflect attention from negative environmental issues.[83] As part of this process, fundamental challenges are left unacknowledged;[84] good news is

78 Jones and Solomon, "Problematising Accounting for Biodiversity"; J.F. Solomon *et al.*, "Impression Management, Myth Creation and Fabrication in Private Social and Environmental Reporting: Insights from Erving Goffman", *Accounting, Organizations and Society* 38 (3) (2013): 195–213.

79 Cf. de Villiers and van Staden, "Can Less Environmental Disclosure Have a Legitimising Effect?"

80 Solomon *et al.*, "Impression Management, Myth Creation and Fabrication in Private Social and Environmental Reporting".

81 Suchman, "Managing Legitimacy: Strategic and Institutional Approaches"; Solomon *et al.*, "Impression Management, Myth Creation and Fabrication in Private Social and Environmental Reporting".

82 B.E. Ashforth and B.W. Gibbs, "The Double-Edge of Organisational Legitimation", *Organization Science* 1 (2) (1990): 177–94.

83 R. Gray, "Is Accounting for Sustainability Actually Accounting for Sustainability…and How Would We Know? An Exploration of Narratives of Organisations and the Planet", *Accounting, Organizations and Society* 35 (1) (2010): 47–62; H. Tregidga *et al.*, "(Re) presenting 'Sustainable Organizations'", *Accounting, Organizations and Society* 39 (6) (2014): 477–94.

84 Tregidga *et al.*, "(Re)presenting 'Sustainable Organizations'".

emphasized while negative reports are either obfuscated or ignored;[85] and common expressions such as "triple-bottom-line", the "sustainability reporting journey" and "integrated thinking" serve as metaphors designed to "blend" ecological and market-orientated discourse to reassure stakeholders that significant steps are being taken to balance environmental concerns with financial performance.[86] In this way, poor environmental reporting can be indicative of a carefully planned strategy to reframe the debate and deflect attention from problems which cannot be readily addressed or which can only be resolved with a significant (and costly) reordering of the business.[87]

When it comes to biodiversity reporting and accounting for pollination services in particular, the scientific research points to a looming environmental crisis which can result in higher food prices, uncertain supply chains and the need to bear the cost of an alternative to natural pollination.[88] Each of these poses significant social and financial challenges which are not easily solved and have the potential to call into question the sustainability practices of food producers and their ability to continue as going concerns. In line with the arguments of critical environmental accounting academics,[89] it is, therefore, no surprise that: environmental disclosures found in almost all of the integrated reports under review are generic and repetitive;[90] business risks and strategies are not context-specific;[91] and there are limited examples of demonstrable measures of environmental metrics, including biodiversity indicators, against key performance measures.[92]

85 Gray, "Is Accounting for Sustainability Actually Accounting for Sustainability"; King, "Comments on: Integrated Reporting and the Integrated Report".

86 C. Higgins and R. Walker, "Ethos, Logos, Pathos: Strategies of Persuasion in Social/Environmental Reports", *Accounting Forum* 36 (3) (2012): 194–208; N. Brennan and D. Merkl-Davies, "Rhetoric and Argument in Social and Environmental Reporting: The Dirty Laundry Case", *Accounting, Auditing & Accountability Journal* 27 (4) (2014): 602–33.

87 J.M. Moneva *et al.*, "GRI and the Camouflaging of Corporate Unsustainability", *Accounting Forum* 30 (2) (2006): 121–37; Tregidga *et al.*, "(Re)presenting 'Sustainable Organizations'".

88 Gess and Gess, *Potential Pollinators of the Cape Group of Crotalarieae*; Allsopp *et al.*, "Valuing Insect Pollination Services with Cost of Replacement"; S. Levy, "What's Best for Bees". *Nature* 479 (November 2011): 164–5; Melin *et al.*, "Pollination Ecosystem Services in South African Agricultural Systems"; Pirk *et al.*, "A Survey of Managed Honey Bee Colony Losses in the Republic of South Africa".

89 J. Bebbington *et al.*, "Seeing the Wood for the Trees: Taking the Pulse of Social and Environmental Accounting", *Accounting, Auditing & Accountability Journal* 12 (1) (1999): 47–52; Gray, "Is Accounting for Sustainability Actually Accounting for Sustainability"; Tregidga *et al.*, "(Re)presenting 'Sustainable Organizations'".

90 Solomon and Maroun, *Integrated Reporting*; Atkins and Maroun, *South African Institutional Investors' Perceptions of Integrated Reporting.*

91 Raemaekers and Maroun, "Trends in Risk Disclosure"; Raemaekers *et al.*, "Risk Disclosures by South African Listed Companies Post-King III".

92 Jones and Solomon, "Problematising Accounting for Biodiversity"; Mansoor and Maroun, "Biodiversity Reporting in the South African Food and Mining Sectors".

The perceived relevance of pollinator services

According to the research on the link between legitimacy theory and environmental reporting, companies react to significant environmental challenges by increasing the extent of information included in their reports to stakeholders. This forms part of a complex process of legitimization in terms of which additional reporting is used to justify a particular outcome or provide reassurance that environmental problems have been addressed or that preventative measures have been taken.[93]

In this context, the fact that South African companies have not reacted to the global decline in bee populations implies that challenges faced by the ecosystem service are not perceived to be a significant threat to stakeholder confidence. As discussed earlier, one explanation is that South African bee populations are relatively healthy when compared with the situation in Europe and the USA[94] or that colony losses are not the direct result of the activities of companies under review. Consequently, there is little risk of companies being held accountable for this. It remains a scientific problem which can be attributed to events and circumstances beyond the organization's control and, accordingly, poses no significant threat to the credibility of companies in the food industry. This perspective is, however, at odds with preliminary results on the state of biodiversity reporting generally which shows that South African companies are not reacting to the risks posed by, for example, climate change, mass extinction or habitat destruction.[95] This is despite the fact that, unlike local bee populations, these concerns have been attributed to industry activity, have been the subject of considerable public debate[96] and are specifically referred to by codes of best reporting practice and governance.[97]

A possible inference is that local companies are aware of these environmental risks but, as found in the South African mining industry, refrain from providing specific disclosures in order to avoid attracting scrutiny or causing panic among

93 de Villiers and van Staden, "Can Less Environmental Disclosure Have a Legitimising Effect?"; Higgins and Walker, "Ethos, Logos, Pathos"; Tregidga *et al.*, "(Re)presenting 'Sustainable Organizations'".

94 Melin *et al.*, "Pollination Ecosystem Services in South African Agricultural Systems"; SANBI, *Crop Agriculture, Pollination and the Honeybee.*

95 Mansoor and Maroun, "Biodiversity Reporting in the South African Food and Mining Sectors".

96 Jones, "Accounting for Biodiversity: A Pilot Study"; Intergovernmental Panel on Climate Change, *Climate Change 2013: The Physical Science Basis*, 2013, accessed 1 June 2015, http://www.ipcc.ch/report/ar5/wg1/.

97 IOD, *The King Code of Governance for South Africa (2009) and King Report on Governance for South Africa (2009) (King-III)*; KPMG, *Carrots and Sticks—Promoting Transparency and Sustainability: An Update on Trends in Voluntary and Mandatory Approaches to Sustainability Reporting*, 2012, accessed 30 June 2013, https://www.globalreporting.org/resourcelibrary/Carrots-And-Sticks-Promoting-Transparency-And-Sustainbability.pdf; GRI, *G4 Sustainability Reporting Guidelines.*

investors.[98] According to O'Dwyer, environmental disclosure can often expose companies to greater demands from stakeholders and can be used as a basis for holding managers accountable for perceived poor environmental performance. For example, environmental reporting—even if not detailed—can be interpreted as a policy statement by the respective organization, which can have the unintended effect of limiting how it reacts to emerging issues if key stakeholders question inconsistencies between current responses and past statements on environmental responsibilities.[99] Added to this is the risk of environmental reporting being interpreted as inconsistent with an organization's actual operations with the result that non-financial reporting, designed to promote the entity, has a significant delegitimizing effect.[100] Consequently, a "self-promotor paradox"[101] leads to the removal of biodiversity reporting as a mechanism for gaining, maintaining or repairing organizational legitimacy.

Alternately, non-disclosure is indicative of a failure to appreciate the relevance of pollinator services and related non-financial reporting requirements. Numerous academics have questioned the sincerity of sustainability reporting.[102] For example, Solomon *et al.*, exploring private social and environmental reporting (SER) in the UK, explain that investors and companies still regard social and environmental disclosure as secondary to traditional financial performance measures.[103] The result is that SER is interpreted simply as a compliance process designed to manage expectations, rather than as an exercise in genuinely reforming business practice and promoting sustainability.[104] Similarly, O'Dwyer notes that managers in environmentally sensitive industries have "little sympathy" with environmental pressure groups and may be avoiding additional non-financial reporting to signal their disapproval or avoid allegations that firms' activities have adverse environmental outcomes.[105]

98 de Villiers and van Staden, "Can Less Environmental Disclosure Have a Legitimising Effect?"

99 B. O' Dwyer, "Managerial Perceptions of Corporate Social Disclosure: An Irish Story", *Accounting, Auditing & Accountability Journal* 15 (3) (2002): 406–36.

100 Suchman, "Managing Legitimacy: Strategic and Institutional Approaches"; O' Dwyer, "Managerial Perceptions of Corporate Social Disclosure".

101 O' Dwyer, "Managerial Perceptions of Corporate Social Disclosure", 424.

102 Gray, "Is Accounting for Sustainability Actually Accounting for Sustainability"; R.L. Burritt, "Environmental Performance Accountability: Planet, People, Profits", *Accounting, Auditing & Accountability Journal* 25 (2) (2012): 370–405.

103 Solomon *et al.*, "Impression Management, Myth Creation and Fabrication in Private Social and Environmental Reporting".

104 Tregidga *et al.*, "(Re)presenting 'Sustainable Organizations'"; J.F. Atkins, "The Emergence of Integrated Private Reporting", *Meditari Accountancy Research* 23 (1) (2015): 28–61.

105 de Villiers and van Staden, "Can Less Environmental Disclosure Have a Legitimising Effect?"

Existing investment models, short-term focus and stakeholder [in]activism

The Code for Responsible Investment in South Africa (CRISA) applies to institutional investors on a comply-or-explain basis, encouraging them to include ESG metrics as part of their investment decision-making.[106] In this way, South Africa is second only to the UK in developing a set of formal principles for the promotion of responsible investment practices in line with the recommendations of the UN-backed Principles for Responsible Investment. This is complemented by the operation of Regulation 28 of the Pensions Funds Act which requires all factors which could materially affect an investment to be taken into account, including relevant ESG issues.[107] The result is an increased awareness of the importance of ESG reporting by both companies and the country's largest institutional investors.[108] This has not, however, translated into widespread reform of investment appraisal and analysis by the country's investment community.

Even in cases where some stakeholders appreciate the relevance of environmental issues, the dominant factor in investor decision-making remains economic performance.[109] Academic research points to the value relevance of non-financial information in the South African capital market but, despite the introduction of Regulation 28 and CRISA, results are far from conclusive.[110] In addition, it is often difficult for qualitative environmental information to be incorporated into quantitative investment models.[111] This is especially true when considering the significant practical difficulties encountered when estimating the expected value of lost biodiversity mass[112] such as replacement cost of pollination services by wild and

106 IOD, *Code for Responsible Investing in South Africa* (Johannesburg, South Africa: Lexis Nexus South Africa, 2011).

107 Deloitte, *The Relationship between CRISA and Regulation 28 of Pension Funds Act and Integrated Reporting*, 2014, accessed 5 September 2015, http://www2.deloitte.com/content/dam/Deloitte/za/Documents/governance-risk-compliance/ZA_TheRelationshipBetweenCRISAAndRegulation28OfPensionFundsAct04042014.pdf.

108 Atkins and Maroun, "Integrated Reporting in South Africa in 2012".

109 Jones, "Accounting for the Environment"; R. Gray, "Back to Basics: What Do We Mean by Environmental (and Social) Accounting and What Is It For? A Reaction to Thornton", *Critical Perspectives on Accounting* 24 (6) (2013): 459–68; Atkins and Maroun, *South African Institutional Investors' Perceptions of Integrated Reporting*; J. Atkins *et al.*, "'Good' News from Nowhere: Imagining Utopian Sustainable Accounting", *Accounting, Auditing & Accountability Journal* 28 (5) (2015): 65170.

110 de Klerk and de Villiers, "The Value Relevance of Corporate Responsibility Reporting".

111 Atkins and Maroun, *South African Institutional Investors' Perceptions of Integrated Reporting*; Mansoor and Maroun, "Biodiversity Reporting in the South African Food and Mining Sectors".

112 Jones and Solomon, "Problematising Accounting for Biodiversity"; Siddiqui, "Mainstreaming Biodiversity Accounting".

managed hives.[113] In this way, limited environmental reporting (including biodiversity-specific disclosures) can be seen as indicative of the challenges encountered when attempting to incorporate these metrics into existing valuation models. From a more critical perspective, however, this problem highlights the limitations imposed on conceptualizations of firm performance by the continuing dominance of financial and economic paradigms.[114] As explained by Gray,[115] "sustainability" is conceptualized in economic terms where:

> conventional financial accounting is a predominantly economic—and not very internally logical—practice which has no substantive conceptual space for environmental or social matters *per se*. It has no space for...market alien values—values such as environmental concern.

Consequently, biodiversity reporting is marginalized, not because issues such as colony collapse disorder are scientifically invalid, but because the existing systems of accounting and accountability are inadequate for reporting the effects of so-called "soft-issues".[116] This goes hand-in-hand with companies viewing the biodiversity loss as a remote or long-term risk which does not warrant current attention.

An established body of international research documents the focus on short-term profitability to the detriment of sustainable business practices,[117] a concern shared by the IIRC and GRI.[118] In a South African context, recent research confirms that non-financial measures are not consistently included in the investment appraisal process which is biased in favour of short-term indicators of financial profitability.[119] Consequently, even though biodiversity loss is a recognized scientific problem, the continuing focus on financial measures of performance, coupled with the difficulty of incorporating environmental performance measures in existing investment analysis systems, limits the extent to which this information is being communicated to investors and other stakeholders.[120]

113 Allsopp *et al.*, "Valuing Insect Pollination Services with Cost of Replacement"; Pirk *et al.*, "A Survey of Managed Honey Bee Colony Losses in the Republic of South Africa".
114 R. Gray, *Integrated Reporting: Integrated With What and For Whom?* 2012, accessed 7 July 2014, https://risweb.st-andrews.ac.uk/portal/en/researchoutput/integrated-reporting-integrated-with-what-and-for-whom(f9797689-51b2-4bb0-b158-e3a6ab0904ca).html; Tregidga *et al.*, "(Re)presenting 'Sustainable Organizations'".
115 Gray, "Back to Basics: What Do We Mean by Environmental (and Social) Accounting and What Is It For?"
116 Gray, "Is Accounting for Sustainability Actually Accounting for Sustainability"; Gray, "Back to Basics: What Do We Mean by Environmental (and Social) Accounting and What Is It For?"; Atkins *et al.*, "'Good' News from Nowhere".
117 Moneva *et al.*, "GRI and the Camouflaging of Corporate Unsustainability"; Solomon *et al.*, "Impression Management, Myth Creation and Fabrication in Private Social and Environmental Reporting".
118 IIRC, *The International Framework: Integrated Reporting*; GRI, *G4 Sustainability Reporting Guidelines*.
119 Atkins and Maroun, "Integrated Reporting in South Africa in 2012".
120 Atkins *et al.*, "'Good' News from Nowhere".

A lack of stakeholder activism adds to the problem. A recent study reports low levels of financial literacy among many South African investors.[121] This is confirmed by a review of investors' views on South African integrated reports which identifies limited awareness of relevant corporate governance principles and poor financial skills, particularly when it comes to some pension fund trustees.[122] Even when deficient skills are not an issue, the continued dominance of finance paradigms, emphasis on short-term profitability and emerging evidence of stakeholder apathy in South Africa means that there is little engagement with local companies on the relevance of significant biodiversity risks, including the risk of the pollination ecosystem service beginning to falter.[123]

Summary

From a purely scientific perspective, the general consensus is that bees, working with other insects, offer a vital service to the South African farmer, contributing millions of dollars in value on an annual basis to the local economy. South Africa is fortunate in that it does not seem to have suffered as severely from the loss of wild and managed bee colonies as the USA. There are, however, reports of initial losses by beekeepers and a risk of inadequate foraging resources to promote the necessary growth in bee populations. Consequently there is a concerted effort by the scientific community to understand better the causes of bee losses in South Africa and how to promote healthy wild and managed colonies.

Disappointingly, the massive monetary value of pollination services—and investment in related scientific research—has gone unnoticed by corporate South Africa. An analysis of 19 of the largest companies in the farming, food and retail sectors yielded only two examples of disclosures dealing with bees. There was no evidence of pollination services being identified as a key natural resource in any of the integrated reports under review or even an initial assessment of the long-term risks posed by declining bee populations.

One explanation is that South African bees are in a healthier position than in the USA and Europe and scientific evidence on local bee populations is not yet conclusive. Nevertheless, there is evidence pointing to significant losses in the country's

121 R. Rensburg and E. Botha, "Is Integrated Reporting the Silver Bullet of Financial Communication? A Stakeholder Perspective from South Africa", *Public Relations Review* 40 (2) (2014): 144–52.

122 Atkins and Maroun, *South African Institutional Investors' Perceptions of Integrated Reporting*; Atkins and Maroun, "Integrated Reporting in South Africa in 2012".

123 Mansoor and Maroun, "Biodiversity Reporting in the South African Food and Mining Sectors".

bee populations in more recent years[124] and, even if this was not the case, South Africa is part of a global food market which is under threat from a disruption of pollination ecosystem services. As such, it is reasonable to expect that organizations committed to providing a complete account of the risks to generating sustainable returns would deal with bee pollination services, at least to some extent, in their integrated or sustainability reports.

A lack of clear reporting guidelines and the time taken to move from shareholder-centric financial reporting to stakeholder-orientated integrated reports may be contributing to limited biodiversity reporting. Adding to this is the slow rate of dissemination of academic research on bees, and their economic relevance, to preparers of corporate reports. From a more critical perspective, however, the neglect of bee-related disclosures calls into question the commitment of large South African companies to engage in transparent risk-management reporting. It also suggests that the investment community is either unaware of the importance of ecosystem services or that short-term profit orientation has resulted in the marginalization of a significant environmental risk which, ironically, may have massive financial implications in the years to come. This is especially worrying given that CRISA and Regulation 28 are aimed specifically at addressing the risk of short-term focus on profitability to the detriment of long-term sustainability and prudent investment choices.

Ultimately, the findings in this chapter highlight the need for immediate action by the environmental accounting research community. This book is part of a first-stage response to the need for more interdisciplinary research which identifies material ESG considerations based on the findings of leading scientific research. More, however, needs to be done to document serious environmental challenges and their implications for the ability of organizations to generate sustainable returns in the short, medium and long term. This will need to be part of a global research agenda which takes cognizance not only of worldwide ecological problems such as colony collapse disorder but also of the specific threats posed to the planet's unique biomes. Researchers will also need to be mindful of the importance of generating valid and reliable results which can still be understood by practitioners and drive meaningful change in reporting practices.

Contemporaneously, more needs to be done to develop a biological-based method of accounting and valuation to complement CRISA, the Principles for Responsible Investment and integrated reporting framework. These provide important principles for describing the objectives of balanced reporting and investment analysis but stop short of giving practitioners a generally accepted method for integrating financial and non-financial information in their investment models. It is unlikely that a viable solution can be developed by relying only on traditional accounting measures. Again, an interdisciplinary approach will be required to understand the

124 Wossler, "South Africa: American Disease That's Wreaking Havoc on the Cape's Honeybee Population".

interconnection between organizations and biodiversity and how best to quantify and communicate the implications of changes in biodiversity.

Bibliography

Agricultural Research Council. *The African Pollinator Initiative,* 2014. The Agricultural Research Council. Accessed 2 January 2015. http://www.arc.agric.za/arc-ppri/Pages/Biosystematics/African-Pollinator-Initiative-(API).aspx.

Aizen, M. A. and L.D. Harder. "The Global Stock of Domesticated Honey Bees Is Growing Slower Than Agricultural Demand for Pollination". *Current Biology* 19 (11) (2009): 915–8.

Allsopp, M. H., W.J. de Lange and R. Veldtman. "Valuing Insect Pollination Services with Cost of Replacement". *PLoS ONE* 3 (9) (2008): e3128.

Ashforth, B. E. and B.W. Gibbs. "The Double-Edge of Organisational Legitimation". *Organization Science* 1 (2) (1990): 177–94.

Atkins, J. and W. Maroun. *South African Institutional Investors' Perceptions of Integrated Reporting.* London: The Association of Chartered Certified Accountants, 2014.

Atkins, J. and W. Maroun, "Integrated Reporting in South Africa in 2012: Perspectives from South African Institutional Investors". *Meditari Accountancy Research* 23 (2) (2015): 197–221.

Atkins, J., B. Atkins, I. Thomson and W. Maroun. "'Good' News from Nowhere: Imagining Utopian Sustainable Accounting". *Accounting, Auditing & Accountability Journal* 28 (5) (2015): 65170.

Atkins, J., E. Barone, D. Gozman, W. Maroun and B. Atkins. "Exploring Rhinoceros Conservation and Protection: Corporate Disclosures and Extinction Accounting by Leading South African Companies". *Meditari Accountancy Research,* 2 July 2015.

Atkins, J. F., A. Solomon, S. Norton and N.L. Joseph. "The Emergence of Integrated Private Reporting". *Meditari Accountancy Research* 23 (1) (2015): 28–61.

Bebbington, J., R. Gray and D. Owen "Seeing the Wood for the Trees: Taking the Pulse of Social and Environmental Accounting". *Accounting, Auditing & Accountability Journal* 12 (1) (1999): 47–52.

Brennan, N. and D. Merkl-Davies. "Rhetoric and Argument in Social and Environmental Reporting: The Dirty Laundry Case". *Accounting, Auditing & Accountability Journal* 27 (4) (2014): 602–33.

Brennan, N., and J. Solomon. "Corporate Governance, Accountability and Mechanisms of Accountability: An Overview". *Accounting, Auditing & Accountability Journal* 21 (7) (2008): 885–906.

Burritt, R. L. "Environmental Performance Accountability: Planet, People, Profits". *Accounting, Auditing & Accountability Journal* 25 (2) (2012): 370–405.

Carels, C., W. Maroun and N. Padia. "Integrated Reporting in the South African Mining Sector". *Corporate Ownership and Control* 11 (1) (2013): 991–1005.

Carvalheiro, L. G., C.L. Seymour, S.W. Nicolson and R. Veldtman "Creating Patches of Native Flowers Facilitates Crop Pollination in Large Agricultural Fields: Mango as a Case Study". *Journal of Applied Ecology* 49 (6) (2012): 1373–83.

Carvalheiro, L. G., R. Veldtman, A.G. Shenkute, G.B. Tesfay, C.W.W. Pirk, J.S. Donaldson and S.W. Nicolson. "Natural and Within-Farmland Biodiversity Enhances Crop Productivity". *Ecology Letters* 14 (3) (2011): 251–9.

Coetsee, D. and N. Stegmann "A Profile of Accounting Research in South African Accounting Journals". *Meditari Accountancy Research* 20 (2) (2012): 92–112.

Conservation South Africa. *Green Choice: Serving Nature's Bounty*, 2011. Accessed 12 May 2015. http://biodiversityadvisor.sanbi.org/wp-content/uploads/2014/07/GreenChoice_brochure-2011_low-res.pdf.

Conservation South Africa and South African National Biodiversity Institute. *Crop Agriculture, Pollination and the Honeybee*, 2014. Accessed 14 March 2016. http://biodiversityadvisor.sanbi.org/wp-content/uploads/2012/11/Crop-Agric-Pollination-and-the-honeybee-Brochure.pdf.

Crous, C. J. and F. Roets. "Realising the Value of Continuous Monitoring Programmes for Biodiversity Conservation". *South African Journal of Science* 110 (11/12) (2014): 7–11.

Cuckston, T. "Bringing Tropical Forest Biodiversity Conservation into Financial Accounting Calculation". *Accounting, Auditing & Accountability Journal* 26 (5) (2013): 688–714.

de Klerk, M. and C. de Villiers. "The Value Relevance of Corporate Responsibility Reporting: South African Evidence". *Meditari Accountancy Research* 20 (1) (2012): 21–38.

de Villiers, C. J. and P. Barnard. "Environmental Reporting in South Africa from 1994 to 1999: A Research Note". *Meditari Accountancy Research* 8 (1) (2000): 15–23.

de Villiers, C. and C.J. van Staden. "Can Less Environmental Disclosure Have a Legitimising Effect? Evidence from Africa". *Accounting, Organizations and Society* 31 (8) (2006): 763–81.

Deloitte. *The Relationship between CRISA and Regulation 28 of Pension Funds Act and Integrated Reporting*, 2014. Accessed 5 September 2015. http://www2.deloitte.com/content/dam/Deloitte/za/Documents/governance-risk-compliance/ZA_TheRelationshipBetweenCRISAAndRegulation28OfPensionFundsAct04042014.pdf.

Department of Environmental Affairs and Tourism. *Convention on Biological Diversity: South African National Report to the Fourth Conference of the Parties Department*. South Africa: Department of Environmental Affairs and Tourism, 1998.

Department of Environmental Affairs and Tourism. *South Africa's Fourth National Report to the Convention on Biological Diversity*. South Africa: Department of Environmental Affairs and Tourism, 2009.

Gess, S. K. and F.W. Gess. *Potential Pollinators of the Cape Group of Crotalarieae (sensu Polhill) (Fabales: Papilionaceae), with Implications for Seed Production in Cultivated Rooibos Tea*. Entomological Society of Southern Africa, 1994.

Grabsch, C., M.J. Jones and J.F. Solomon. *Accounting for Biodiversity in Crisis: A European Perspective*. Working Paper, London: Kings College, 2012.

Gray, R. "Is Accounting for Sustainability Actually Accounting for Sustainability…and How Would We Know? An Exploration of Narratives of Organisations and the Planet". *Accounting, Organizations and Society* 35 (1) (2010): 47–62.

Gray, R. *Integrated Reporting: Integrated With What and For Whom?* 2012. Accessed 7 July 2014. https://risweb.st-andrews.ac.uk/portal/en/researchoutput/integrated-reporting-integrated-with-what-and-for-whom(f9797689-51b2-4bb0-b158-e3a6ab0904ca).html.

Gray, R. "Back to Basics: What Do We Mean by Environmental (and Social) Accounting and What Is It For? A Reaction to Thornton". *Critical Perspectives on Accounting* 24 (6) (2013): 459–68.

Global Reporting Initiative (GRI). *G4 Sustainability Reporting Guidelines*, 2014. Accessed 10 February 2015. https://www.globalreporting.org/reporting/g4/Pages/default.aspx.

Higgins, C. and R. Walker. "Ethos, Logos, Pathos: Strategies of Persuasion in Social/Environmental Reports". *Accounting Forum* 36 (3) (2012): 194–208.

Hindley, T. and P. Buys. "Integrated Reporting Compliance With The Global Reporting Initiative Framework: An Analysis Of The South African Mining Industry". *International Business & Economics Research Journal* 11 (11) (2012): 1249–60.

International Integrated Reporting Council (IIRC). *The International Framework: Integrated Reporting*, 2013. Accessed 1 October 2013. http://www.theiirc.org/wp-content/uploads/2013/12/13-12-08-THE-INTERNATIONAL-IR-FRAMEWORK-2-1.pdf.

Intergovernmental Panel on Climate Change. *Climate Change 2013: The Physical Science Basis*, 2013. Accessed 1 June 2015. http://www.ipcc.ch/report/ar5/wg1/.

Institute of Directors in Southern Africa (IOD) *The King Report on Corporate Governance in South Africa—2002 (King-II)*. Johannesburg, South Africa: Lexis Nexus South Africa, 2002.

IOD. *The King Code of Governance for South Africa (2009) and King Report on Governance for South Africa (2009) (King-III)*. Johannesburg, South Africa: Lexis Nexus South Africa, 2009.

IOD. *Code for Responsible Investing in South Africa*. Johannesburg, South Africa: Lexis Nexus South Africa, 2011.

Integrated Reporting Committee of South Africa (IRCSA). "Framework for Integrated Reporting and the Integrated Report", 2011. Accessed 5 June 2012. www.sustainabilitysa.org.

Jones, M. J. "Accounting for Biodiversity: A Pilot Study". *The British Accounting Review* 28 (4) (1996): 281–303.

Jones, M. J. "Accounting for Biodiversity: Operationalising Environmental Accounting". *Accounting, Auditing & Accountability Journal* 16 (5) (2003): 762–89.

Jones, M. J. "Accounting for the Environment: Towards a Theoretical Perspective for Environmental Accounting and Reporting". *Accounting Forum* 34 (2) (2010): 123–38.

Jones, M. J. and J.F. Solomon. "Problematising Accounting for Biodiversity". *Accounting, Auditing & Accountability Journal* 26 (5) (2013): 668–87.

King, M. "Comments on: Integrated Reporting and the Integrated Report". *International Corporate Governance Conference, Johannesburg, South Africa. 23 October 2012.*

KPMG. *Carrots and Sticks—Promoting Transparency and Sustainability: An Update on Trends in Voluntary and Mandatory Approaches to Sustainability Reporting*, 2012. Accessed 30 June 2013. https://www.globalreporting.org/resourcelibrary/Carrots-And-Sticks-Promoting-Transparency-And-Sustainbability.pdf.

Levy, S. "What's Best for Bees". *Nature* 479 (November 2011): 164–5.

Loate, B., N. Padia and W. Maroun. "Acid Mine Drainage in South Africa: A Test of Legitimacy Theory". *Journal of Governance and Regulation* 4 (2) (2015): 26–40.

Mansoor, H. and Maroun, W. "Biodiversity Reporting in the South African Food and Mining Sectors". In*: 27th International Congress on Social and Environmental Accounting Research, 27 August 2015, Egham, United Kingdom*. United Kingdom: Centre for Social and Environmental Accounting Research.

Maroun, W. and C. Jonker. "Critical and Interpretive Accounting, Auditing and Governance Research in South Africa". *Southern African Journal of Accountability and Auditing Research* 16 (2014): 51–62.

Melin, A., M. Rouget, J.J. Midgley and J.S. Donaldson. "Pollination Ecosystem Services in South African Agricultural Systems". *South African Journal of Science* 110 (11/12) (2014): 25–33.

Merkl-Davies, D., N. Brennan and P. Vourvachis. "Text Analysis Methodologies in Corporate Narrative Reporting Research". In *23rd CSEAR International Colloquium, 2011, St Andrews, United Kingdom*.

Meyer, J. W. and B. Rowan. "Institutionalized Organizations: Formal Structure as Myth and Ceremony". *American Journal of Sociology* 83 (2) (1977): 340–63.

Moneva, J. M., P. Archel and C. Correa. "GRI and the Camouflaging of Corporate Unsustainability". *Accounting Forum* 30 (2) (2006): 121–37.

O'Donovan, G. "Environmental Disclosures in the Annual Report". *Accounting, Auditing & Accountability Journal* 15 (3) (2002): 344–71.

O' Dwyer, B. "Managerial Perceptions of Corporate Social Disclosure: An Irish Story". *Accounting, Auditing & Accountability Journal* 15 (3) (2002): 406–36.

Patten, D. M. "The Relation between Environmental Performance and Environmental Disclosure: A Research Note". *Accounting, Organizations and Society* 27 (8) (2002): 763–73.

Pick n Pay. *Pick n Pay: Integrated Annual Report: 2012*. Accessed 1 March 2015. http://www.picknpayinvestor.co.za/financials/annual_reports/2012/downloads/integrated_ar/pnp_integrated_ar.pdf.

Pick n Pay. “Where have all the bees gone?” Accessed 1 March 2015. http://www.picknpay.co.za/green-news/where-have-all-the-bees-gone.

Pirk, C. W. W., H. Human, R.M. Crewe and D. vanEngelsdorp. “A Survey of Managed Honey Bee Colony Losses in the Republic of South Africa—2009 to 2011”. *Journal of Apicultural Research* 53 (1) (2014): 35–42.

Raemaekers, K. and W. Maroun. “Trends in Risk Disclosure: Practices of South African Listed Companies”. *Accounting Perspectives in Southern Africa* 2 (1) (2014): 8–15.

Raemaekers, K., W. Maroun, and N. Padia. “Risk Disclosures by South African Listed Companies Post-King III”. *South African Journal of Accounting Research* (2015): 1–20.

Rensburg, R. and E. Botha. “Is Integrated Reporting the Silver Bullet of Financial Communication? A Stakeholder Perspective from South Africa”. *Public Relations Review* 40 (2) (2014): 144–52.

Rimmel, G. and K. Jonäll. “Biodiversity Reporting in Sweden: Corporate Disclosure and Preparers' Views”. *Accounting, Auditing & Accountability Journal* 26 (5) (2013): 746–78.

Samkin, G. “Changes in Sustainability Reporting by an African Defence Contractor: A Longitudinal Analysis”. *Meditari Accountancy Research* 20 (2) (2012): 134–66.

Samkin, G., A. Schneider. and D. Tappin. “Developing a Reporting and Evaluation Framework for Biodiversity”. *Accounting, Auditing & Accountability Journal* 27 (3) (2014): 527–62.

South African National Biodiversity Institute (SANBI). *Crop Agriculture, Pollination and the Honeybee*. Accessed 14 March 2016. http://biodiversityadvisor.sanbi.org/wp-content/uploads/2012/11/Crop-Agric-Pollination-and-the-honeybee-Brochure.pdf.

SANBI. *Honeybees in South Africa—What Landowners Can Do to Help*. Accessed 8 May 2015. http://www.sanbi.org/sites/default/files/documents/documents/honbeybees-sa-brochure-28-may-eng-proof.pdf.

SANBI. *Biodiversity Information Policy Document- Policy Series: Principle and Guidelines*. South Africa: SANBI, 2010.

SANBI. “SANBI Biodiversity for Life”. Accessed 15 August 2014. http://www.sanbi.org/.

Siddiqui, J. “Mainstreaming Biodiversity Accounting: Potential Implications for a Developing Economy”. *Accounting, Auditing & Accountability Journal* 26 (5) (2013): 779–805.

Solomon, J. *Corporate Governance and Accountability*, 3rd edition. Chichester, UK: John Wiley and Sons Ltd, 2010.

Solomon, J. and W. Maroun. *Integrated Reporting: The New Face of Social, Ethical and Environmental Reporting in South Africa?* London: The Association of Chartered Certified Accountants, 2012).

Solomon, J. F., A, Solomon, N.L. Joseph and S.D. Norton. “Impression Management, Myth Creation and Fabrication in Private Social and Environmental Reporting: Insights from Erving Goffman”. *Accounting, Organizations and Society* 38 (3) (2013): 195–213.

Suchman, M. C. “Managing Legitimacy: Strategic and Institutional Approaches”. *The Academy of Management Review* 20 (3) (1995): 571–610.

Tregidga, H., M. Milne and K. Kearins. “(Re)presenting ‘Sustainable Organizations’”. *Accounting, Organizations and Society* 39 (6) (2014): 477–94.

Turpie, J. K. “The Existence Value of Biodiversity in South Africa: How Interest, Experience, Knowledge, Income and Perceived Level of Threat Influence Local Willingness to Pay”. *Ecological Economics* 46 (2) (2003): 199–216.

van Liempd, D. and J. Busch. “Biodiversity Reporting in Denmark”. *Accounting, Auditing & Accountability Journal* 26 (5) (2013): 833–72.

Watmough, R. H. "The potential of Megachile gratiosa Cameron, Xylocopa caffra (Linnaeus) (Hymenoptera: Megachilidae and Anthophoridae) and other solitary bees as pollinators of alfalfa, *Medicago sativa* L. (Fabaceae), in the Oudtshoorn District, South Africa". Entomological Society of Southern Africa, 1999.

West, A. "A Commentary on the Global Position of South African Accounting Research". *Meditari Accountancy Research* 14 (1) (2006): 121–37.

Woolworths. *Woolworths Holdings Limited: 2011 Good Business Journey Report.* Accessed 1 March 2015. http://www.woolworthsholdings.co.za/downloads/2011_good_business_journey_report.pdf.

Woolworths. *Woolworths Holding Limited: 2012 Integrated Report.* Accessed 1 March 2015. http://www.woolworthsholdings.co.za/downloads/2012/2012_integrated_report.pdf.

World Conservation Monitoring Centre. *Global Biodiversity: Status of the Earth's Living Resources.* 1992. Accessed 1 March 2015. http://www.biodiversitylibrary.org/item/97636#page/5/mode/1up.

Wossler. "South Africa: American Disease That's Wreaking Havoc on the Cape's Honeybee Population". *All Africa,* January 2015. Accessed 22 September 2015. http://allafrica.com/stories/201508020001.html.

Wynberg, R. "A Decade of Biodiversity Conservation and Use in South Africa: Tracking Progress from the Rio Earth Summit to the Johannesburg World Summit on Sustainable Development: Review Article". *South African Journal of Science* 98 (5 & 6) (2002): 233–43.

14

Corporate bee accountability among Swedish companies

Kristina Jonäll
Gothenburg University, Sweden

Gunnar Rimmel
Jönköping University, Sweden

Imagine billions of workers who work without a break, in silence and without pay. This is the reality for bees. For a long time no one saw any value in the work they do. It has taken a tragedy for us to understand their economic value. In the US a large part of the natural wild bee population has died off; the same thing has happened in Europe.[1]

Pollinating insects are vital for the ecosystem to function, for the global economy, for modern consumer culture and for human survival. Without bees and other pollinators, more than 50% of the food we consume would disappear or sharply rise in price.[2] Primary vegetables, fruit and berries are pollinated by bees. Meat and dairy products are dependent on pollinators, since large proportions of livestock fodder such as clover or alfalfa require pollination. Cotton is also pollinated by insects, as well as rubber trees. So without bees humans would have to live without jeans, T-shirts or sneakers, as well as refreshing fruits or a cup of coffee in the morning.

1 D. van Engelsdorp et al., "Colony Collapse Disorder: A Descriptive Study", *PLoS ONE* 4(8) (2009): e6481, doi: 10.1371/journal.pone.0006481.

2 J. Sass, *Busy as a Bee: Pollinators Put Food on the Table* (National Resources Defense Council, 2015); N. Holland, "The Economic Value of Honeybees", *BBC News*, 23 April 2009.

Most of us would survive solely on corn, rice and wind-pollinated grains, but we would probably suffer from deficiency diseases such as scurvy.

In landscapes where wild pollinators are decreasing, honey bees promote the maintenance of plant species; therefore honey bee losses are of great concern. Current honey bee colony losses worldwide are caused by colony collapse disorder, the mite *Varroa destructor* and pesticides.[3] One of the first alarms raised concerning mass bee death came from a beekeeper in Florida in November 2006 who discovered that his bees had disappeared.[4] Further reports showed that the phenomenon could be found in the USA, Canada and it was also detected in Germany, France, Holland and Italy.

Although much is still to be researched and explained about massive bee death, most researchers agree that the answer is to be found in a combination of the following possible explanations. Genetically manipulated crops, inbreeding, chemical pesticides, parasites, stress from modern industrial bee management, where bee colonies are shipped on trucks between huge mono crops that require fertilization, are all part of the problem.[5] The issue most frequently identified is the use of chemicals in modern farming. In particular, the use of **neonicotinoid** pesticides can be linked to the mass death of bees. This pesticide is spread in plant tissues and is deadly to insects throughout the growing season, including during flowering when honey bees consume their pollen. Neonicotinoids affect insects' central nervous system. Some studies show how neonicotinoids affected the bees' ability to navigate.[6] Another reaction is that even very small amounts of neonicotinoids deteriorate the reproductive ability of bumblebees, another important pollinating insect.[7]

In Sweden gains and losses in bee stock have been documented for almost 100 years. Sweden has not suffered from mass death of bees, which is reported in other parts of the world. In Sweden the bees, in recent years, have hibernated relatively well due to warm winters. There is also a completely different pressure on the usage of pesticides and chemicals in agriculture in Sweden compared to other countries, as many Swedish beekeepers fight mites using organic methods. It is illegal to use substances such as neonicotinoids in crops that are attractive to bees or other pollinators. So even if pesticides are a problem for pollinators internationally, this is

3 T.R. Pedersen et al., Massdöd av bin - samhällsekonomiska konsekvenser och möjliga åtgärder (Rapport 2009:24) (Jönköping: Jordbruksverket, 2009).

4 D. Engelsdorp and J.S. Pettis, "Colony Collapse Disorder", in *Bee Health and Veterinarians*, edited by W. Ritter (Paris: World Organisation for Animal Health, 2014).

5 Pedersen et al., Massdöd av bin - samhällsekonomiska konsekvenser och möjliga åtgärder.

6 J.P. van der Sluijs et al., "Neonicotinoids, Bee Disorders and the Sustainability of Pollinator Services", *Current Opinion in Environmental Sustainability* 5 (2013): 293–305.

7 C. Lu et al., "Sub-lethal Exposure to Neonicotinoids Impaired Honey Bees Winterization before Proceeding to Colony Collapse Disorder", *Bulletin of Insectology* 67 (2014): 125–30.

less of an issue in Sweden. The spread of the varroa mite is a far greater problem for Swedish bees.[8]

Last year's media attention on the mass death of bees has affected both people and companies. Disasters often trigger human reaction to prevent further deterioration or extinction. The mass death of bees is linked to money for most of us, as we are nearly all invested in stock markets through pension funds and other forms of institutional investment, and stock markets are clearly affected.[9] This financial connection also creates a great opportunity for marketing and sales of products that in one way or another can be associated with honey or pollinating bees. As a reaction several companies donated money to bee research: for example the cosmetics chain The Body Shop donated 20 Swedish kronor per sold product of one of their make-up series for three weeks in summer 2011 to the Swedish Beekeepers Association.[10]

Many companies are not satisfied with simply donating money or demonstrating concern for the bees in their advertising. Instead, some companies try to get involved in other ways. One way they are involving themselves in bee decline is by introducing beehives close to their corporate buildings, such as in courtyards, on terraces or on rooftops. This type of initiative is a new and increasingly common phenomenon in Sweden. In this chapter, we discuss this corporate use of beehives as a means of demonstrating sustainable development; we explore corporate accountability in relation to bee populations and the impact of these initiatives on urban biodiversity and environment.

The value of pollination

There are many signs that the threat to life on the planet and the loss of biodiversity has to be taken seriously. There are about 1,900 species of pollinating bees and bumblebees in Europe, 20% of them are endemic and many of these are threatened with extinction, according to the IUCN Red List.[11] In Germany, for instance, there are 560 bee species and 289 of them are on the red list. Every loss of pollinating species means a step nearer an approaching collapse of ecosystems that depend on pollination, which threatens our food production. Many experts speak of a global

8 Pedersen et al., Massdöd av bin - samhällsekonomiska konsekvenser och möjliga åtgärder.

9 See Chapters 1, 11, 12, 13 and 16.

10 B. Johansson, Make up med honung för mångfalden. *Bitidningen.* Sveriges Biodlares Riksförbund, Biodlarna, 2011.

11 A. Nieto et al., *European Red List of Bees* (IUCN Global Species Programme, IUCN European Union Representative Office, 2015).

pollination crisis.[12] About 75% of all crop species require pollination by animals of some sort. Recent studies suggest that about one-third of pollination is delivered by honey bees, the rest being carried out by a range of wild insects, flies, butterflies, birds or even bats.[13] If vital pollinators cannot survive or if there are not enough pollinators, farmers will be forced to hand-pollinate crops. This is now happening in a province of China, where hired workers use brushes to hand-pollinate pear trees.[14] Such measures are possible for a limited number of high-value crops, but there are not enough humans in the world to pollinate all of our crops by hand.

The pollination service of bees has for a long time been economically invisible. In recent years the commercial value of pollination has been calculated. The global annual economic value of insect pollination was estimated to be €153 billion during 2005 (i.e. 9.5% of the total economic value of world agricultural output considering only crops that are used directly for human food.[15] This means that the value of refined products such as pickled cucumbers or tomatoes used for ketchup are not included. The corresponding figure for the value of bee pollination for agriculture in Sweden 2011 has been calculated as approximately €26–47 million. In Sweden, the value of pollination is estimated as 3% of the total contribution of agriculture to the Swedish GDP.[16] In these calculations the value of non-commercial farming such as gardening, pollination of wild plants and berries was not included. The value of honey bee pollination of wild flora is difficult to estimate but probably it is as important as it is for commercial farming. Due to the pollinating crisis and the worrying decline in the number of specialized wild pollinators, humans have become dependent on a super-generalist honey bee visiting a large number of plant species.[17] Knowing that bees' pollinating service is worth €153 billion, we may now look at them in a different way, give them a little more attention. There is so much value in a beehive but we humans have not even thought about it.

Pollination in Swedish agriculture

The term **mass death of bees** is used collectively for unusually large winter losses and colony collapse disorder (CCD). CCD is quite a new concept, primarily an American and a Canadian phenomenon, but similar symptoms have been

12 S. Kluser and P. Peduzzi, *Global Pollinator Decline: A Literature Review* (United Nations Environment Programme, 2007).
13 T.D. Breeze et al., "Pollination Services in the UK: How Important are Honeybees?" *Agriculture, Ecosystems & Environment* 142 (2011): 137–43.
14 U.M.A. Patrap et al., "Pollination Failure in Apple Crop and Farmers' Management Strategies in Henduan Mountains, China", *Acta Hortic (ISHS)* (2001): 225–30.
15 S.G. Potts et al., "Global Pollinator Declines: Trends, Impacts and Drivers", *Trends in Ecology and Evolution* 25 (2010): 345–53.
16 T.R. Pedersen, *Värdet av honungsbins pollinering av grödor i Sverige* (Jönköping: Jordbruksverket, 2011).
17 M. Rundlof et al., "Seed Coating with a Neonicotinoid Insecticide Negatively Affects Wild Bees", *Nature* 521 (2015): 77–80.

recorded in other countries. There is no official registration of the CCD in Sweden, although beekeepers have reported similar symptoms.[18] The Israeli acute paralysis virus, which is one of the possible causes of CCD, has not been found in Sweden. Chemicals legislation and the procedure for pesticides approval are considerably stricter in Sweden than in other countries. The whole structure of beekeeping is also different in comparison with countries such as the United States. Beekeeping in Sweden is rather immobile unlike in the United States where beehives are moved around the crops. However, in Sweden there is poor control of the presence of viral diseases and there is a lack of planning to prevent the spread of viruses and mites. Normally the level of winter losses of Swedish bees has been between 5 and 10% but statistics show that winter losses are increasing and the number of bee colonies has declined in recent years. This could be an effect of many factors, for example a decrease in colony vitality that makes it more vulnerable to parasites, diseases, agricultural poisons and breeding among others.[19]

In 2009, the Swedish Board of Agriculture initiated a project to investigate why honey bees are dying and what can be done to prevent this from happening. The study showed that the varroa mite and associated viruses are the biggest threats to honey bees in Sweden. Several other threats were also identified including: lack of pollen and nectar plants; reduced genetic variation within bee populations because of modern bee breeding; and pesticides such as neonicotinoids. Several projects have started to tackle the threats that have emerged.

One of these projects is "Diversity on the plain". In this project farmers in areas with intensive agricultural production are saving plants that honey bees, bumblebees and other bees thrive in. The decline in farmland biodiversity is often said to be a result of agricultural intensification and structural changes in the agricultural landscape.[20] Contemporary agricultural landscapes often lack forage resources for pollinators. The intensification of agriculture with larger fields, efficient and diverse cultivation and denser crops are some explanations for the disappearance of pollinators.[21] Flower strips, field islets and ditch banks are examples of methods to manage the decline of pollinators. The aim of the project was to encourage farmers in the plains area to implement simple and inexpensive measures to protect biodiversity, as far as is possible, while even improving agricultural profitability.[22]

18 Pedersen et al., Massdöd av bin - samhällsekonomiska konsekvenser och möjliga åtgärder.

19 *Ibid.*

20 M. Rundlöf and H.G. Smith, "The Effect of Organic Farming on Butterfly Diversity Depends on Landscape Context", *Journal of Applied Ecology* 43 (2006): 1121–7.

21 R. Bommarco et al., "Drastic Historic Shifts in Bumble Bee Community Composition in Sweden", *Proceedings of the Royal Society B. Biological Sciences* 279 (2012): 309–15.

22 S. Eriksson and M. Rundlöf, Pollinatörer i insådda ettåriga blomremsor - en fältundersökning av förekomsten av blombesökande insekter i insådda blommande remsor i tre slättbygdsområden i Sverige 2011-12. Hushållningssällskapet, 2013.

Sweden also has a national programme to improve the production and marketing of apiculture products. In 2013, the Swedish Board of Agriculture introduced The National Honey Program. Two out of the programme's four objectives are connected to increasing the bee population and decreasing winter death. The average number of bee colonies should increase by 2% over a period of three years from the 2013 level for at least two-thirds of the counties. The decrease in winter mortality during the current programme period will be lower than the corresponding winter mortality for the previous period.

A project called the **pollination pool** started by the "Swedish Professional Beekeepers" makes it possible for plant growers and beekeepers to rent and let bee colonies for pollination assignments. With the help of pollination services plant growers can increase the harvest without increasing the amount of fertilizer or the area of cultivated land. Swedish research has shown that there can be a yield increase of up to 20% by adding two beehives per hectare.[23]

Another project will develop a new contingency plan for mites and other bee pests, as well as examining the impacts of neonicotinoids on honey bees, bumblebees and solitary bees under field conditions in Sweden. This is an extension of the 2009 study on the economic consequences of and possible interventions in colony losses in honey bees.[24] The current project is continuing to study the threats of plant protection products and exotic pests.

Bees in Swedish cities

In addition to national Swedish agricultural programmes, there has been a growing interest among firms in adopting beehives and placing them in cities. A new type of "employment agency" is becoming more common, which rents bees instead of a human workforce. Instead of "man power" these agencies provide "bee power".

Contrary to what we might expect, bees are doing well in the city. Cities often provide a huge diversity of sites: gardens, meadows and nature reserves. All of these habitats can add up to a really special resource for pollinators. There is a wide variety of flowers and other plants on balconies, terraces and allotment gardens and lots of flowers. This ensures that bees will find pollen during a greater part of the year. Within cities there is more varied bee food. Hence, bees will be more efficient and can produce more honey in each hive. Moreover, it tends to be slightly warmer in cities than in the countryside. This makes it easier for bees to over-winter. Studies show that bees do not seem to be affected significantly by the exhaust gases from vehicles and in town we use fewer agricultural poisons compared with use in

23 Biodlingsföretagarna, Öka skörden med pollineringspoolen. Hallvigs tryckeri, 2010.
24 Pedersen et al., Massdöd av bin - samhällsekonomiska konsekvenser och möjliga åtgärder.

the countryside. Samples have been taken from both urban bees and their honey to investigate contamination. However, there have been no findings of contamination in the honey and the bees showed no raised levels of toxins. Nonetheless, cities are often bereft of bees and wild pollinators. Therefore, an increase in urban beekeepers could both serve as insurance for the bees' survival and for pollination of plants.

Research methods

We started by searching all companies on the Stockholm Stock Exchange, OMX, for bee-related disclosures. Due to its voluntary nature, bee-related disclosure may appear anywhere in corporate communications. Therefore, all website sections were examined, not just the sustainability sections. That process required examination of archives, presentations, news announcements, company brochures and reports in electronic form, as well as corporate activity in social media. Companies' websites, online information and documents were searched for words related to: bee(s), honey, pollination and beehives. Since we only got a few hits the search was broadened to all Swedish companies. This resulted in the finding that it seems to be a new trend for companies to engage in hosting beehives on rooftops or company premises. The study therefore focuses on companies that are hosting beehives.

This study comprises 32 companies from the following sectors: (17) property and/or municipal housing; (3) transportation (airport, railway); (5) hotel and conference; (1) architecture; (1) energy; (1) marketing; (2) culture (beehives were placed on top of the cities' opera houses); (1) food production; and (1) recycling.

Four years of annual and sustainability reports (2011 to 2014), if available, were analysed for all companies. In addition, the content on companies' websites, blogs and Facebook pages or other social media were studied.

We conducted two stages of analysis. First we performed a content analysis to extract themes from the bee-related disclosures. Second we looked at bee-related initiatives to find out why companies engaged with bees; this could later be included in the themes.

In this study we applied a range of different content analysis such as meaning oriented and interpretive content analysis. This implies that we are looking for the underlying themes in the texts as an interpretative content analysis assuming that words derive their meanings when they are used in specific situations. By looking at words and phrases in their context (paragraph or whole text) we can discover themes that exist independently from the interpreter.[25]

25 M. Smith and R.J. Taffler, "The Chairman's Statement: A Content Analysis of Discretionary Narrative Disclosures", *Accounting Auditing & Accountability Journal* 13 (2000): 624–46.

Bees disclosure by Swedish companies

In reports and websites of Swedish listed companies, just a few companies disclosed information about bees. Searching more widely on Swedish companies and the link to bees, we found that there was great interest on the part of companies to invest in bees and beehives as part of their sustainability work. Companies in the real estate business were most devoted to this type of activity; 17 of the 32 companies in this study belong to this group. Several companies that let and manage hives for large companies have, in recent years, entered the market. The hives are placed on the roofs of company buildings, preferably in the middle of large cities or nearby places where there are high levels of emissions (e.g. in areas of high density traffic).

Companies that invested in beehives were mainly to be found in large cities: 18 of the companies are located in Stockholm, nine in the area of Gothenburg, three in Malmö, one in Karlstad and one in Västerås. We can also see that the companies are focused in a few industries, mainly the property industry and municipal housing as noted above (17 of 32).

The review of corporate information regarding bee-related disclosure has shown that the information is available at various locations within the corporate communications (see Table 14.1). Out of the 32 studied companies, 14 provided the information in the company's annual report, 10 featured bee disclosure in the sustainability report. Three of these companies disclosed the information in both the annual and the sustainability report; one of the studied firms had an integrated report. All companies provided information on the website, one had a blog and another had a Facebook page; we also looked at Twitter accounts.

The information disclosed by companies that rent or own hives varies between companies. In most cases there is information on the company's website where they disclose their environmental commitment in general, adding information about the beehives and the purpose of keeping bees. Some of the companies link the information to more than an ecological value and also mention an economic value.

After studying 32 company websites, annual reports, sustainability reports and other documents it can be noticed that the available corporate disclosure concerning involvement in bees and beehives can be categorized into six themes. These themes are described below.

Contributing to biodiversity

Out of the 32 analysed companies, 17 write that they want to **contribute to biological diversity**. Above all it is the bee's role as pollinator the companies are writing about. Bees play an important role for both biodiversity and a sustainable society. The pollination of the neighbourhood, especially in cities, is a factor that contributes to biodiversity. There are also disclosures about the importance of spreading knowledge about the bee's vital role in the ecosystem. If the companies are committed to biodiversity and sustainability they say that they "take the environmental efforts to the next level". By investing in hives they put things in a "larger perspective", which "inspires others", and increases understanding of the importance of biodiversity. This applies particularly in urban environments. "Together we will spread awareness of the importance of increased biodiversity and the bee's role in our ecosystem".[26]

Corporate social responsibility

Investing in beehives could be one part of a company's **CSR** work. It is a way of portraying themselves as respecting the environment and showing local commitment. Some use the bees as a way to be target-oriented and goal-oriented with respect to environmental issues. Since honey and other products from the bees are sold, companies can also, as a part of their CSR work, donate all surpluses to non-profit organizations or for charitable purposes. Research shows the importance of corporate philanthropy as a mechanism to support economic prosperity and growth, especially when it comes to urgent social purposes. Surplus donations to charity fit well with this view.

Responsible property ownership

Most of the studied real estate companies say that they are taking care of all tenants, bees as well as humans, big as well as small. One writes that "our latest, and maybe most important tenants, are smaller than what we're used to—they are bees"[27] another that "we like to take care of all our tenants humans as well as bees".[28] Sometimes bees are mentioned as part of the staff. "Our new staff, the worker bees, makes sure we get a lot of fruit in autumn".[29] Taking care of the tenants is what many of the companies refer to as **responsible property ownership**. Similarities between bees and companies in the real estate business are highlighted. Among several of the property companies, colonies of bees are compared with human society. Similarities include the fact that bees are very sociable; they undergo various stages of maturity akin to humans (i.e. children and adults); and they have different types of tasks or work. One example of this is when a company writes "Bees take care of their children in a similar ways as humans".[30] Or, "bees, like humans

26 Wihlborgs fastigheter, Hållbara höjdpunkter, Sustainability Report 2014
27 Aspholmen fastigheter, Miljöarkiv, 2013, accessed 13 February 2015, Vår binäring – honungsbin http://www.aspholmenfastigheter.se/hallbarhet/miljoarkiv/2014
28 AFA fastigheter, Våra minsta hyresgäster surrar mest, accessed 13 February 2015, http://www.afafastigheter.se/Om_oss/Bikupa
29 Brostaden press release, 28 June 2013, accessed 13 February 2015, http://www.brostaden.se/om-brostaden/nyheter-och-media/nyheter/2013/30-000-nya-arbetsplatser-pa-taket/
30 Aspholmen fastigheter, 2013, accessed 13 February 2015, http://www.aspholmenfastigheter.se/hallbarhet/miljoarkiv/2013-11-28-30-000-nya-honungsproducerande-och-miljoframjande-hyresgaster-i-kopparlunden.html

are working, make sure that there is enough food in the larder and help each other to keep warm in the winter".[31] There is some discussion in the corporate discourse of bees as community builders, which are very diligent and concerned with health and safety. One property company wrote that "...to become a sponsor for a beehive is an excellent way to bring to life an environmental policy and sustainability initiatives for both employees and tenants."[32] The more a company can add ecological value, by planting vegetation on the roofs, establishing plantations, beehives, wetlands or other initiatives, the more it increases the value of the environmental project and the more sustainable the company will be perceived. The purpose is often talked about as a willingness to be involved and take responsibility for society, a way of giving back to nature.

Raising awareness of bee decline and the pollination crisis

With small and simple but valuable contributions companies hope that they could contribute to bee survival and help to **raise awareness of global bee deaths** and the **pollination crisis**. Bees are in a precarious position around the world. Hence, spreading knowledge is of utmost importance. By raising the issue and spreading knowledge about the work bees are doing through pollination, people could become aware and influence decision-makers to take action against global bee death and the pollination crisis.

Measuring performance

Especially for companies in the transportation sector, **bees are used as performance measurers**. Malmö Airport has an ongoing project with bees and bee products, which serves as an environmental indicator for assessing air quality around the airport. Honey bees are considered to be good indicators of chemical pollution in the environment in two ways. First, bees experience high mortality rates when in contact with pesticides and second their bodies and products accumulate pollutants, which can be measured in laboratories. Comparative analysis between the bees on and far from the airport has shown no significance difference between air pollution levels between measuring points. Also the honey and beeswax produced in hives near the airport were analysed to show which chemical substances are present in the bees' environment. Having beehives on airport runways could also been seen as a way of neutralizing CO_2 emissions and reducing the carbon footprint. This could be known as a "Bees as 21st century canaries" theme as they are being used in the same way that canaries in cages were used to detect gas in mines.

31 *Ibid.*
32 Wihlborgs fastigheter, Hållbara höjdpunkter, Sustainability Report, 2014

Education

Many of the beehives are also used for **education**. The hives are placed on the ground and are built so that anyone is able to look in behind secure glass to see how the bees live and work. Some of the companies are also sponsoring activities for schools; the bee rental companies, hiring the hives to other companies, are educating young children in school about the life of bees and the value of pollinators.

Discussion

A large number of studies[33] use legitimacy theory in an attempt to explain CSR disclosures in annual reports. Most of such disclosures focus on general sustainability and most companies provide little if any information on their impact on ecosystems.[34] A previous study shows that less than one-third of Swedish companies report such information.[35] Legitimacy theory assumes that the legitimacy of a firm to operate in society depends on a social contract between the firm and the society.[36] Legitimacy theory also assumes that companies will adopt disclosure strategies to conform to society's expectations.[37] According to legitimacy theory, a company needs to have legitimacy in the sense of a social "licence to operate".[38] Without this "licence" a company won't access the necessary resources to successfully conduct business. If society perceives that a company is not operating in an

33 See, for example, J. Guthrie and L.D. Parker, "Corporate Social Disclosure Practice: A Comparative International Analysis", *Advances in Public Interest Accounting* 3 (1990): 159–75; D.M. Patten, "Exposure, Legitimacy and Social Disclosure", *Journal of Accounting Public Policy* 10 (1991): 297–308; D.M. Patten, "Intra-industry Environmental Disclosures in Response to the Alaskan Oil Spill: A Note on Legitimacy Theory", *Accounting, Organizations and Society* 17 (1992): 471–5; C. Deegan and M. Rankin, "Do Australian Companies Report Environmental News Objectively? An Analysis of Environmental Disclosures by Firms Prosecuted Successfully by the Environmental Protection Authority", *Accounting, Auditing & Accountability Journal* 9 (1996): 50–67; and D.J. Campbell, "Legitimacy Theory or Managerial Reality Construction? Corporate Social Disclosure in Marks & Spencer Corporate Reports, 1969–1997", *Accounting Forum* 24 (2000): 80–100.

34 M.J. Jones and J.F. Solomon, "Problematising Accounting for Biodiversity", *Accounting, Auditing & Accountability Journal* 26 (2013): 668–87.

35 G. Rimmel and K. Jonäll, "Biodiversity Reporting in Sweden: Corporate Disclosure and Preparers' Views", *Accounting, Auditing and Accountability Journal* 26 (2013): 746–78.

36 J. Guthrie and L.D. Parker, "Corporate Social Reporting: A Rebuttal of Legitimacy Theory", *Accounting and Business Research* 19 (1989): 343–52.

37 C. Deegan, "The Legitimising Effect of Social and Environmental Disclosures: A Theoretical Foundation", *Accounting, Auditing, & Accountability Journal* 15 (2002): 282–311.

38 *Ibid.*

acceptable way, legitimacy will be potentially threatened. Companies use disclosure to enhance their "corporate image" and strengthen their "corporate identity".[39]

The companies studied demonstrate through the act of investing in beehives that they are engaging in what is for humanity an important and urgent matter. By investing in hives and additionally disclosing information about their engagement with bees, the firms demonstrate that the issue of pollination is taken seriously. Previous research[40] has shown that companies report to their stakeholders in order to legitimize corporate activities. In this study, however, it seems that action itself is more important than reporting. Therefore, since the information is disclosed rather vaguely, it is the specific action that the companies report, where they are putting the bees in focus. When it comes to the property and municipal housing companies we studied, there is a link between their own business, their activities and beekeeping. They compare the beehives to their own houses, the colonies to the residents and the bees' work to the employees' work. They also write that beekeeping is a part of corporate sustainability initiatives, especially linked to the local area; by investing in a hive the firms show local commitment to pollination and diversity.

Hahn and Kühnen conclude in their summary of previous research that companies want to signal good performance; this implies a positive effect on reporting. They also conclude that companies with poorer sustainability performance may face greater stakeholder pressure. Consequently, companies may be more actively engaged in reporting to mitigate legitimacy threats. This implies a negative relation between performance and sustainability reporting.[41]

What has been observed in this study is that the amount of disclosure from the companies regarding bee-related information is not very large in the reports to shareholders or investors. Most information is given on the corporate website or in the companies' magazines aimed at the residents. When it comes to the property and municipal housing companies, it is the residents who are targeted. The content of the disclosure is linked to housing and community. Companies in this sector often compare beehives with their residential properties and bees are also mentioned as new tenants moving in, tenants who will secure biodiversity. Although there is relatively little reporting of bees, the engagement in comparison is relatively

39 R. Hooghiemstra, "Corporate Communication and Impression Management: New Perspectives Why Companies Engage in Corporate Social Reporting", *Journal of Business Ethics* 27 (2000): 55–68.

40 See for example R.M. Haniffa and T.E. Cooke, "The Impact of Culture and Governance on Corporate Social Reporting", *Journal of Accounting and Public Policy* 24 (2005): 341–430.

41 R. Hahn and M. Kühnen, "Determinants of Sustainability Reporting: A Review of Results, Trends, Theory, and Opportunities in an Expanding Field of Research", *Journal of Cleaner Production* 59 (2013): 5–21.

large. We can assume that companies do not face legitimacy threats regarding this type of disclosure. The disclosure focuses on positive effects through investing in beehives, highlighting the bees' work and proving to be responsible property owners who take care of all tenants, including bees.

For the studied companies in the transportation sector, bees are used as performance measurers. In this case there seems to be a stronger link to corporate legitimacy and focus on showing that the operation meets society's expectations that the companies work for a cleaner environment. Making use of bees as a measure of emissions can be linked to how stakeholders perceive the company. To ensure that emissions are so small that they do not affect sensitive animals also shows that the demands for cleaner vehicles and less pollution are taken seriously and thus contribute to strengthening the company's legitimacy.

In society today there is a call to protect nature and to ensure that diversity is maintained. Biodiversity is a term that includes all variations and all the interactions between plants, animals and their environment. It is important to preserve biological diversity for several reasons. Among other things, functioning ecosystems perform numerous ecological services that we often take for granted. Biodiversity loss is accelerating and this represents one of today's most serious environmental issues. Loss of species affects ecosystems and food security on Earth. From an ecosystems perspective, taking an integrated approach to bee decline, the loss or severe decline in one particular species can have catastrophic and unknown consequences on other species and on nature as a whole. The demand for organic food increases gradually as consumer awareness increases. The honey produced in companies' beehives is often packaged in a way that conveys the message that the company's brand stands for sustainable development and biodiversity. The companies' disclose their beekeeping in a way that gets people to start talking about what happens in nature around us. There are many companies working proactively to reduce their environmental impact and find business opportunities in the conservation of biodiversity and ecosystem services. That companies contribute to biodiversity or at least not destroy it is one of the conditions for companies to gain legitimacy. The bees bring environmental benefits and companies that place hives on their plots or roofs receive goodwill.

The importance of pollination for human food production has been a hot topic, especially after the threats of extinction of bees have been highlighted in the media. As an extended discussion of biodiversity many of the companies in the study intend to increase awareness of global bee deaths and the pollination crisis. This ensures that information and knowledge will be spread that can contribute to bee conservation.

Some reflections and possibilities

Through the interpretative content analysis we found six major themes in the bee-related disclosures. Companies were quite eager to communicate that they contributed to biological diversity by hosting bees. Beekeeping was also a part of the companies' CSR work and also a way of portraying themselves as respecting the environment and showing local commitment. The majority of the studied companies were in the property and/or municipal housing sector. These firms compared the beehives to human society and observed many similarities between bees and humans.

Among companies in the central parts of Stockholm and other big cities, having their own hives has become the latest way to communicate sustainability and ecological awareness among clients and competitors. Companies will in this way spread knowledge and information about bees and biodiversity and how they relate to sustainable urban development and human well-being.

For decades, companies have used disclosures about sustainability, climate change and ecosystems to create a picture of being a "good company". Some of these words have been overused and have become outdated; companies need something new to lean on. Using bees engages the public in a natural way since many are aware of the problems with diminishing numbers of pollinators. Installing beehives near company properties is fantastic for building public awareness about both the company and nature without using words.

Table 14.1 **Overview of Swedish company bee-related disclosure**

		Information in		
Company	**Business**	**Annual report**	**Sustainability report**	**Website**
Swedavia Malmö	Airport	0	2011/2013	X
Swedavia Landvetter	Airport	0	0	X
Jernhusen	Railway	2014	0	X
White arkitekter	Arcitect	2013	0	X
Borås energi	Energy	2014	0	Blog
Ibis Arlanda	Hotel	0	2012	X
Clarion Arlanda	Hotel	0	0	X
Hotel Hilton Slussen	Hotel	0	0	X
Radisson Blue Arlanda	Hotel	0	0	Facebook
Svenska mässam	Hotel conference	2014	0	X
Bee production	Marketing content företag	0	0	X
Familijebostäder	Municipal housing	0	0	X
Bostadsbolaget	Municipal housing	2013/2014	2013	X
Stockholmshem	Municipal housing	2013/2014	0	X

Company	Business	Information in		
		Annual report	Sustainability report	Website
Poseidon	Municipal housing	0	0	X
Wihlborgs fastigheter	Property	0	2014	X
Brostaden (Castellum)	Property	0	2013/2014	X
Axfast (Axel Johnson koncernen)	Property	2013	0	X
Fabege	Property	2012	2012	X
Huge	Property	2012	0	X
Einar Mattsson	Property	0	0	X
Riksbyggen	Property	0	0	X
HSB	Property	2013	2012	X
Chalmersfastigheter	Property	2014	0	X
Aspholmen fastigheter (Castellum)	Property	0	2013/2014	X
AFA fastigheter	Property	0	0	X
AMF fastigheter	Property investment and development	0	2013	X
Skanska	Property and construction	2013	0	X
Löfbergs Lila	Food production	0	2013/2014	X
Operahuset i Malmö	Operahouse	0	0	Facebook/Twitter
Göteborgs Operan	Operahouse	2012	0	X
IK Recycling	Recycling	2013	Integrated	Blog

References

AFA fastigheter. Våra minsta hyresgäster surrar mest. Accessed 13 February 2015. http://www.afafastigheter.se/Om_oss/Bikupa

Aspholmen fastigheter. Miljöarkiv. Accessed 13 February 2015. Vår binäring – honungsbin http://www.aspholmenfastigheter.se/hallbarhet/miljoarkiv/Var-binaring–honungsbin.html

Aspholmen fastigheter, 30 000 nya honungsproducerande och miljöfrämjande hyresgäster i Kopparlunden.,2013. Accessed 13 February 2015. http://www.aspholmenfastigheter.se/hallbarhet/miljoarkiv/2013-11-28-30-000-nya-honungsproducerande-och-miljoframjande-hyresgaster-i-kopparlunden.html

Biodlingsföretagarna. Öka skörden med pollineringspoolen. Hallvigs tryckeri, 2010.

Bommarco, R., O. Lundin, H.G. Smith and M. Rundlöf. "Drastic Historic Shifts in Bumble Bee Community Composition in Sweden". *Proceedings of the Royal Society B. Biological Sciences* 279 (2012): 309–15.

Breeze, T. D., A.P. Bailey, K.G. Balcombe and S.G. Potts. "Pollination Services in the UK: How Important are Honeybees?" *Agriculture, Ecosystems & Environment* 142 (2011): 137–43.

Brostaden press release. 28 June 2013. Accessed 13 February 2015. http://www.brostaden.se/om-brostaden/nyheter-och-media/nyheter/2013/30-000-nya-arbetsplatser-pa-taket/

Campbell, D. J. "Legitimacy Theory or Managerial Reality Construction? Corporate Social Disclosure in Marks & Spencer Corporate Reports, 1969–1997". *Accounting Forum* 24 (2000): 80–100.

Deegan, C. "The Legitimising Effect of Social and Environmental Disclosures: A Theoretical Foundation". *Accounting, Auditing, & Accountability Journal* 15 (2002): 282–311.

Deegan, C. and M. Rankin. "Do Australian Companies Report Environmental News Objectively? An Analysis of Environmental Disclosures by Firms Prosecuted Successfully by the Environmental Protection Authority". *Accounting, Auditing & Accountability Journal* 9 (1996): 50–67.

Engelsdorp, D. and J.S. Pettis. "Colony Collapse Disorder". In *Bee Health and Veterinarians*, edited by W. Ritter. Paris: World Organisation for Animal Health, 2014.

Eriksson, S. and M. Rundlöf. Pollinatörer i insådda ettåriga blomremsor - en fältundersökning av förekomsten av blombesökande insekter i insådda blommande remsor i tre slättbygdsområden i Sverige 2011-12. Hushållningssällskapet, 2013.

Guthrie, J. and L.D. Parker. "Corporate Social Reporting: A Rebuttal of Legitimacy Theory". *Accounting and Business Research* 19 (1989): 343–52.

Guthrie, J. and L.D. Parker. "Corporate Social Disclosure Practice: A Comparative International Analysis". *Advances in Public Interest Accounting* 3 (1990): 159–75.

Hahn, R. and M. Kühnen. "Determinants of Sustainability Reporting: A Review of Results, Trends, Theory, and Opportunities in an Expanding Field of Research". *Journal of Cleaner Production* 59 (2013): 5–21.

Haniffa, R. M. and T.E. Cooke. "The Impact of Culture and Governance on Corporate Social Reporting". *Journal of Accounting and Public Policy* 24 (2005): 341–430.

Holland, N. "The Economic Value of Honeybees". *BBC News*, 23 April 2009.

Hooghiemstra, R. "Corporate Communication and Impression Management: New Perspectives Why Companies Engage in Corporate Social Reporting". *Journal of Business Ethics* 27 (2000): 55–68.

Johansson, B. Make up med honung för mångfalden. *Bitidningen*. Sveriges Biodlares Riksförbund, Biodlarna, 2011.

Jones, M. J. and J.F. Solomon. "Problematising Accounting for Biodiversity". *Accounting, Auditing & Accountability Journal* 26 (2013): 668–87.

Kluser, S. and P. Peduzzi. *Global Pollinator Decline: A Literature Review*. United Nations Environment Programme, 2007.

Lu, C., K.M. Warchol and R.A. Callahan. "Sub-lethal Exposure to Neonicotinoids Impaired Honey Bees Winterization before Proceeding to Colony Collapse Disorder". *Bulletin of Insectology* 67 (2014): 125–30.

Nieto, A., S. P. M. Roberts, J. Kemp, P. Rasmont, M. Kuhlmann, M. García Criado, J.C. Biesmeijer, P. Bogusch, H.H. Dathe, P. De La Rúa, T. De Meulemeester, M. Dehon, A. Dewulf, F.J. Ortiz-Sánchez, P. L'homme, A. Pauly, S.G. Potts, C. Praz, M. Quaranta, V.G. Radchenko, E. Scheuchl, J. Smit, J. Straka, M. Terzo, B. Tomozii, J. Window and D. Michez. *European Red List of Bees*. IUCN Global Species Programme, IUCN European Union Representative Office, 2015.

Patrap, U. M. A., T.E.J. Patrap and H.E. Yonghua. "Pollination Failure in Apple Crop and Farmers' Management Strategies in Henduan Mountains, China". *Acta Hortic (ISHS)* (2001): 225–30.

Patten, D. M. "Exposure, Legitimacy and Social Disclosure". *Journal of Accounting Public Policy* 10 (1991): 297–308.

Patten, D. M. "Intra-industry Environmental Disclosures in Response to the Alaskan Oil Spill: A Note on Legitimacy Theory". *Accounting, Organizations and Society* 17 (1992): 471–5.
Pedersen, T. R. *Värdet av honungsbins pollinering av grödor i Sverige*. Jönköping: Jordbruksverket, 2011.
Pedersen, T. R., R. Bommarco, K. Ebbersten, A. Falk, I. Fries, P. Kristiansen, P. Kryger, H. Nätterlund and M. Rundlöf. Massdöd av bin - samhällsekonomiska konsekvenser och möjliga åtgärder (Rapport 2009:24). Jönköping: Jordbruksverket, 2009.
Potts, S.G., J.C. Biesmeijer, C. Kremen, P. Neumann, O. Schweiger and W.E. Kunin. "Global Pollinator Declines: Trends, Impacts and Drivers". *Trends in Ecology and Evolution* 25 (2010): 345–53.
Rimmel, G. and K. Jonäll. "Biodiversity Reporting in Sweden: Corporate Disclosure and Preparers' Views". *Accounting, Auditing and Accountability Journal* 26 (2013): 746–78.
Rundlöf, M. and H.G. Smith. "The Effect of Organic Farming on Butterfly Diversity Depends on Landscape Context". *Journal of Applied Ecology* 43 (2006): 1121–7.
Rundlof, M., G.K.S. Andersson, R. Bommarco, I. Fries, V. Hederstrom, L. Herbertsson, O. Jonsson, B.K. Klatt, T.R. Pedersen, J. Yourstone and H.G. Smith. "Seed Coating with a Neonicotinoid Insecticide Negatively Affects Wild Bees". *Nature* 521 (2015): 77–80.
SASS, J. *Busy as a Bee: Pollinators Put Food on the Table*. National Resources Defense Council, 2015.
Smith, M. and R.J. Taffler. "The Chairman's Statement: A Content Analysis of Discretionary Narrative Disclosures". *Accounting Auditing & Accountability Journal* 13 (2000): 624–46.
van der Sluijs, J. P., N. Simon-Delso, D. Goulson, L. Maxim, J-M. Bonmatin and L.P. Belzunces. "Neonicotinoids, Bee Disorders and the Sustainability of Pollinator Services". *Current Opinion in Environmental Sustainability* 5 (2013): 293–305.
van Engelsdorp, D., J.D. Evans, C. Saegerman, C. Mullin, E. Haubruge, B.K. Nguyen and J.S. Pettis. "Colony Collapse Disorder: A Descriptive Study". *PLoS ONE* 4(8) (2009): e6481. doi: 10.1371/journal.pone.0006481.
Wihlborgs fastigheter. Hållbara höjdpunkter. Sustainability Report 2014.

15

Bees and accountability in Germany

A multi-stakeholder perspective

Christoph F. Biehl
Henley Centre for Governance, Accountability and Responsible Investment, UK

Martina N. Macpherson
Sustainable Investment Partners Ltd, UK

In this chapter we focus on the roles of the main stakeholders involved in the bee controversies in Germany. We focus on Germany as it is not only home to two of the largest European beekeepers' associations, an eco-conscious civil society and government, but also to industrial farmers and some of the largest chemical producers worldwide. The main aspects analysed in the chapter are the decline in the number of beekeepers, the impact of the different stakeholder actions on bee health and the impact of renewable energy land use on bees' natural habitat.

Sustainable social and economic development, in line with the multifaceted principles and values of liberty, respect, dignity, freedom, cultural diversity and integrity, have been ingrained in German society post-1945.[1]

Constitutionally, these values are guaranteed through freedom of speech and association and socially through access to (free) education, care and welfare—key "pillars" of modern German society. Hence, from a young age, individuals are encouraged to participate in social and economic life through schools, clubs, and a range of public associations and engagement programmes.

1 Jürgen Hartmann, "Fundamental Pillars of Democracy", 2012, https://www.deutschland.de/en/topic/politics/germany-europe/fundamental-pillars-of-democracy.

Civil activism—the study and practice of non-violent conflict and resistance—has become a well-established way to disrupt political and corporate systems in Germany. Politicians, public authorities and companies are mandated to provide access to (objective) information, (political) decision-making, new regulation and their impact on citizens and consumers alike; for example, via media channels, activist and interest groups.[2]

Over the last 50 years, key areas of focus for many civil activist groups in Germany have been the natural environment and animal welfare ("Natur-" und "Tierschutzorganisationen").[3] Since the early 1980s, respecting, preserving and protecting the global ecosystem, biodiversity, soil, natural habitat and animals—above and beyond their economic utility—has gradually become an educational and political theme as well as an "intrinsic value" within German society.[4]

Over the years, law and regulations have followed suit. First, in the form of the Federal Nature Conservation Act (Bundesnaturschutzgesetz, BNatSchG), now in its amended version of 21 September 1998, which reflects on the protection, preservation and conservation of the natural environment, the natural landscape and natural species and also focuses on the involvement of citizens' action groups.

Other regulatory frameworks include the Federal Forestry Act (Bundeswaldgesetz), introduced on 2 May 1975, which contains environmental law provisions as they relate to forests, the Federal Soil Protection Act (Bundesbodenschutzgesetz, BBodSchG), introduced on 17 March 1998, and the Animal Protection Act (Tierschutzgesetz), which is now in its amended version of 25 May 1998. The German Civil Code no longer considers animals as inanimate objects, but determines that the law pertaining to inanimate objects should apply *mutatis mutandis* to animals if no other provision is made.[5]

However, regardless of activist and environmental groups' educational and political campaigns and more and more regulatory frameworks for nature conservation and preservation—including bees and pollinators—scientists, activist groups and

2 Mark Engler and Paul Engler, "History Didn't Bring Down the Berlin Wall—Activists Did", http://fpif.org/history-didnt-bring-berlin-wall-activists/.

3 Dieter Rucht and Jochen Roose, "The Transformation of Environmental Activism in Berlin" (Social Science Research Center Berlin, 2011), accessed August 13, 2015, http://ecpr.eu/filestore/paperproposal/05262578-28eb-46a6-9c93-8c4da44d2d98.pdf.

4 However, recent controversial corporate activities in Germany, such as the Volkswagen AG scandal, have not only caused reputational and economic damage to the organizations involved, but also raised significant questions in relation to corporate and stakeholder governance, transparency, disclosure and "intrinsic values".

5 Peter-Christoph Storm, "Environmental Laws: Introduction", http://www.iuscomp.org/gla/literature/envirmt.htm.

policy-makers have made limited progress to explain, analyse and reverse the significant losses of insect pollinator colonies[6] (in particular bees) in Germany.

What are the causes? Much of the work to date has focused on the decline of the honey bee, *Apis mellifera*—and the most important factors so far identified relate to parasitic, bacterial and viral disease, as well as to wider agricultural practices that can impact at many points in the life-cycle of bees. In short, the decline is undoubtedly the product of multiple factors—both known and unknown—acting singly or in combination.[7]

But what about the impact? Without insect pollination more than 70% of our crops will probably suffer some decrease in productivity.[8] In Europe alone, 84% of the 264 crop species are animal-pollinated and 4,000 vegetable varieties exist thanks to pollination by bees.[9]

Undoubtedly, the most nutritious and interesting crops in our diet, including key fruits and vegetables, together with some crops used as fodder in meat and dairy production, can be badly affected by a decline in insect pollinators; in particular, the production of apples, strawberries, tomatoes and almonds would suffer.

Besides crop plants, most wild plants (around 90% of them) need animal-mediated pollination to reproduce, and thus other ecosystem services and the wild habitats providing them also depend—directly or indirectly—on insect pollinators.[10]

And what are the financial costs? The most recent estimate of the global economic benefit of pollination amounts to some €265 billion,[11] assessed as the value of crops dependent on natural pollination.[12] This is not a "real" value, of course, as

6 Neil Carter, *The Politics of the Environment Ideas, Activism, Policy* (Cambridge: Cambridge University Press, 2007).

7 Dharam P. Abrol, "Decline in Pollinators", in *Pollination Biology*, ed. D.P. Abrol (Dordrecht: Springer, 2012), 545–601; Yves Le Conte, Marion Ellis, and Wolfgang Ritter, "Varroa Mites and Honey Bee Health: Can Varroa Explain Part of the Colony Losses?", *Apidologie* 41(3) (2010): 353–63, doi: 10.1051/apido/2010017.

8 UNFAO, "Protecting the Pollinators", *Spotlight*, 2005, http://www.fao.org/ag/magazine/0512sp1.htm. Reyes Tirado, Gergely Simon, and Paul Johnston, "A Review of Factors That Put Pollinators and Agriculture in Europe at Risk", *Greenpeace Research Laboratories Technical Report (Review)*, vol. 01/2013 (Amsterdam: Greenpeace International, 2013), 5 ff.

9 Everything Connects, "Pollinator Decline", 2014, http://www.everythingconnects.org/pollinator-decline.html.

10 IRGC, "Risk of Loss of Pollination Services", 2008, 1-28, www.irgc.org/IMG/doc/IRGCS_TC8Sept08_11_DraftCN_Pollination.doc.

11 Sven Lautenbach *et al.*, "Spatial and Temporal Trends of Global Pollination Benefit", *PLoS One* 7, no. 4 (2012): 1-16, doi:10.1371/journal.pone.0035954.

12 Stéphane Kluser *et al.*, "Global Honey Bee Colony Disorder and Other Threats to Insect Pollinators", *UNEP Emerging Issues* (UNEP, 2010): 1-14.

it hides the fact that, should natural pollination be severely compromised or end, it might prove impossible to replace—effectively making its true value infinitely high.[13]

With so much at stake, including food security, policy-makers are eager to take action. One of the European Commission's (EC) main objectives is halting the loss of biodiversity by 2020, a move likely to benefit all kinds of pollinators. The EC has also designated a reference laboratory for bee health, a decision meant to improve the quality of collected data and to harmonize surveillance. Risk assessment procedures for plant protection products are also being revised.[14]

In their recent report,[15] the European Food Safety Authority has also noted that, at an overall European level, there are significant gaps in knowledge of the multiple stressors affecting both wild and cultured pollinators, in particular the impact of mixtures of pesticides. The report notes inter alia the need to devise, through coordinated research, common methods of monitoring bees and of hazard identification for different classes of chemicals.[16]

On the basis of these scientific assessments by EFSA, the European Commission enforced an EU-wide ban of three pesticides belonging to a class of chemicals known as neonicotinoids in December 2013. The ban received strong support from multiple European member states.[17]

However, two producers of these pesticides, Bayer AG and Syngenta, sought to overturn the ban[18] while, in turn, environmental and consumer organizations such as Greenpeace International, Bee Life European Beekeeping Coordination,

13 Elizabeth Grossman, "Declining Bee Populations Pose A Threat to Global Agriculture", *Yale Environment 360*, 2013, http://e360.yale.edu/feature/declining_bee_populations_pose_a_threat_to_global_agriculture/2645/; Tirado, Simon, and Johnston, "A Review of Factors That Put Pollinators and Agriculture in Europe at Risk", 3, 18; Nicola Gallai *et al.*, "Economic Valuation of the Vulnerability of World Agriculture Confronted with Pollinator Decline", *Ecological Economics* 68, no. 3 (2009): 810–21, doi: 10.1016/j.ecolecon.2008.06.014; Richard Black, "Bees and Flowers Decline in Step", *BBC News*, 2006, http://news.bbc.co.uk/2/hi/science/nature/5201218.stm.

14 Ettore Capri and Alexandru Marchis, "Bee Health in Europe: Facts & Figures", *Bridging Science and Policy* (Piacenza: Opera Research Centre, 2013), 14 ff., http://www.pollinator.org/PDFs/OPERAReport.pdf.

15 EFSA, "Towards an Integrated Environmental Risk Assessment of Multiple Stressors on Bees: Review of Research Projects in Europe, Knowledge Gaps and Recommendations", *EFSA Journal* 12(3) (2014): 3594.

16 *Ibid.*

17 European Commission, "Bees & Pesticides: Commission Goes Ahead with Plan to Better Protect Bees" (European Commission, 2013), http://ec.europa.eu/food/archive/animal/liveanimals/bees/neonicotinoids_en.htm.

18 Helga Einecke, "Bayer Klagt Gegen Pestizid-Verbot Nach Bienensterben", 2013, http://www.sueddeutsche.de/wirtschaft/chemiekonzern-bayer-klagt-gegen-pestizid-verbot-nach-bienensterben-1.1756210.

Pesticides Action Network Europe, ClientEarth, Buglife and SumOfUs applied to intervene at the European Court of Justice to defend the ban.[19]

The political, academic and industry-led debate about the links between bee decline and the usage of pesticides is still ongoing—in particular in Germany where a range of chemical companies are headquartered which fund research into bee health and lobby governments.

Most recently, in August 2015, the President of the European Beekeepers' Association, Walter Haefeker, openly criticized the lack of transparency and independent research in Germany while outlining the links between pesticides and bee decline in a public TV interview produced by MDR/ARD.[20]

In this chapter, we aim to assess three key factors that are contributing to the losses of insect pollinators, in this case the honey bees, in Germany:

- The decline of the bee keepers,
- Key issues related to bee health,
- Renewable energy-specific land use practices

We will also assess the involvement and influence of multiple stakeholder groups including beekeepers such as the German Beekeepers' Association (Deutscher Imkerbund e.V., DIB), German Professional Beekeepers' Association (Deutscher Berufs- u. Erwerbsimkerbund e.V., DBIB), environmental and consumer-led activist groups, scientists, farmers (Deutscher Bauernverband, DBV), the German Government and other policy-makers as well as corporates—and their role in this process.

Bee decline in Germany: beekeepers

Global bee decline has been caused by a variety of different factors, as outlined above. While some of them are vastly complex, for example the impact of neonicotinoid-based pesticides on the bees' immune system, others show a more direct impact, such as the decline in numbers of beekeepers. In the following section, we will analyse the situation in Germany and the development of the number of beekeepers and beehives over recent years.

19 Greenpeace International, "NGOs and Beekeepers Take Legal Action to Defend EU Ban of Bee-Killing Pesticides against Syngenta and Bayer" (Greenpeace International, 2013), http://www.greenpeace.org/eu-unit/en/News/2013/NGOs-and-beekeepers-take-legal-action-to-defend-EU-ban-of-bee-killing-pesticides-against-Syngenta-and-Bayer.

20 Daniel Baumbach, "Sie Dürfen Nicht Sterben! - Bienen in Not", *Exakt - Die Story* (Mitteldeutscher Rundfunk, 2015), http://www.mdr.de/exakt/die-story/video290090.html.

Stakeholders

The two main beekeeping organizations in Germany are the German Beekeepers' Association (DIB) and the German Professional Beekeepers' Association (DBIB). Ninety per cent of all German beekeepers are organized within DIB, Germany's amateur beekeeper's association. DBIB represents Germany's professional beekeepers, i.e. beekeepers who rely on the income from beekeeping for their livelihood. It is important to explore their respective roles in the following sections.

Over the last few years, nationwide initiatives have been developed that are fundamental in raising awareness of the work of beekeepers. We will discuss the initiative "Deutschland summt" as the most recent example.

DIB

Germany's Beekeepers' Association (DIB) is the largest national association of beekeepers within Europe. Founded over a hundred years ago,[21] it counted 97,400 members at the end of 2014.

The association has a federal structure made up by 19 regional associations (roughly matching the 16 German federal states). The regional associations are made up of more than 3,000 local chapters; each one is organized as an association (e.V., eingetragener Verein) and forms a link between the national association and the beekeepers.[22]

The advantage of this structure is that while the small local associations enable information exchange on specific topics relevant to the local beekeeper community, the power of all beekeepers combined supports the national association of beekeepers to target large-scale problems (such as bee health), develop best practice recommendations and interact with multiple national and international stakeholders.

In terms of funding the DIB unfortunately lacks transparency and disclosure; there is almost no publicly accessible information available on funding structure, sources of funding, overall income and outgoings. Having said that, the DIB discloses the annual budget, as well as the detailed income and outgoings, to the regional chapters. On request, the DIB press office stated that the four sources of income are membership fees, licensing income from the DIB honey jars and labels, sale of information material and the *Apidologie* journal, and the advertisement

21 Petra Friedrich, "Honigbienen Nutzen Mensch Und Natur", 2015, http://www.bauernverband.de/bienen.

22 Deutscher Imkerbund e.V., "Jahresbericht 2013/2014 Des Deutschen Imkerbundes E. V.", vol. 2013/2014 (Wachtberg: Deutscher Imkerbund e.V., 2014), 5.

membership fee (to be spent on advertisement only).[23] The membership fee is €3.58 for each member, in addition to a "PR contribution fee" of €0.26 per hive. Based on these fees alone the annual income of DIB exceeds €500,000.[24]

In Germany, approximately 90% of all beekeepers are members of the national beekeepers' association and one of its associated local chapters.[25] Due to the high level of membership, reliable data is available, including but not limited to the overall number of German beekeepers, demographic characteristics of German beekeepers, and the number of colonies. From a 2014 report, the average DIB beekeeper keeps bees as a hobby, has 6.9 colonies, is male and 56.6 years old.[26]

It is crucial to elaborate further on these statistics, as they are key in order to better understand the situation of beekeeping in Germany. Just 1% of all German beekeepers keep bees for a living which means that the vast majority of beekeepers see beekeeping as a hobby. As a consequence, 96% of the beekeepers have 20 or fewer colonies.

While the number of beekeepers with 150+ colonies is slowly increasing, the number of beekeepers with 21–150 colonies is decreasing.[27] This shows a very early and slow start of professionalization in German beekeeping, and, at the same time, the vanishing of the mid-sized, semi-professional beekeepers.

According to Figure 15.1, DIB had 97,524 members by the end of 2014. Meanwhile, the differences between the regional associations are striking: for example, while Bavaria is home to more than 25,000 beekeepers, less than 2,300 members are based in Thuringia.

23 Email from DIB press office 17 August 2015.

24 Deutscher Imkerbund e.V., "Werbefonds 2013" (http://deutscherimkerbund.de/download/0-340: Deutscher Imkerbund e.V., 2013).

25 Deutscher Imkerbund e.V., "Imkerei in Deutschland", 2015, http://deutscherimkerbund.de/160-Die_deutsche_Imkerei_auf_einen_Blick.

26 Deutscher Imkerbund e.V., "Jahresbericht 2013/2014 Des Deutschen Imkerbundes E. V.", 13–14.

27 *Ibid.*, 13.

Figure 15.1 **Number of DIB beekeepers by federal state**

Source: Raw data provided by the DIB press office and DESTATIS

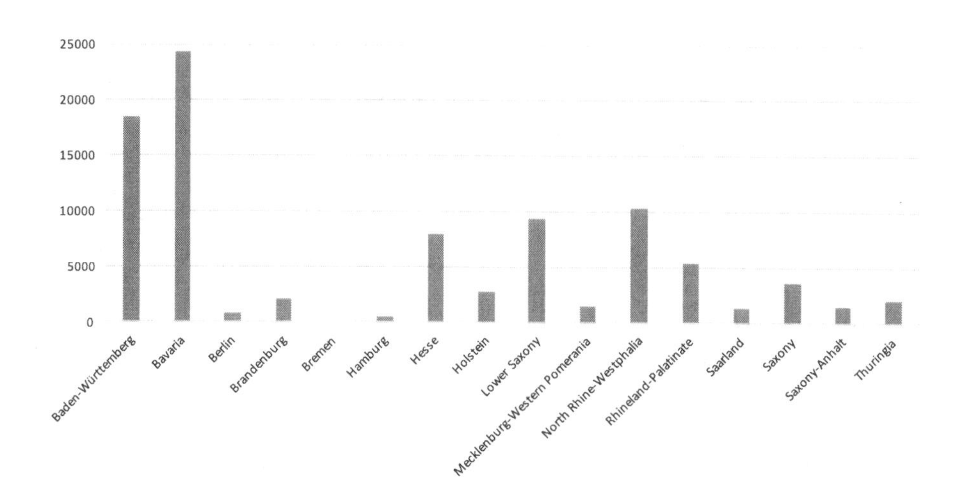

In an international context, the decline in the number of beekeepers has been mentioned as one of the key problems leading to the decline of bee populations.

For the younger generation, beekeeping is no longer perceived as an attractive hobby, which means that many beehives have disappeared along with the older generations of beekeepers. This is clearly a problem which can be attributed to the aforementioned fact that over 90% of German beekeepers keep bees as a hobby.

The situation in Germany is displayed in Figure 15.2: whereas a decrease in the number of beekeepers can be observed from 1951 to 2006,[28] the number has increased ever since.

28 The sharp increase in 1990 is due to the German reunification and change in statistical methods.

Figure 15.2 **Number of DIB beekeepers**

Source: Raw data provided by the DIB press office

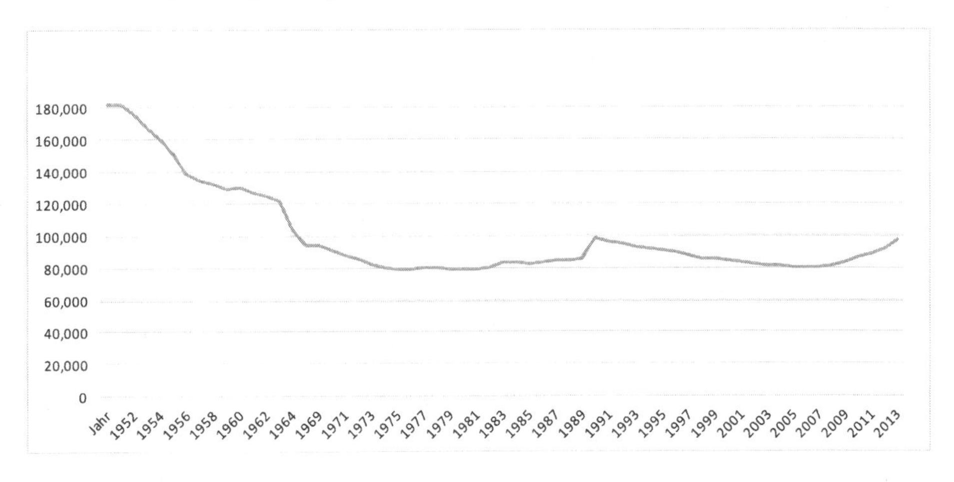

It is important to put this positive development into context. The aforementioned decline in semi-professional beekeeping (21–150 colonies) and the consequential reduction in colonies per beekeeper has led to an almost continuous decline in the total number of bee colonies in Germany since 1951 (see Fig. 15.3(a) and (b)). Data shows that in 1991 there were 3.2 colonies to pollinate each km^2; by 2013 this number had fallen to just 1.96 colonies per km^2.[29]

Hence, even though the number of beekeepers has continuously increased since 2006, the total number of colonies and therefore the number of bees in Germany is decreasing.

29 *Ibid.*, 13.

Figure 15.3 **(a) Average number of beehives per DIB beekeeper; (b) total number of beehives**

Source: Raw data provided by the DIB press office

(a)

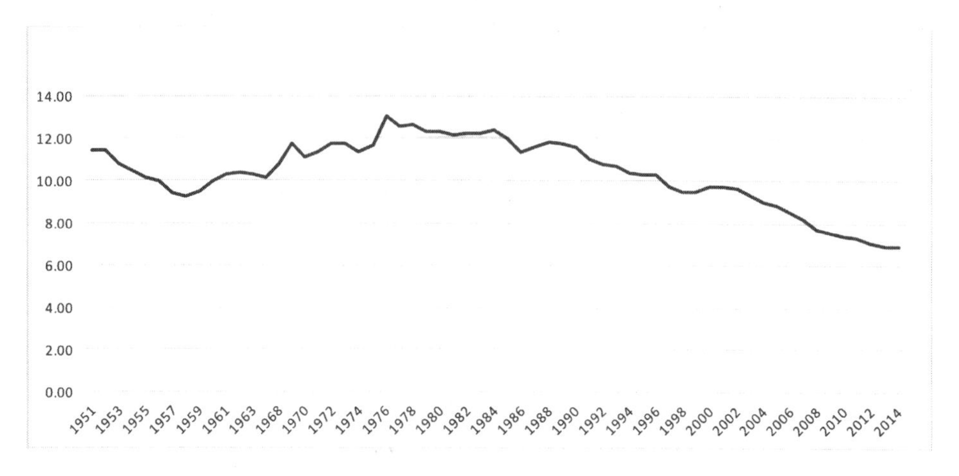

(b)

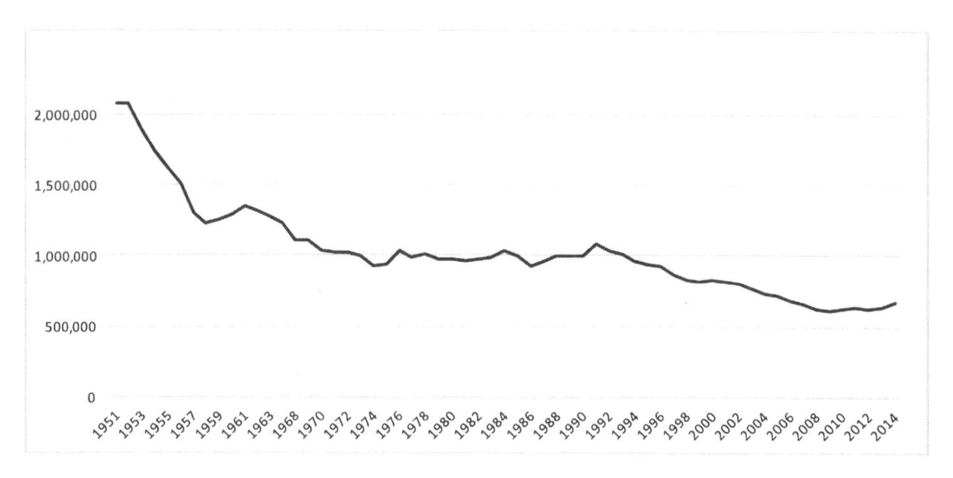

The possible effect of this trend for pollination in Germany becomes clear when looking at the individual federal states, because the level of agriculture and therefore the required pollination varies significantly among the states. Figure 15.4 shows the number of DIB bee colonies per square kilometre of agricultural land.[30]

30 Figure 15.4 does not include the German city states Berlin, Bremen and Hamburg. These city states are treated as outliers, because only a very small amount of land is used for agriculture.

Figure 15.4 **Number of DIB colonies per square kilometre of agricultural land**

Source: Raw data provided by the DIB press office and DESTATIS

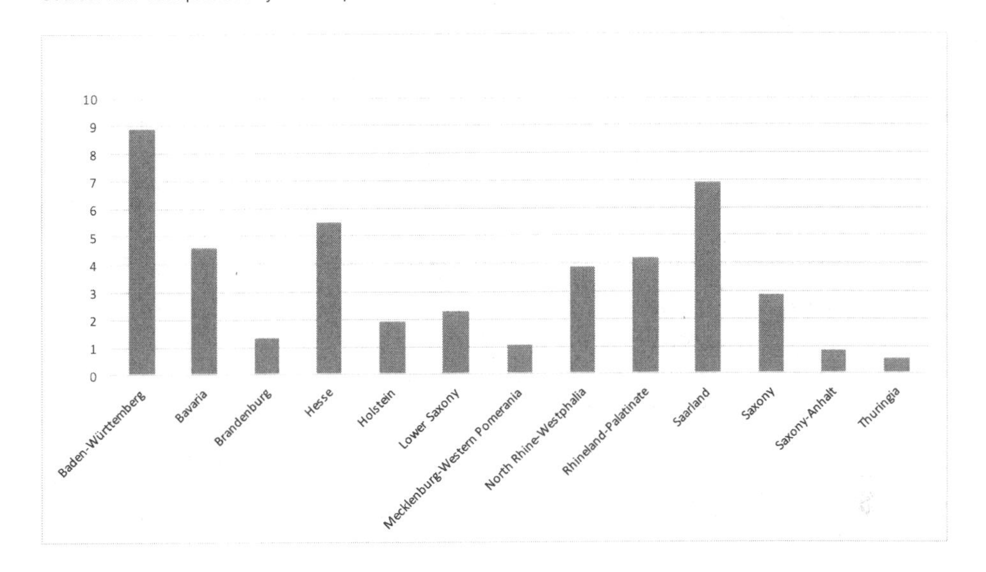

In order to fight this trend, the beekeeper associations are trying to make beekeeping more attractive. Many of the local chapters welcome new beekeepers by providing them with a free "colony starter set". In addition, DIB and its regional and local chapters are organizing free events and seminars all over Germany to raise awareness of the importance of bees and beekeeping.

DBIB

In order to become an ordinary member of the German Association of Professional Beekeepers (Deutsche Berufs- und Erwerbsimkerbund e.V., DBIB) it is mandatory to keep bees for a living.[31] While DBIB only consists of approximately 800 members, and therefore only represents a small fraction of beekeepers in Germany, these professional beekeepers own around 50% of all bee colonies in Germany, according to DBIB's president Manfred Hederer. As DBIB's members rely on bees for their livelihood rather than a hobby, they observe changes to their bees carefully and analytically. Their approach becomes apparent when analysing DBIB's position regarding bee health.

While there is no (annual) report available that reveals DBIB's budget, funding sources, membership statistics, and so on, DBIB's vision, mission and position in the bee health debate is communicated through press releases and publications.

31 Deutscher Berufs und Erwerbsimkerbund e.V., "Satzung" (DBIB e.V., 2013), http://www.berufsimker.de/index.php/satzung.

DBIB is mainly represented by its president, Manfred Hederer, and its vice-president, Walter Haefeker, who is also President of the European Professional Beekeepers' Association (EPBA).

During each year, DBIB organizes events to enable (professional) beekeepers to expand their knowledge and expertise. The main event each year, which draws international attention, is the international bee fair in Donaueschingen.[32]

"Deutschland summt" ("Germany hums")

The initiative "Deutschland summt" was started in 2010 by Dr Corinna Hölzer and Cornelis Hemmer, as part of their work for the Stiftung Mensch und Umwelt (Foundation for Humans and the Environment).[33] The initiative was started after securing funding from the Bundeskulturstiftung (Federal Culture Foundation) and Haus der Kulturen (House of Cultures), and gained strong momentum in 2013 when the German President's partner became its patron.[34]

The mission's initiative is to raise public awareness of the decline in bees. In order to achieve this aim, the initiative employs a variety of strategies, inter alia, education of adults and kids, national media campaigns and visible public campaigns in major German cities.

In order to educate adults, a book has been published that describes how "ordinary people" can help bees by deploying simple measures. In terms of education for children, two educational kits have been developed for different ages. The aim is to teach children in nurseries and kindergarten about the importance of bees and about the role they play in their daily lives.[35]

To support outreach and awareness activities an educational campaign on national television and online media (ProSieben) as well as on social media has been established to promote the initiative.[36]

In addition to these centralized efforts, there is also a decentralized part of the initiative: sub-initiatives started in eight cities across Germany, including "Berlin

32 Deutscher Berufs und Erwerbsimkerbund e.V., "Süddeutsche Berufsimkertage in Donaueschingen" (Deutscher Berufs und Erwerbsimkerbund e.V., 2015), http://www.berufsimker.de/index.php/berufsimkertage/donaueschingen-sueddeutsche-berufsimkertage/veranstaltung.

33 In their day jobs both work for a "green" media agency (Green Media Net) founded by Dr Corinna Hälzer.

34 Stiftung für Mensch und Umwelt, "Die Initiative—Deutschland Summt", 2015, http://www.deutschland-summt.de/die-initiative-deutschland.html; Stiftung für Mensch und Umwelt, "Daniela Schadt - Schirmherrin von 'Deutschland Summt'" 2015, http://www.deutschland-summt.de/unsere-schirmherrin.html.

35 Stiftung für Mensch und Umwelt, "Unser Ratgeber 'Wir Tun Was Für Bienen - Bienengarten, Insektenhotel Und Stadtimkerei,'" 2015, http://www.deutschland-summt.de/unsere-schirmherrin.html; Stiftung für Mensch und Umwelt, "Umweltbildung Leicht Gemacht", 2015, http://www.bienenkoffer.de.

36 ProSiebenSat.1 Digital GmbH, "Deutschland Summt", 2015, http://www.prosieben.de/tv/greenseven/deutschland-summt.

summt", 'Frankfurt summt" and "München summt". In order to raise awareness, beehives are being placed visibly on landmarks, supported by campaigns in the respective cities.[37]

The initiative is continuously growing, especially through decentralized activities. Unfortunately it is not clear how this growth is funded. The initiative's funding/ financial statement is very vague, mentioning that the initiative and its different projects are funded by a mix of foundations, private donors and sponsors. Given the fact that the initiative is involved in educational activities, more transparency and disclosure would be desirable.[38]

When analysing the aforementioned data it becomes apparent that changes in beekeeping have led to a decline in colonies and therefore in the number of bees. This seems to be mainly due to a reduction in the number of colonies per beekeeper.

The advanced average age of German beekeepers also poses a problem for the future: if the hives are not passed on to younger beekeepers then the colonies will die with their owners. Looking at the demographics this could lead to a severe decline of bee colonies in Germany in the very near future.

As a result of this problem the educational roles especially of DBIB and the initiative "Deutschland summt" are essential and a step in the right direction. "Deutschland summt" was able to transform an environmental niche topic into a matter of national interest. The positive consequences of this effort for the future of bees in Germany cannot be overstated.

DIB, DBIB and "Deutschland summt" are involved in educational, public outreach and political activities; hence we would like to see a higher degree of transparency and public disclosure when it comes to income/revenue streams and funding sources.

Bee decline in Germany: bee health

The health of bee colonies can be affected by a variety of factors. The potential impact of pesticides has been debated widely in European industry and academia, especially since the EFSA supported a ban of neonicotinoid-based pesticides by

37 Stiftung für Mensch und Umwelt, "Partnerstädte", 2015, http://www.deutschland-summt.de/startseite.html.

38 Stiftung für Mensch und Umwelt, "Wer Sind Die Förderer von 'Deutschland Summt!'?", 2015, http://www.deutschland-summt.de/faq-leser/items/wer-sind-die-foerderer-von-deutschland-summt.html.

the European Commission in 2013.[39] Two chemical companies and producers of pesticides, Bayer AG and Syngenta, have since sought to overturn this ban.[40]

However, various international environmental or consumer-led campaigns continue to fight for the ban and raise awareness of bee health issues linked to (neonicotinoid-based) pesticides.[41] In the following section we aim to assess how key stakeholder groups in Germany assess and respond to these bee health concerns. We will also discuss a range of other factors that can influence bees' life and their natural habitat in Germany. Following the same format as in the previous section, we start with an introduction of the main stakeholders in relation to bee health before assessing their impact.

Stakeholders

The situation among the group of key stakeholders in Germany is complex and without consensus when it comes to bee health.

In the following sections we discuss the role of amateur and professional beekeeper associations, farmers, producers of pesticides, research institutions and NGOs. We will also introduce two key German bee health initiatives as well as the "bee-friendly certified" label supported by DBIB.

DIB

Beekeepers are the main stakeholders when it comes to the health of beehives. As described above, beekeepers are privately organized and often members of Germany's DIB (German Beekeepers' Association). DIB publishes an annual report, which summarizes inter alia its actions and general position on bee health:

Regarding **pesticides**, DIB mentions that it sees pesticides as a risk for bee health and that it welcomes the EU's ban of certain neonicotinoid-based pesticides. DIB asks for additional research to be carried out during the time of the ban, in order to encourage well-informed decision-making about whether or not to extend the ban.

39 EFSA, "EFSA Identifiziert Risiken Durch Neonicotinoide Für Bienen", 2013, http://www.efsa.europa.eu/de/press/news/130116.

40 Einecke, "Bayer Klagt Gegen Pestizid-Verbot Nach Bienensterben."

41 Greenpeace International, "NGOs and Beekeepers Take Legal Action to Defend EU Ban of Bee-Killing Pesticides against Syngenta and Bayer"; 38 Degrees, "Keep the Ban on Bee-Killing Pesticides", 2015, https://speakout.38degrees.org.uk/campaigns/ban-bee-killing-pesticides-for-good-937d4563-7694-41a8-a642-65e6b0e51453; AVAAZ, "3 Million to Save the Bees", 2013, http://www.avaaz.org/en/save_the_bees_global/; Buglife, "Pesticide Approval Strikes Blow for Bees", 2015, https://www.buglife.org.uk/news-%26-events/news/pesticide-approval-strikes-blow-for-bees; Friends of the Earth, "Let's Help Britain Bloom for Bees", *The Bee Cause*, https://www.foe.co.uk/what_we_do/the_bee_cause_home_map_39371.

In addition, it states that it has engaged on the national as well as on the European level in order to guarantee stricter authorization procedures for pesticides.[42]

When it comes to detailed information about bee health preservation activities, DIB sources generally refer to the FitBee and DeBiMo (German Bee Monitoring) projects. Both are research projects that aim to monitor and analyse a range of factors that might have an impact on bees' livelihoods and habitat in Germany and factors that specifically impact bee health (please see below for more information).

In addition to generic statements, DIB's annual report includes a short report on bee health by the chairman of the academic advisory group, Dr von der Ohe.[43] Even though the topic is being extensively discussed and assessed in the research community, Dr von der Ohe's report does not highlight the status quo of the international bee health research discourse. The fact that bee viruses might impact bee health is just mentioned once and very briefly in Dr von der Ohe's report.[44]

The most prominent topic in the academic advisory council's section of the DIB report is clearly the link between varroa mites and winter losses. The main focus is on avoiding and fighting a varroa infection.

In a recent public TV documentary about bee decline, German beekeepers and the president of the European Professional Beekeepers' Association demanded more independent research into bee health and decline and drew a clear link between pesticides and bee decline in Germany.[45]

DBIB "Certified Bee Friendly" label

Since 2011, DBIB and its European parent organization, the European Professional Beekeepers' Association (EPBA), have worked with a broad range of international food, textile and energy producers in promoting a "certified bee friendly" label for their products. The label was introduced at the 42nd International Apicultural Congress "Apimondia" in Buenos Aires and promotes pollinator-friendly systems. It assesses 27 criteria with a focus on the traceability of crop rotations, use of pesticides, direct or indirect use of genetically modified organisms (GMOs), areas of biodiversity conservation, mortality rates of pollinators during harvesting

42 Deutscher Imkerbund e.V., "Jahresbericht 2013/2014 Des Deutschen Imkerbundes e.V.", 4–5, 8, 18, 25, 30–31.

43 Dr von der Ohe works for the Bee Institute Celle, which forms part of the FITBee and DeBiMo projects. Please see the relevant sections below for additional information about these projects, including inter alia information regarding funding structure and project set up.

44 *Ibid.*, 42.

45 Baumbach, "Sie Dürfen Nicht Sterben! - Bienen in Not"; MDR, "Interview Zur Sendung: Walter Haefeker", *Exakt - Die Story* (Mitteldeutscher Rundfunk, 2015), http://www.mdr.de/exakt/die-story/video209232.html.

operations, partnership initiatives with beekeepers' associations and support for research projects.[46]

One of the first "bee-friendly" consumer products, "Sternenfair milk", was developed in cooperation with Milchvermarktungs (MVS) GmbH in Bavaria. Since 2013, a range of German supermarkets have stocked this bee-friendly milk whose producers are committed to abandoning genetic engineering and pesticides as well as other controversial environmental practices.[47]

The participating farmers are willing to accept these additional requirements in order to obtain a competitive advantage and ultimately more customers among beekeepers and their clients.[48]

DBV, Association of German Farmers

Germany's farmers are another important stakeholder group in this debate. The farmers are represented by the Association of German Farmers (Deutscher Bauernverband, DBV), which consists of 18 regional chapters, which, in turn, are divided into c. 300 district chapters. In total, DBV represents more than 300,000 German farmers.[49]

DBV states that bees are an important factor in the pollination process and therefore essential for common agricultural practices. In addition, DBV highlights the economic importance of honey bees in agriculture: they estimate the value of bee pollination in Germany to be €2 billion.[50]

Having identified the essential role of bees in agriculture, DBV also emphasises the importance of bees' health and cooperates with DIB to address concerns about bee health.

This cooperation has led to a better mutual understanding of the concerns and challenges of beekeepers and farmers. However, a central cooperation strategy at national level is still missing. Instead, DBV emphasises that farmers and beekeepers cooperate locally on a voluntary basis.[51]

When it comes to specific details on bee health, DBV follows the "varroa narrative", stating that bee colony losses are caused by varroa mites. However, at the

46 Bee Friendly Association, "Bee Friendly Specifications Synthesis", http://www.certifiedbeefriendly.org/wp-content/uploads/2014/09/BEE-FRIENDLY_Specifications_Synthesis.pdf, 1-2.

47 Sternenfair, "Sternenfair: Die 5-Sterne-Milch", 2015, http://www.sternenfair.de.

48 Walter Haefeker, "Bienenfreundlich Hergestellt— 'Certified Bee Friendly'-Siegel, Weltweit", 2014, http://ev.mellifera.de/fix/doc/Bienenfreundlich Schreiben 2014.pdf, 1.

49 Deutscher Bauernverband, "Die Deutschen Bauern: Veränderung Gestalten, Geschäftsbericht Des Deutschen Bauernverbandes 2014/2015", vol. 2014/2015 (Berlin: Deutscher Bauernverband, 2014), 60.

50 Friedrich, "Honigbienen Nutzen Mensch Und Natur."

51 Deutscher Bauernverband, "Faktencheck Landwirtschaft 2014" (Berlin: Deutscher Bauernverband, 2014), http://www.bauernverband.de/faktencheck, 2.

same time the association of farmers also encourages further research into the effects of pesticides on bees.[52]

Chemical producers

Bayer AG and Syngenta are two of the main producers of pesticides, in general, and of neonicotinoid-based pesticides, in particular, worldwide.

Bayer AG and Syngenta are currently engaged in separate legal actions against the EU over the 2013 ban of certain neonicotinoid-based pesticides.[53]

Analysing the latest annual company reports, sustainability reports and/or integrated reports of 30 companies listed on the German stock exchange, DAX, we found that of all these companies only Bayer AG explicitly refers to the situation of pollinators and bees (see Appendix 15.1 for more details) in its reports.

oekom research's Controversies Monitor reports that Bayer AG, Syngenta and BASF SE have all been involved in key "sustainable chemicals management" controversies over the last 12–18 months that have allegedly caused the decline of bee populations (see "oekom Controversies Monitor Cases, 2014–15", Appendix 15.2 screenshots). The controversy has been analysed and included in oekom's Corporate Ratings[54] for these companies.

oekom's Corporate Ratings Report 2015 cites the following key areas of concern regarding Bayer AG's "weaknesses" in its "product and substance risk management and assessment" ("oekom Corporate Rating Report, Bayer"; see Appendix 15.2), "strategy to reduce and substitute substances of concern" ("oekom Corporate Rating Report, Bayer"; see Appendix 15.2) and regarding its involvement in the bee health controversy: "B.2.2.5. Major environmental controversies, fines or settlements relating to sustainable chemicals management" ("oekom Corporate Rating Report, Bayer"; see Appendix 15.2).

In its publications, Bayer AG frequently refers to "studies" to underpin the credibility of the company's research, products, corporate strategy and position.

On request, Bayer Bee Care provided the authors with nine publications that reflect their position and views; 56% of the publications provided pre-date 2013 and therefore do not include recent findings regarding the impact of neonicotinoid-based pesticides on the physiology and metabolism of bees. Furthermore, 33% of the studies are co-authored by Bayer AG employees.[55]

52 Deutscher Bauernverband, "Geschäftsbericht Des Deutschen Bauernverbandes 2013/2014", vol. 2013/2014 (Berlin: Deutscher Bauernverband, 2013), 76; Deutscher Bauernverband, "Faktencheck Landwirtschaft 2014" (Berlin: Deutscher Bauernverband, 2014), http://www.bauernverband.de/faktencheck, 2.

53 Einecke, "Bayer Klagt Gegen Pestizid-Verbot Nach Bienensterben".

54 oekom research, Corporate Ratings 2015 (Scale = A+ / Best to D- / Worst) for Bayer AG = C+ (NOT PRIME), Syngenta = C- (NOT PRIME), BASF SE = B- (PRIME); PRIME sector threshold for chemicals = B-) (oekom, "oekom corporate rating", accessed 17 March 2016, http://www.oekom-research.com/index_en.php?content=corporate-rating).

55 Email from Bayer Bee Care with key studies from 9June 2015.

The key message concerning bee health in the Bayer AG publications that have been analysed—reports, websites, brochures—is the following: "[B]ee health is correlated with the presence of the varroa mites, viruses and many other factors, but not with the use of insecticides".[56]

DeBiMo

The German "DeBiMo" ("German bee monitoring") project is the largest bee-monitoring project in the world. The project was started and coordinated by a project committee consisting of members from several bee institutes,[57] including DIB, DBIB (Association of Professional German Beekeepers), DBV, BMVEL (German Ministry of Consumer Protection, Nutrition and Agriculture), BASF SE, Bayer CropScience AG, Bayer HealthCare AG and Syngenta.[58]

The project includes over 100 beekeepers from different parts of Germany. Each beekeeper designates ten colonies to be included in the project giving a total of more than 1,000 colonies.

The DeBiMo project is now solely funded by the German Ministry of Agriculture and Nutrition.[59] However, the project design, set-up and the first stage set-up (pre-2010) were funded (up to 50%) by Germany's IVA (Industrieverband Agrar) consisting inter alia of BASF SE, Bayer CropScience AG, Bayer HealthCare AG and Syngenta.[60]

The declared aim of DeBiMo is primarily to monitor the status of bee health. However, based on the data provided several articles have been published, which try to explain causal relationships. In the light of current discussions they comment inter alia on the effects of pesticides on bee health or the causes of bee colony losses.[61]

56 Bayer AG, "Neonicotinoids", accessed 3 July 2015, http://beecare.bayer.com/home/what-to-know/pesticides/neonicotinoids.

57 In their article Hoppe and Safer mention nine bee institutes. However, at this point in time the DeBiMo website only lists seven: Bayerische Landesanstalt für Weinbau und Gartenbau; Fachzentrum Bienen, Dienstleistungszentrum ländlicher Raum Westerwald-Osteifel; Fachzentrum Bienen und Imkerei, Länderinstitut für Bienenkunde Hohen Neuendorf e.V.; Landesanstalt für Bienenkunde, Universität Hohenheim; Landesbetrieb Landwirtschaft Hessen, Bieneninstitut Kirchhain; LAVES Institut für Bienenkunde; Nationales Referenzlabor für Bienenkrankheiten, Friedrich-Löffler-Institut.

58 Peter P. Hoppe and Anton Safer, "Das Deutsche Bienenmonitoring-Projekt: Anspruch Und Wirklichkeit - Eine Kritische Bewertung", 2011, http://www.bund.net/fileadmin/bundnet/pdfs/chemie/20110125_chemie_bienenmonitoring_studie.pdf, 1; Peter Rosenkranz, "Erwiderung Zur Presseerklärung Des NABU Vom 25.01.2011", 2011, http://www.ag-bienenforschung.de/debimo_kritik.html.

59 *Ibid.*

60 Hoppe and Safer, "Das Deutsche Bienenmonitoring-Projekt", 1, 11–13.

61 See, for example, DeBiMo, "Schlussbericht, Deutsches Bienenmonitoring - 'DeBiMo'" (Stuttgart: DeBiMo, 2013), https://www.uni-hohenheim.de/fileadmin/einrichtungen/bienenmonitoring/Dokumente/DEBIMO-Bericht-2011-2013.pdf; Werner von der

DeBiMo's position on causes of bee colony losses follows the varroa narrative, mentioning repeatedly infections with varroa mites as a key cause.

FitBee

The "FitBee" project is a cooperative effort created to analyse the synergy effects between a single bee and its colony, taking into account bee health and environmental influences.[62] The project consists of seven bee institutes[63] and seven industry partners, including inter alia Bayer and Syngenta.[64]

The project's mission has three stages: in stage one the project aims to analyse synergy effects between a single bee, its colony, bee health and environmental factors. Based on these conditions, the factors for "healthy bees" are established in stage two. Finally, in stage three the project aims to improve the conditions for bees through targeted measures.[65]

In order to achieve the aforementioned aims seven modules were set up, analysing different aspects of "bee health". Module five has received particular attention as the DeBiMo report states that it will conduct a multi-factor analysis including the effects of sub-lethal doses of pesticides, bee viruses and varroa mite infections.[66]

A recent publication based on the FitBee project suggests that the neonicotinoid-based pesticide thiacloprid has no negative effect on bees.[67] Due to the modular structure of this project it is impossible to find a dominant narrative. Unfortunately, so far no publications can be found based on module 5 which could be of great interest for all stakeholders.[68]

Ohe and Dieter Martens, "Das Deutsche Bienenmonitoring, Pflanzenschutzmittel-Rückstände Im Bienenbrot", *ADIZ* 10/2011 (2011): 8–9; Annette Schroeder, "Das Deutsche Bienenmonitoring", *Rheinische Bienenzeitung* 21 (2012).

62 FIT BEE, "Was Ist 'FIT BEE'?", accessed 13 August 2015, http://fitbee.net.

63 Bayerische Landesanstalt für Weinbau und Gartenbau, Fachzentrum Bienen; Dienstleistungszentrum ländlicher Raum Westerwald-Osteifel; Institut für Bienenkunde Oberursel; Institut für Biologie, Martin-Luther-Universität Halle-Wittenberg; Landesanstalt für Bienenkunde, Universität Hohenheim; Landesbetrieb Landwirtschaft Hessen, Bieneninstitut Kirchhain; LAVES Institut für Bienenkunde Celle.

64 Bayer CropScience AG, BioSolutions Halle GmbH, Interactive Network Communications GmbH, IP SYSCON GmbH, IS Insect Services GmbH, Lechler GmbH, Syngenta Agro GmbH.

65 FIT BEE, "Ziele", accessed 13 August 2015, http://fitbee.net/ziele.

66 DeBiMo, "Schlussbericht, Deutsches Bienenmonitoring - 'DeBiMo'", 17.

67 Lena Faust and Reinhold Siede, "Insektizide Im Raps", *Deutsches Bienen Journal* 5/2014 (2014): 6–7.

68 The last update that can be found on the FitBee website regarding module 5 dates back to 2012.

Discussion

The variety and diversity of the stakeholders involved in the bee health debate leads to a high level of complexity. This complexity is only partly caused by the scientific complexity of the underlying research questions. A large part of the complexity seems to be caused by extra-scientific incentives, scientific cross-links and economic interests of some stakeholders that we have introduced in the section above.

Varroa mite and viruses

The discussion about bee colony losses in Germany is dominated by the narrative that infections with varroa mites lead to the death of bee colonies.

In the recent discussion this message is backed up by early research results from DeBiMo: Around 2011, DeBiMo publications state that the varroa mite infection was 3.6 per 100 bees in the case of surviving colonies and 15.8 per 100 bees in the case of colonies that did not survive. Based on these findings the publication concludes that fighting varroa mite infections is the key to fighting winter losses.[69] The logic of this line of argumentation can be seen in Figure 15.5.

Stage 1 (Early publications)

Figure 15.5 **Correlation and causality between varroa mites and bee colony losses**

More recent reports acknowledge the existence of multiple factors which might contribute to the losses of bee colonies. It is stated that in colonies that did not survive the winter, the deformed wing virus and the acute bee paralysis virus have been found, in addition to varroa mites. In the report the role of varroa mites as a carrier of these viruses is emphasized, concluding that fighting the varroa mite will prevent winter losses.[70] The logic of this line of argumentation can be seen in Figure 15.6(a) and (b).

69 Annette Schroeder, "Struktur Und Ergebnisse Des DeBiMo", *ADIZ* 9/2011 (2011): 8–9.
70 Schroeder, "Das Deutsche Bienenmonitoring", 28–29.

Stage 2 (Recent publications)

Figure 15.6 **(a) Correlation and causality between varroa mites, viruses and bee colony losses; (b) varroa mites, viruses and bee colony losses and the "no varroa mites, no bee colony losses" conclusion**

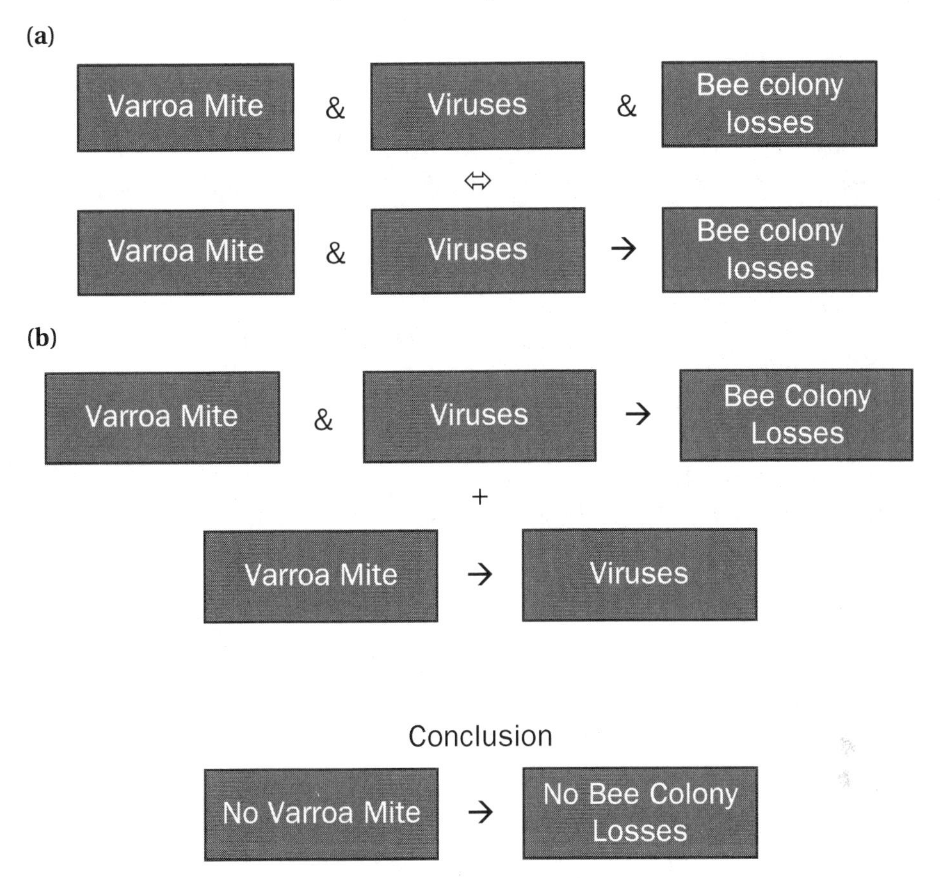

The main problem with both interpretations of DeBiMo's research results is the existence of a "cum hoc, ergo propter hoc" (false cause) fallacy; a correlation between A (varroa mite) and B (colony winter loss) can be observed. This correlation between the research parameters is interpreted as causality.[71] DeBiMo's and some of the aforementioned Bayer AG publications completely neglect the fact that additional known or unknown factors could be the cause of the bee colony losses or could act as a catalyst.[72]

71 Hoppe and Safer, "Das Deutsche Bienenmonitoring-Projekt", 5–11.

72 Le Conte, Ellis, and Ritter, "Varroa Mites and Honey Bee Health: Can Varroa Explain Part of the Colony Losses?"; D J T Sumpter and S J Martin, "The Dynamics of Virus Epidemics in Varroa-Infested Honey Bee Colonies", *Journal of Animal Ecology* 73, no. 1 (2004): 51–63, doi: 10.1111/j.1365-2656.2004.00776.x.

The simplistic mono-causal interpretation of the correlation between research parameters forms the basis of several publications (e.g. DeBiMo reports and articles) and of a vast amount of public information material (e.g. print and online publications). Through the introduction of an "easy-to-understand" causal link, a simplistic narrative has been created to explain the bee colony (winter) losses.

A key weakness of this "varroa narrative" is that it does not explain the increasing summer losses. DeBiMo admits in their report that summer losses have not been analysed robustly. In response to a critical article it is stated that summer losses were reported. However, at the same time it is stated that this assessment is not part of the project's aim and design.[73] The latest interim report as well as the final report mention singular varroa related losses during summer, but do not seem to analyse summer losses systematically.[74]

Pesticides[75]

In 2013, the EU Commission voted for a ban of several neonicotinoid-based pesticides following the recommendation of the European Food and Safety Agency (EFSA).[76] This decision sparked controversy between pesticide producers and a variety of "pro-ban" stakeholders (e.g. beekeepers, researchers). As a consequence of the ban Bayer, among others, decided to sue the European Commission with the declared aim for the ban to be completely revoked.[77]

The debate regarding the impact of pesticides on bee health in general and neonicotinoid-based pesticides in particular is still in full swing. Although a full review is beyond the scope of this publication, we would like to highlight the different positions in the debate. As the current debate is primarily about the recent ban of neonicotinoid-based pesticides, we will focus on the effect of these types of pesticides.

73 Rosenkranz, "Erwiderung Zur Presseerklärung Des NABU Vom 25.01.2011".

74 DeBiMo, "Schlussbericht, Deutsches Bienenmonitoring - 'DeBiMo'", 17; DeBiMo, "Zwischenbericht 2014, Deutsches Bienenmonitoring - 'DeBiMo'" (Stuttgart: DeBiMo, 2014), https://www.uni-hohenheim.de/fileadmin/einrichtungen/bienenmonitoring/Dokumente/DEBIMO-Bericht-2011-2013.pdf.

75 We do acknowledge the differences between pesticides, herbicides and fungicides. However, in this chapter, we use pesticides as an umbrella term to include substances from all three groups.

76 EFSA, "Towards an Integrated Environmental Risk Assessment of Multiple Stressors on Bees: Review of Research Projects in Europe, Knowledge Gaps and Recommendations", 3594.

77 Deutscher Imkerbund e.V., "Jahresbericht 2013/2014 Des Deutschen Imkerbundes e.V.", 31.

Essentially, there are two sides: on one side, the varroa narrative argues that the colony losses can be stopped by fighting varroa and that there are no links to pesticides. These views are supported inter alia by pesticide producers such as Bayer and Syngenta, but also by initiatives such as DeBiMo and FitBee.[78]

The other side argues that neonicotinoid-based pesticides form one factor in a multi-causal effect chain, which cannot be identified through mere observation. This side consists inter alia of (independent) German research institutions and professional beekeepers (DBIB).

In the current public debate in Germany the varroa narrative argument is often based on the DeBiMo and FitBee findings. As described above DeBiMo as well as FitBee analyse samples from roughly 1,200 beehives from 120 beekeepers selected by the partnering bee institutes. The samples are taken from bees, beebread,[79] honey and honey dome.[80] The pesticide analyses are then carried out by the LUFA institute in Speyer.

The results reported by DeBiMo show that in most cases only traces of pesticides could be found in the samples. However, in 2014 the researchers conducted tests for 401 substances and found a total of 77,[81] with a maximum of 28 substances in one sample and an average across all samples of approximately five substances. While the researchers acknowledge that the mixture of many different substances will cause a debate about interaction effects and sub-lethal effects, they conclude that based on their findings a link between the substances found and bee colony winter losses does not exist. The main reason for bee colony winter losses is infection with varroa mites.[82]

The focus of DeBiMo lies on samples drawn directly from the hive. While this is a very useful way of determining inter alia the rate of varroa or virus infection, it does not take recent findings concerning pesticides into consideration.

Inter alia, Krupke *et al.*[83] show that neonicotinoid-based pesticides can be found not only on the treated plant itself, but also in the soil and neighbouring plants. Taking samples from beehives does not do justice to the complexity of the chain of effects: bees are not only exposed to pesticides when they are on the blossoms, which is what is analysed in the samples taken by DeBiMo, but also inter alia

78 This is based on the FitBee module 5 publications which were available when this chapter was written.

79 Beebread is the pollen that is stored in the beehive, i.e. it is a mixture of pollen and bee saliva.

80 Honey dome refers to the dome-shaped cells around the queen that are used to feed the queen.

81 65 above the limit of measurement and 11 above the limit of detection

82 DeBiMo, "Zwischenbericht 2014, Deutsches Bienenmonitoring - 'DeBiMo'", 58, 60–64.

83 Christian H. Krupke *et al.*, "Multiple Routes of Pesticide Exposure for Honey Bees Living Near Agricultural Fields", *PLoS One* 7, no. 1 (2012), doi: 10.1371/journal.pone.0029268.t001.

through contact with the soil. The concentration found in the bee products in the bee hives might give an indication, but it falls short of explaining the full exposure of bees to pesticides.

There is doubt about the mono-causal link suggested by the varroa narrative with regard to summer as well as winter losses. Henry *et al.*[84] analyse the ability of bees to find their way back to their hives. They find that this ability is impacted by neonicotinoid-based pesticides. Bees not being able to find their way back could potentially be a reason for summer losses, which were not included in the project design of DeBiMo.[85]

But even when it comes to winter losses the mono-causal varroa narrative does not seem to hold. DiPrisco *et al.*[86] find that neonicotinoid-based pesticides can act as a catalyst for bee viruses. The study analyses the effects of clothianidin on the immune system of bees. The authors conclude that there is a negative impact on the bees' immune defence, which leads to an increase in virus infections.

Based on these findings it seems that the mono-causal varroa narrative does not hold. An updated logical structure based on the findings by DeBiMo can be found in Figure 15.7.

Figure 15.7 **Correlation and causality between varroa mites, viruses, neonicotinoid-based pesticides and bee colony losses**

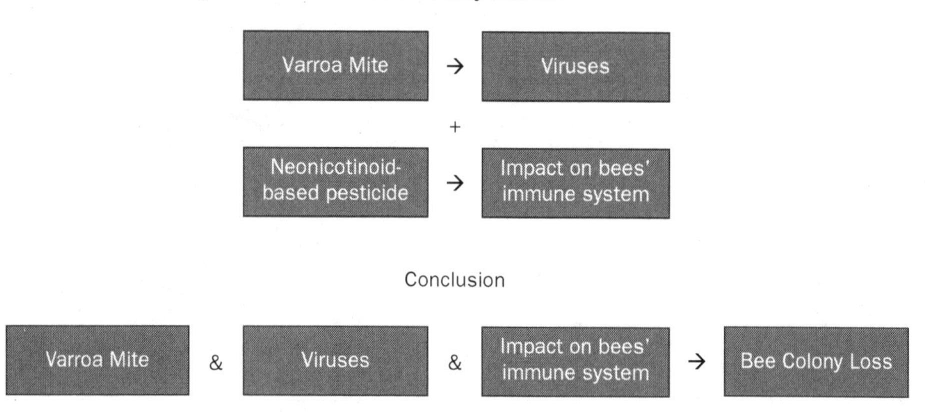

84 M Henry *et al.*, "A Common Pesticide Decreases Foraging Success and Survival in Honey Bees", *Science* 336 (6079) (2012): 348–50.

85 Hoppe and Safer, "Das Deutsche Bienenmonitoring-Projekt"; Rosenkranz, "Erwiderung Zur Presseerklärung Des NABU Vom 25.01.2011".

86 G Di Prisco *et al.*, "Neonicotinoid Clothianidin Adversely Affects Insect Immunity and Promotes Replication of a Viral Pathogen in Honey Bees", *Proc Natl Acad Sci USA* 110, no. 46 (2013): 18466–71, doi: 10.1073/pnas.1314923110.

However, this structure is still too simplistic. In order to answer the question about the role of pesticides for bee colony losses, the projects must account for physiological effects on bees which simply cannot be analysed using DeBiMo's research set-up.[87]

Flügel[88] suggests the existence of an unreasonable focus on the varroa narrative among German bee researchers. Wenzel[89] confirms this by stating: "The scientific research for detecting those sub-lethal effects of NN [neonicotinoids] has been done mostly in European countries, while German bee experts are still sticking to work on the varroa mite".

The existence of the mono-causal varroa narrative can be seen as a mystery, given the evidence which points towards multi-factor causality.

It has been observed that many (research) projects into bee health or decline, that are (co-)funded by companies such as Bayer AG and Syngenta, do confirm the varroa narrative. In an interview, the president of the European Professional Beekeeper Association, Walter Haefeker, states that the varroa narrative is used by the chemical industry to distract from the role of pesticides. In addition, he argues that chemical companies have "shaped" the German research landscape through the funding of certain projects. He sees the varroa narrative as the result of the chemical industry's well-funded PR machinery.[90,91]

87 Campact e.V., Deutscher Berufs und Erwerbsimkerbund e.V., and Mellifera e.v., "Antwortschreiben Auf Stellungnahme von Aigner Zur Petition" (https://www.google.de/url?sa=t&rct=j&q=&esrc=s&source=web&cd=1&cad=rja&uact=8&ved=0CCIQFjAAahUKEwi45fqz7L7HAhUo9HIKHXY1AwA&url=http%3A%2F%2Fwww.berufsimker.de%2Findex.php%2Fdownloads%2Fcategory%2F2-allgemein%3Fdownload%3D45%3Aantwortschreiben-auf-stellung: Deutscher Berunfs und Erwerbsimkerbund e.V., 2013); Deutscher Imkerbund e.V. *et al.*, "Offener Brief Zum Thema Bienen Und Verbot Der Neonicotinoide" (http://www.buckfastnrw.de/offener-brief-zum-thema-bienen-und-verbot-der-neonicotinoide/: Landesverband Nordrhein-Westfälischer Buckfastimker, 2013); Hoppe and Safer, "Das Deutsche Bienenmonitoring-Projekt", 11–13.

88 Hans-Joachim Flügel, "Von Columella Bis CCD – Das Bienensterben Im Wandel Der Zeit", *Entomologische Zeitschrift* 125/1 (2015): 27–40.

89 Klaus-Werner Wenzel, "Neonikotinoid-Insektizide Als Verursacher Des Bienensterbens", *Entomologische Zeitschrift* 2, no. 125 (2015).

90 Baumbach, "Sie Dürfen Nicht Sterben! - Bienen in Not"; MDR, "Interview Zur Sendung: Walter Haefeker".

91 Haefeker, "Bienenfreundlich Hergestellt – 'Certified Bee Friendly'-Siegel, Weltweit".

Bee decline in Germany: renewable energy

Germany's renewable energy sector is one of the most innovative and successful sectors worldwide.

By some estimates, renewables will provide about 14% of Germany's gross electricity consumption by the end of 2015, and in July 2015 the German Government increased its targets for renewable energy to 27% of electricity by 2020 (up from 20%) and at least 45% by 2030.[92]

Two key sources of renewable energy are relevant in the context of the insect pollinators' loss of natural habitat: biomass monocultures and solar energy-specific land use practices.

In the following section, we follow our usual structure and start by introducing the main stakeholders in the renewable energy debate, discussing their positions and conclude by assessing their impact on bees.

Stakeholders

In this section we discuss the role of the key stakeholders in Germany such as the German Government, farmers, beekeeper associations and NGOs in relation to renewable energy land use practices. We will also outline the risks and opportunities linked to renewable energy such as biomass and solar energy, and touch on the importance of DBIB's "bee-friendly certified" label in the context of biomass monocultures.

The German Government

The German Government was one of the first to decide to invest heavily in renewable energy (see above). Germany's role in energy from biomass cannot be overstated: in 2013 two-thirds of renewable energy (electricity, heat and fuel) was created from biomass.[93]

The renewable energy sector benefited when the German Green Party joined the federal government between 1998 and 2005. Support for renewable energy

92 Bruno Burger, "Electricity Production from Solar and Wind in Germany in 2014", *Fraunhofer Institute Publication*, 2014, 268, http://www.ise.fraunhofer.de/en/downloads-englisch/pdf-files-englisch/data-nivc-/electricity-production-from-solar-and-wind-in-germany-2014.pdf; Janet L Sawin, "Germany Leads Way on Renewables, Sets 45% Target by 2030" (Worldwatch Institute, 2013), http://www.worldwatch.org/node/5430; C. Winter, "Germany Reaches New Levels of Greendom, Gets 31 Percent of Its Electricity From Renewables", *Bloomberg Business* (Bloomberg, 2014), http://www.bloomberg.com/bw/articles/2014-08-14/germany-reaches-new-levels-of-greendom-gets-31-percent-of-its-electricity-from-renewables.

93 Deutscher Bauernverband, "Situationsbericht 2014/2015 Des Deutschen Bauernverbandes: Ressourcenschutz in Der Landwirtschaft" (Berlin: Deutscher Bauernverband, 2014), 46.

continued under all following governments, regardless of composition, including the current CDU/CSU and SPD coalition government that came to power in 2013.[94]

The renewable energy sector was aided especially by the Renewable Energy Sources Act (Erneuerbare-Energien-Gesetz, EEG, established in 2000) that promotes renewable energy and outlines the strategic approach of the current German Government with regard to renewable energy: for example energy from solar and biomass.

The latest version of the law was passed in 2014 (EEG 2014)[95] and includes subsidies for electricity from renewable energy such as biomass (Sections 44–46 EEG); these subsidies have been confirmed by the European Union. The law also states that an annual increase of 100 MW in electricity from biomass is set as the target (Section 3 EEG).

The German Government also introduced a special "(wild)flower bonus" in the amendment of the EEG Act. Following extensive lobbying by DBIB, energy producers receive 2 cents more per kWh if wildflowers instead of corn are processed in a biogas plant.[96]

Within this legal framework, the German Government also introduced restrictions on renewable energy sources: for example on solar electricity generation, there is currently a 10-megawatt limit on solar arrays to avoid conflict with other productive types of land use. But in 2016, these restrictions are to be relaxed to include relatively unproductive agricultural land.[97]

In 2012, the Leopoldina Academy published a report on the limits of biomass energy (bioenergy). It highlighted that the demand for bioenergy leads to monocultures of silo maize and rapeseed. In addition, the report argues that the land use for agricultural activities in Germany is at its limit. The desired and incentivized increase of bioenergy through the German Government is therefore not possible without trade-offs. Therefore, decisions need to be made regarding either land use (i.e. converting land into agriculturally usable land) or types of crop plants.

However, the Leopoldina Academy argues that there is no land available to be transformed into agriculturally usable land. This means that trade-offs regarding the types of crops planted on the existing agricultural land have to be made. In any case this will lead to a shortage in whatever crop is replaced, which in turn will have to be imported.[98]

94 Matthias Lang and Annette Lang, "CDU/CSU and SPD Present Coalition Agreement – 55% to 60% Renewables by 2035 and More", *German Energy Blog*, 2013, http://www.germanenergyblog.de/?p=14825.

95 We will refer to the law as EEG. The full name of the law is as follows: Gesetz für den Ausbau erneuerbarer Energien (Erneuerbare-Energien-Gesetz-EEG 2014).

96 Haefeker, "Bienenfreundlich Hergestellt—'Certified Bee Friendly'-Siegel, Weltweit".

97 Craig Morris, "German Government Announces New Rules for Solar", 2015, http://energytransition.de/2015/01/german-government-announces-new-rules-for-solar/.

98 Deutsche Akademie der Naturforscher Leopoldina, "Bioenergy: Möglichkeiten Und Grenzen, Bioenergy - Chances and Limits" (Halle: Nationale Akademie der Wissenschaften, 2012), 6–13.

DBV

The largest stakeholder group in this debate is made up of the farmers. As mentioned above, the farmers are represented by the Association of German Farmers (Deutscher Bauernverband, DBV).

The obvious reason for the importance of the farmers is that they are producing the crops which are being used for the production of bioenergy.[99] Currently 20% of cropland is used for renewable energy crops; 86.5% of those renewable energy crops are either maize (corn) or rapeseed.[100]

In addition to producing the crops needed for the production of bioenergy, German farmers have invested c. €20 billion in biogas as well as solar power facilities between 2009 and 2013. This has led to the astonishing fact that in Germany three-quarters of all biogas facilities are owned by farmers.[101] The farmers are therefore not only the producers of biomass, but also the suppliers of bioenergy.

Finally, they are also opponents of certain solar energy-specific land use practices—DBV strictly rejects the construction of solar plants on arable or grassland and rather mandates solar arrays on sealed surfaces as well as on building roofs.[102]

DIB

DIB (German Beekeepers' Association; see above) has three main areas of concern in relation to monocultures: the starvation of bees due to a lack of food, the negative effects on the immune system, and the exposure to pesticides.

When assessing the problem of bee starvation, "maize desert" has become the term summarizing the main concerns over the increasing production of silo maize. Bees are not able to find food in maize fields, which can lead to their starvation. Beekeepers need to provide food for the bees so that the colony does not starve during the summer months.[103]

The problem of starvation, however, also relates to rapeseed production. Large areas of arable land are used to produce rapeseed and rapeseed stops blossoming in mid-May. However, bees require food from April until August and therefore

99 We use bioenergy as an umbrella term to include biofuels and electricity from biomass.

100 Deutscher Bauernverband, "Situationsbericht 2014/2015 Des Deutschen Bauernverbandes: Ressourcenschutz in Der Landwirtschaft", 45–46.

101 *Ibid.*, 45.

102 Deutscher Bauernverband, "Keine Fotovoltaik-Anlagen Auf Landwirtschaftlichen Flächen" (Deutscher Bauernverband, 2014), http://www.bauernverband.de/keine-fotovoltaik-anlagen-auf-landwirtschaftlichen-flaechen.

103 Baumbach, "Sie Dürfen Nicht Sterben! - Bienen in Not"; Deutscher Imkerbund e.V., "Jahresbericht 2013/2014 Des Deutschen Imkerbundes e.V."; Jaboury Ghazoul, "Buzziness as Usual? Questioning the Global Pollination Crisis", *Trends in Ecology & Evolution* 20, no. 7 (2005): 367–73, doi: 10.1016/j.tree.2005.04.026.

struggle to find food during the summer months. If they are not fed by the beekeepers, they are under threat of starvation.[104]

The second problem caused by monocultures relates to bee nutrition and physiology: recent studies have shown that it is essential for bees to get a mix of food. A mono-source diet has been shown to impact negatively on the bees' immune system.[105] The consequences of a reduction in the immune system's activity have also been highlighted in the pesticides section, above.

The final problem relates to exposure to pesticides. The DeBiMo project showed that seven out of nine pesticides found were used during the production of rapeseed. For example the pesticide clothianidin is used especially frequently on rapeseed.[106] As mention above this pesticide is neonicotinoid-based and it has been shown in studies that it reduces the activity of the bees' immune system.[107] Please see the section on bee health above for more details regarding the negative consequences of pesticide exposure for bees.

DBIB

Since 2011, DBIB and its European parent organization, the European Professional Beekeepers' Association (EPBA), have worked with a broad range of international food, textile and energy producers to mandate and promote a "certified bee-friendly" label for these products and services.[108] One of the main objectives is to address the aforementioned problem of bee starvation during the summer months—as a consequence of maize and rapeseed monocultures.[109]

DBIB publically criticizes Germany's current political views and Angela Merkel's, (Germany's Chancellor) position regarding pesticides, genetically modified crops (see above) and monocultures[110].

In 2013, DBIB and Bund für Umwelt und Naturschutz Deutschland (BUND), a key environmental organization in Germany, launched a joint educational campaign

104 Günter Friedmann, "Imker Und Wissenschaftler Schlagen Erneut Alarm: Honigbienen Verhungern Mitten Im Sommer. Die Landschaft Ernährt Ihre Insekten Nicht Mehr", 2009, http://www.demeter.de/presse/imker-und-wissenschaftler-schlagen-erneut-alarm-honigbienen-verhungern-mitten-im-sommer-die.

105 Flügel, "Von Columella Bis CCD – Das Bienensterben Im Wandel Der Zeit".

106 DeBiMo, "Schlussbericht, Deutsches Bienenmonitoring - 'DeBiMo'", 44–46, 61.

107 Di Prisco *et al.*, "Neonicotinoid Clothianidin Adversely Affects Insect Immunity and Promotes Replication of a Viral Pathogen in Honey Bees".

108 Haefeker, "Bienenfreundlich Hergestellt—'Certified Bee Friendly'-Siegel, Weltweit".

109 Walter Haefeker, "Agrarpolitik Ist Bienenpolitik - Die EU-Agrarreform Und Die Zukunft Der Imkerei", in *Der Kritische Agrarbericht 2013*, ed. Manuel Schneider, Andrea Fink-Kessler, and Friedhelm Stodieck (München: ABL Verlag, 2013): 178–81; NABU Landesverband Thüringen *et al.*, "Situation Der Bestäubenden Insekten" (https://thueringen.nabu.de/imperia/md/content/thueringen/positionspapiere/position_imker_und_nabu.pdf: NABU Thüringen).

110 Deutscher Berufs und Erwerbsimkerbund e.V., "Der Schwarze Pinsel", 2013, http://www.berufsimker.de/index.php/aktuelles-2/7-website?start=24.

for farmers to outline the risks of pesticides and their impact on bees in the context of rapeseed monocultures for biomass production.[111]

"Wild statt mono"

The negative impact of monocultures has been widely recognized and analysed in Germany, *inter alia* by Bayerische Landesanstalt für Weinbau und Gartenbau (LWG), a Bavarian governmental agency in charge of viticulture and horticulture. To mitigate the impact, this agency started an initiative called "Wild statt mono" ("wild[flowers] instead of mono[cultures]").

The aim of the initiative is to create a mix of ecologically valuable wildflowers which can also be harvested and used to create energy from biomass. This can provide the benefits of biomass energy, without the disadvantages, i.e. the negative impact on biodiversity in general and bees in particular.

The project started in 2008 when LWG set out to develop eight different mixes of wildflower seeds, each one optimized for different water and soil characteristics. Since then, tests have been carried out in order to analyse the ecological as well as economic outcomes, especially in comparison with silo maize.[112]

These "wildflower into biomass" initiatives are subsidized by the German Government (see above) and supported by environmental groups and farmers alike. Support for them is growing from a social and economic perspective: a key example here is the recent Energiebündel and Flower Power[113] initiative in Baden-Wuerttemberg.[114]

The positive effects of wildflowers on biodiversity and their important role for bees have been confirmed in previous studies.[115] However, the "Wild statt mono" sets out to also analyse the economic viability. Final results are pending and the initial results are mixed: it can be shown that in two out of three cases the wild flower biomass yields exceed the biomass yield of maize. This effect is especially strong in very dry and very wet conditions, where the respectively optimized wildflower seed mixes outperform maize in terms of yield.[116,117]

111 Thomas Brückmann, "Gestiegene Gefahren Durch Neuartige Pestizide – Neonikotinoide Sind Nervengifte" (Berlin: Bund für Umwelt und Naturschutz Deutschland e.V., 2013).
112 Birgit Vollrath, Werner Kuhn, and Antje Werner, *"Wild" Statt "mono"—Neue Wege Für Die Biogaserzeugung* (Veitshöchheim: Bayerische Landesanstalt für Weinbau und Gartenpflege, 2010): 1–6.
113 Werner Kuhn and Stefan Zeller, "Energiebuendel & Flower Power", http://www.energiebuendel-und-flowerpower.de/
114 Jürgen Jonas, "Biomasse Aus Wildblumen", *Schwäbisches Tagblatt*, 2015.
115 Vollrath, Kuhn, and Werner, *"Wild" Statt "mono"—Neue Wege Für Die Biogaserzeugung*, 1–6.
116 So far one trial with one repetition.
117 Vollrath, Kuhn, and Werner, *"Wild" Statt "mono"—Neue Wege Für Die Biogaserzeugung*, 6.

When looking at the methane yield, wildflowers show a 14% lower result compared with silo maize. However, the initiative states that the methane yield can be improved by optimizing the seed mixes.

At this stage, wildflowers are not yet on par with silo maize in terms of the economic performance. However, there are cases when it could be economically favourable to use wildflowers: wildflowers require only a fraction of the time and effort, fertilizer and pesticides needed for the production of silo maize. In addition they perform very well in very dry and very wet conditions. This means that wildflowers are an economically viable solution under the aforementioned circumstances.[118]

It is also very important to keep in mind that the ecological advantages of wildflowers are overwhelming. In addition to providing an essential habitat for many species and therefore positively contributing to biodiversity, using wildflowers instead of silo maize can also lead to a reduction in the use of fertilizers and pesticides, and to an improvement in soil quality because of the superior production of humus soil.

Biomass used for the production of biogas and biofuels is one of Germany's most important sources of renewable energy as it forms two-thirds of Germany's renewable energy supply.

However, biomass has become one of the main drivers of monocultures through the cultivation practices for corn and ploughing of valuable grassland. As stated in German literature, Germany used nearly 2.1 million hectares of its arable land for energy crops in 2013. This figure is equivalent to 12.6% of the 16.7 million hectares of farmland in Germany. The upper limit for biomass used for bioenergy is 4 million hectares by 2020. To prevent further losses of arable land and its effects, Germany's government introduced (2000) and revised the Renewable Energy Sources Act (EEG in 2014) that now limits the amount of corn and grain eligible for "special compensation" such as bioenergy production.[119]

Another key source of renewable energy in Germany is solar energy, generated through photovoltaics (PVs). The country has been the world's top PV installer for several years and still leads in terms of the overall installed capacity, which amounted to 38,754 megawatts (MW) by the end of May 2015, ahead of China, Japan, Italy and the United States. Germany has an official target of 2,500 megawatts of newly built PV annually.[120]

118 *Ibid.*, 6; Birgit Vollrath and Antje Werner, "Wildpflanzen Zur Biogasgewinnung – Eine Ökonomische Alternative Zur Silomais" (Veitshöchheim: Bayerische Landesanstalt für Weinbau und Gartenpflege, 2012): 1–6.

119 BMWi, "Bioenergie" (Bundesministerium für Wirtschaft und Energie, 2015): 18 ff., http://www.erneuerbare-energien.de/EE/Navigation/DE/Technologien/Bioenergie/bioenergie.html; Craig Morris and Martin Pehnt, "Energy Transition", ed. Dorothee Landgrebe and Rebecca Bertram, *The German Energiewende* (Berlin: Heinrich Böll Stiftung, 2015).

120 Morris, "German Government Announces New Rules for Solar".

In many countries, gigantic solar PV arrays are being built on otherwise unusable land, but in Germany, the land used for solar plants generally consists of either farmland or forest.[121]

Key stakeholders in the renewable energy debate in Germany are moving in different directions, none of them with a focus on natural or wildlife protection or conservation practices *per se*. The German Government, on the one hand, is focused on meeting its renewable energy and climate change targets while ensuring that these targets are protected through the new EEG legal framework. The German Farmers' Association (DBV), on the other hand, is focused on protecting the economic benefits of agricultural products and hence farmland. At the same time, the German farmers have an invested interest in renewable energy and are now facing a potential conflict of interest, as opponents to, investors in or supporters of renewable energy-led land use practices.[122]

Economic benefits and stakeholder interests aside, biomass monocultures and solar energy-specific land use practices can have a large impact on the insect pollinators' natural habitat, feeding practices, nesting abilities and ultimately on their health.[123]

Intensification of agriculture and monocultures as well as new forms of land use practices have prompted the loss and fragmentation of valuable natural to semi-natural perennial habitats for pollinators, such as agroforestry systems, grasslands, old fields, shrub lands, forests and hedgerows.[124]

On the other hand, land use systems that work with biodiversity and without chemicals, such as ecological farming systems, can benefit pollinator communities, both managed and wild. As stated by Tirado, Simon and Johnston[125] sustainable land use practices at scale can also benefit the pollinators.

Further scientific research, regulations and subsidies, best industry practices (e.g. Belectric's "Big60 Million" solar park project in the UK)[126] and labels such as "bee-friendly certified" by DBIB[127] are important to make room for wildlife and environment protection and preservation strategies on Germany's renewable

121 *Ibid.*

122 Christian Troost, Teresa Walter, and Thomas Berger, "Climate, Energy and Environmental Policies in Agriculture: Simulating Likely Farmer Responses in Southwest Germany", *Land Use Policy* 46 (2015): 50–64, doi: 10.1016/j.landusepol.2015.01.028.

123 Clara I. Nicholls and Miguel A. Altieri, "Plant Biodiversity Enhances Bees and Other Insect Pollinators in Agroecosystems: A Review", *Agronomy for Sustainable Development* 33, no. 2 (2012): 257–74, doi: 10.1007/s13593-012-0092-y.

124 Tomás E. Murray *et al.*, "Local-Scale Factors Structure Wild Bee Communities in Protected Areas", *Journal of Applied Ecology* 49, no. 5 (2012): 998–1008, doi: 10.1111/j.1365-2664.2012.02175.x.

125 Tirado, Simon, and Johnston, "A Review of Factors That Put Pollinators and Agriculture in Europe at Risk".

126 Belectric, "Big60 Million", accessed 17 March 2106, http://www.belectric.co.uk/portfolio/honey-bees/.

127 Haefeker, "Bienenfreundlich Hergestellt—'certified bee friendly'"-Siegel, Weltweit".

energy (Energiewende) agenda and to avoid potential (economic) conflicts of interest between corporate, governmental, NGO and academic stakeholders (see details above).

Conclusion

In this chapter on "Bees and Accountability in Germany" we have outlined that the steep decline in bee populations in Germany and the issues linked with their health, colony collapse disorder and other phenomena, are the product of a range of known and unknown causality factors that can act singly or in combination.[128]

We have focused on three factors that drive this multi-factor causality in Germany: 1) beekeeping practices, 2) bee health issues, and 3) issues related to renewable energy-specific land use practices.

In assessing these factors, we focused on the role of key stakeholders such as amateur and professional beekeeper associations, activist groups, scientists, farmers, the German Government and other policy-makers as well as corporates—and their role in contributing to, accelerating or reversing this process.

We conclude that investors can also play a key role in addressing bee health and colony decline through divesting from or engaging with companies that are linked to "bee controversies".

After assessing the complexity of the issues and stakeholders involved, we see a demand for more independent, publicly funded research, results-driven awareness campaigns, consumer-oriented labels and certifications as well as for the implementation of best practices involving key stakeholder groups[129] to successfully address the bee decline and colony losses in Germany. Given the potential conflicts of interest of these groups, stakeholder governance and oversight also remain key.

Key findings

Factor 1 focused on beekeeping practices

In Germany, beekeeping is declining and that is linked to the fact that beekeeping nowadays is mainly pursued as a hobby: 80% of DIB beekeepers keep just

128 G.R. Williams *et al.*, "Colony Collapse Disorder in Context", *Bioessays* 32, no. 10 (2010): 845–46, doi: 10.1002/bies.201000075; Le Conte, Ellis, and Ritter, "Varroa Mites and Honey Bee Health: Can Varroa Explain Part of the Colony Losses?"

129 For a recommendation regarding "alternative management practices, research, regulation and policies and a multifaceted approach to bee colony decline", see D.A. Mangosing and D.M. Gute, "Bee-Ware: Investigating Bee Colony Decline and its Ecological Effects on Human Health", 2014, accessed 22 August 2015, http://www.academia.edu/8715722/Bee-Ware_Investigating_Bee_Colony_Decline_and_its_Ecological_Effects_on_Human_Health.

1–20 colonies, 18% keep 21–50 colonies and only about 2% keep more than 50 colonies.[130]

As outlined above, better beekeeping and state-of-the-art equipment can lead to healthier bees, higher quality and increased volumes of bee products, easier data collection and improved disease treatment. That said, even the most experienced beekeeper may not be able to protect bees from other threats such as pests and diseases as well as pesticides.

Factor 2 looked at the issues related to bee health

Bees often suffer from diseases and parasites that weaken and kill them. Most of these diseases and parasites are invasive species that cannot be fought through the natural adaptation of native bees or emergence of resistance.

In particular, the invasive parasitic mite, *Varroa destructor*, is a serious threat to apiculture globally. Other parasites, such as *Nosema ceranae*, have been found to be highly damaging to honey bee colonies in some southern European countries. Other new viruses and pathogens are likely to put further pressure on bee colonies.

In Germany, as outlined above, a varroa narrative has been developed and promoted especially by the chemical industry.[131] While varroa mites do pose a risk to bee colonies, the causality between the varroa infection and the death of bees is more complex than the varroa narrative suggests.

The ability of bees to resist diseases and parasites seems to be influenced by a number of factors, particularly their nutritional status and their exposure to toxic chemicals. Bee-harming pesticides (insecticides, herbicides and fungicides) can have a significant impact on pollinator health and seem to weaken honey bees that then are becoming more susceptible to infection and parasitic infestation.

However, these chemicals are still used in Germany, and at high concentrations have been shown to acutely affect bees but also other pollinators. Observed effects include impairment of foraging ability (bees getting lost when coming back to the hive after foraging, and an inability to navigate efficiently), impairment of learning ability (olfactory or smelling memory, essential in a bee's behaviour), increased mortality, and dysfunctional development, including in larvae and queens.[132]

130 Deutscher Imkerbund e.V., "Jahresbericht 2013/2014 Des Deutschen Imkerbundes E. V.", 13–14.

131 Baumbach, "Sie Dürfen Nicht Sterben! - Bienen in Not".

132 Tirado, Simon, and Johnston, "A Review of Factors That Put Pollinators and Agriculture in Europe at Risk", 29 ff; Johannes Fischer *et al.*, "Neonicotinoids Interfere with Specific Components of Navigation in Honeybees", *PLoS One* 9, no. 3 (2014): e91364, doi: 10.1371/journal.pone.0091364.

Factor 3 assessed the impact of Germany's Energiewende and the impact of renewable energy-led land use practices

Bees are losing their natural habitat for a diversity of reasons, mostly related to industrial agriculture and monocultures and controversial land use practices that remove field margins (borders, hedges etc.) that usually hold a diversity of plants around farms.

Any progress in transforming the current destructive, chemical-intensive agricultural system into an ecological farming system[133] will have many associated (co-)benefits on other dimensions of the environment and on human food security, quite apart from clear benefits to global pollinator health. In the short to medium term, there are specific issues that society can begin to address in order to benefit global pollinator health.

Many land use practices, including those for renewable energy, that increase plant diversity, at different scales, can improve the flower resources available to pollinators, in both space and time.

Labels such as "certified bee-friendly"[134] can help to provide more clarity for consumers about GMO- and pesticide-free products and, at the same time, can help to encourage bee-friendly land use practices.

Finally, the recent expansion of organic agriculture, together with the application of innovative techniques that reduce and/or eliminate chemical pesticides (i.e. integrated pest management), demonstrates that farming and other land use without pesticides is entirely feasible, consumer-oriented, economically profitable and environmentally safe.

A final note: many of the predicted consequences of climate change (which Germany aims to tackle through a multitude of renewable energy investments), such as increasing temperatures, changes in rainfall patterns and more erratic or extreme weather events, clearly can affect pollinators, individually and ultimately their communities, becoming reflected in higher extinction rates of the insect pollinator species. Climate change can also modify flowering patterns, displace plants that were major sources of food for bees in a given area, or cause "season creep", where flowering no longer coincides with the emergence of bees in the spring.[135] Unfortunately, we are unable to focus on this topic at length but will leave room for thought to continue the discussion in this direction.

133 Vollrath, Kuhn, and Werner, *"Wild Statt mono"—Neue Wege Für Die Biogaserzeugung.*
134 Haefeker, "Bienenfreundlich Hergestellt—'Certified Bee Friendly'-Siegel, Weltweit".
135 John Bryden *et al.*, "Chronic Sublethal Stress Causes Bee Colony Failure", *Ecology Letters* 16, no. 12 (2013): 1463–69, doi: 10.1111/ele.12188; Kluser *et al.*, "Global Honey Bee Colony Disorder and Other Threats to Insect Pollinators".

Bibliography

Abrol, D. "Decline in Pollinators". In *Pollination Biology*, edited by D.P. Abrol, 545–601. Dordrecht, Netherlands: Springer, 2012.

AVAAZ. "3 million to Save the Bees". Accessed 13 August 2015. http://www.avaaz.org/en/save_the_bees_global/.

Baumbach, D. 2015. "Sie dürfen nicht sterben! - Bienen in Not. Exakt - Die Story". Mitteldeutscher Rundfunk.

Bayer AG. "Neonicotinoids". Accessed 3 July 2015. http://beecare.bayer.com/home/what-to-know/pesticides/neonicotinoids.

Bee Friendly Association. "Bee Friendly Specifications Synthesis". Accessed 22 August 2015. http://www.certifiedbeefriendly.org/wp-content/uploads/2014/09/BEE-FRIENDLY_Specifications_Synthesis.pdf.

Belectric, "Big60 Million". Accessed 17 March 2016. http://www.belectric.co.uk/portfolio/honey-bees/.

Black, R. "Bees and Flowers Decline in Step". *BBC News*, 2006. Accessed 13 August 2015. http://news.bbc.co.uk/2/hi/science/nature/5201218.stm.

BMWI. "Bioenergie". Bundesministerium für Wirtschaft und Energie. Accessed 13 August 2015. http://www.erneuerbare-energien.de/EE/Navigation/DE/Technologien/Bioenergie/bioenergie.html.

Brückmann, T. *Gestiegene Gefahren durch neuartige Pestizide—Neonikotinoide sind Nervengifte*. Berlin: Bund für Umwelt und Naturschutz Deutschland e.V, 2013.

Bryden, J., Gill, R. J., Mitton, R. A. A., Raine, N. E. and Jansen, V. A. A. "Chronic Sublethal Stress Causes Bee Colony Failure". *Ecology Letters* 16 (2013): 1463–69.

Buglife. "Pesticide Approval Strikes Blow for Bees". Accessed 13 August 2015. https://www.buglife.org.uk/news-%26-events/news/pesticide-approval-strikes-blow-for-bees.

Burger, B. "Electricity Production from Solar and Wind in Germany in 2014". *Fraunhofer Institute Publication*, 2014. Accessed 13 August 2015. http://www.ise.fraunhofer.de/en/downloads-englisch/pdf-files-englisch/data-nivc-/electricity-production-from-solar-and-wind-in-germany-2014.pdf.

Campact E.V., Deutscher Berufs und Erwerbsimkerbund E.V. and Mellifera E.V. "Antwortschreiben auf Stellungnahme von Aigner zur Petition". Berlin: Deutscher Berunfs und Erwerbsimkerbund e.V, 2013. Available at https://www.google.de/url?sa=t&rct=j&q=&esrc=s&source=web&cd=1&cad=rja&uact=8&ved=0CCIQFjAAahUKEwi45fqz7L7HAhUo9HIKHXY1AwA&url=http%3A%2F%2Fwww.berufsimker.de%2Findex.php%2Fdownloads%2Fcategory%2F2-allgemein%3Fdownload%3D45%3Aantwortschreiben-auf-stellungnahme-von-aigner-zur-petition&ei=ZIzZVbijN6joywP26gw&usg=AFQjCNEFiJxgdtt23W6nMIX7hfBUjJls3g.

Capri, E. and Marchis, A. "Bee Health in Europe – Facts & Figures, Bridging Science and Policy". Piacenza: Opera Research Centre, 2013. Accessed 13 August 2015. http://www.pollinator.org/PDFs/OPERAReport.pdf.

Carter, N. *The Politics of the Environment Ideas, Activism, Policy*. Cambridge, Cambridge University Press, 2007.

DeBiMo. *Zwischenbericht 2014, Deutsches Bienenmonitoring - "DeBiMo"*. Stuttgart: DeBiMo, 2014.

DeBiMo. *Schlussbericht, Deutsches Bienenmonitoring - "DeBiMo"*. Stuttgart: DeBiMo, 2013.

Deutsche Akademie der Naturforscher Leopoldina. *Bioenergy: Möglichkeiten und Grenzen* [Bioenergy: Chances and limits]. Halle: Nationale Akademie der Wissenschaften, 2012.

Deutscher Bauernverband. *Die deutschen Bauern: Veränderung gestalten, Geschäftsbericht des Deutschen Bauernverbandes 2014/2015*. Berlin: Deutscher Bauernverband, 2014.

Deutscher Bauernverband. *Faktencheck Landwirtschaft 2014*. Berlin: Deutscher Bauernverband, 2014.

Deutscher Bauernverband. *Situationsbericht 2014/2015 des deutschen Bauernverbandes: Ressourcenschutz in der Landwirtschaft.* Berlin: Deutscher Bauernverband, 2014.
Deutscher Bauernverband. "Keine Fotovoltaik-Anlagen auf landwirtschaftlichen Flächen". Deutscher Bauernverband, 2014. Accessed 13.08.2015. http://www.bauernverband.de/keine-fotovoltaik-anlagen-auf-landwirtschaftlichen-flaechen.
Deutscher Bauernverband. *Geschäftsbericht des Deutschen Bauernverbandes 2013/2014.* Berlin: Deutscher Bauernverband, 2013.
Deutscher Berufs und Erwerbsimkerbund E.V. "Süddeutsche Berufsimkertage in Donaueschingen". Deutscher Berufs und Erwerbsimkerbund e.V, 2015. Accessed 22 August 2015. http://www.berufsimker.de/index.php/berufsimkertage/donaueschingen-sued deutsche-berufsimkertage/veranstaltung.
Deutscher Berufs und Erwerbsimkerbund E.V. "Satzung". Deutscher Berufs und Erwerbsimkerbund e.V, 2013. Accessed 22 August 2015. http://www.berufsimker.de/index.php/satzung.
Deutscher Berufs und Erwerbsimkerbund E.V. "Der Schwarze Pinsel". Deutscher Berufs und Erwerbsimkerbund e.V, 2013. Accessed 22 August 2015. http://www.berufsimker.de/index.php/aktuelles-2/7-website?start=24.
Deutscher Imkerbund E.V. *Imkerei in Deutschland.* Accessed 13 August 2015. http://deutscherimkerbund.de/160-Die_deutsche_Imkerei_auf_einen_Blick.
Deutscher Imkerbund E.V. *Jahresbericht 2013/2014 des Deutschen Imkerbundes e. V.* Wachtberg: Deutscher Imkerbund e.V., 2014.
Deutscher Imkerbund E.V. *Werbefonds 2013.* Wachtberg: Deutscher Imkerbund e.V., 2013. Available at http://deutscherimkerbund.de/download/0-340.
Deutscher Imkerbund E.V., Deutscher Berufs und Erwerbsimkerbund E.V., Bioland E.V., Demeter E.V., Mellifera E.V., Gemeinschaft der Europäischen Buckfastimker and European Professional Beekeepers' Association. „Offener Brief zum Thema Bienen und Verbot der Neonicotinoide". Landesverband Nordrhein-Westfälischer Buckfastimker, 2013. Available at http://www.buckfastnrw.de/offener-brief-zum-thema-bienen-und-verbot-der-neonicotinoide/.
Di Prisco, G., Cavaliere, V., Annoscia, D., Varricchio, P., Caprio, E., Nazzi, F., Gargiulo, G. and Pennacchio, F. "Neonicotinoid Clothianidin Adversely Affects Insect Immunity and Promotes Replication of a Viral Pathogen in Honey Bees". *Proc Natl Acad Sci USA* 110 (2013): 18466–471.
EFSA. "Towards an Integrated Environmental Risk Assessment of Multiple Stressors on Bees: Review of Research Projects in Europe, Knowledge Gaps and Recommendations". *EFSA Journal* 12(3) (2014): 1–102.
EFSA. *EFSA identifiziert Risiken durch Neonicotinoide für Bienen,* 2013. Accessed 13 August 2015. http://www.efsa.europa.eu/de/press/news/130116.
Einecke, H. *Bayer klagt gegen Pestizid-Verbot nach Bienensterben,* 2013. Accessed 13 August 2015. http://www.sueddeutsche.de/wirtschaft/chemiekonzern-bayer-klagt-gegen-pestizid-verbot-nach-bienensterben-1.1756210.
Engler, M. and Engler, P. *History Didn't Bring Down the Berlin Wall—Activists Did,* 2014. Accessed 13 August 2015. http://fpif.org/history-didnt-bring-berlin-wall-activists/.
European Commission. "Bees & Pesticides: Commission Goes Ahead with Plan to Better Protect Bees". European Commission, 2013. Accessed 13 August 2015. http://ec.europa.eu/food/archive/animal/liveanimals/bees/neonicotinoids_en.htm.
Everything Connects. *Pollinator Decline,* 2014. Accessed 23 November 2015. http://www.everythingconnects.org/pollinator-decline.html.
Faust, L. and Siede, R. "Insektizide im Raps". *Deutsches Bienen Journal* 5 (2014): 6–7.
Fischer, J., Müller, T., Spatz, A.-K., Greggers, U., Grünewald, B. and Menzel, R. "Neonicotinoids Interfere with Specific Components of Navigation in Honeybees". *PLoS ONE* 9 (2014): e91364.
FIT BEE. "Was ist 'FIT BEE'?" Accessed 13 August 2015. http://fitbee.net.

FIT BEE. "Ziele", 2014. Accessed 13 August 2015. http://fitbee.net/ziele.

Flügel, H.-J. "Von Columella bis CCD – das Bienensterben im Wandel der Zeit". *Entomologische Zeitschrift* 125/1 (2015): 27–40.

Friedmann, G. *Imker und Wissenschaftler schlagen erneut Alarm: Honigbienen verhungern mitten im Sommer. Die Landschaft ernährt ihre Insekten nicht mehr*, 2009. Accessed 13 August 2015. http://www.demeter.de/presse/imker-und-wissenschaftler-schlagen-erneut-alarm-honigbienen-verhungern-mitten-im-sommer-die.

Friedrich, P. *Honigbienen nutzen Mensch und Natur*. Accessed 13 August 2015. http://www.bauernverband.de/bienen.

Friends of the Earth. "Let's Help Britain Bloom for Bees". Accessed 13 August 2015. https://www.foe.co.uk/what_we_do/the_bee_cause_home_map_39371.

Gallai, N., Salles, J.-M., Settele, J. and Vaissière, B. E. "Economic Valuation of the Vulnerability of World Agriculture Confronted with Pollinator Decline". *Ecological Economics* 68 (2009): 810–21.

Ghazoul, J. "Buzziness as Usual? Questioning the Global Pollination Crisis". *Trends in Ecology & Evolution* 20 (2005): 367–73.

Greenpeace International. "NGOs and Beekeepers Take Legal Action to Defend EU Ban of Bee-Killing Pesticides against Syngenta and Bayer". Greenpeace International, 2013. Accessed 13 August 2015. http://www.greenpeace.org/eu-unit/en/News/2013/NGOs-and-beekeepers-take-legal-action-to-defend-EU-ban-of-bee-killing-pesticides-against-Syngenta-and-Bayer.

Grossman, E. "Declining Bee Populations Pose A Threat to Global Agriculture". *Yale Environment 360* (2013). Accessed 13 May 2015. http://e360.yale.edu/feature/declining_bee_populations_pose_a_threat_to_global_agriculture/2645/.

Haefeker, W. Bienenfreundlich hergestellt - „CERTIFIED BEE FRIENDLY"-Siegel, weltweit, 2014. Accessed 22 August 2015. http://ev.mellifera.de/fix/doc/Bienenfreundlich%20Schreiben%202014.pdf.

Haefeker, W. "Agrarpolitik ist Bienenpolitik - Die EU-Agrarreform und die Zukunft der Imkerei". In *Der kritische Agrarbericht 2013*, edited by M. Schneider, A. Fink-Kessler and F. Stodieck. München: ABL Verlag, 2013.

Hartmann, J. *Fundamental pillars of democracy*, 2012. Accessed 13 August 2015. https://www.deutschland.de/en/topic/politics/germany-europe/fundamental-pillars-of-democracy.

Henry, M., Beguin, M., Requier, F., Rolline, O., Odoux, J.-F., Aupinel, P., Aptel, J., Tchamitchian, S. and Decourtye, A. "A Common Pesticide Decreases Foraging Success and Survival in Honey Bees". *Science* 336 (6079) (2012): 348–50.

Hoppe, P. P. and Safer, A. "Das Deutsche Bienenmonitoring-Projekt: Anspruch und Wirklichkeit Eine kritische Bewertung", 2011. Accessed 13 August 2015. http://www.bund.net/fileadmin/bundnet/pdfs/chemie/20110125_chemie_bienenmonitoring_studie.pdf.

IRGC. "Risk of Loss of Pollination Services", 2008. Accessed 13 August 2015. www.irgc.org/IMG/doc/IRGCS_TC8Sept08_11_DraftCN_Pollination.doc.

Jonas, J. "Biomasse aus Wildblumen". *Schwäbisches Tagblatt*, 14 August 2015.

Kluser, S. P., Neumann, P., Chauzat, M.-P. and Pettis, J. S. *Global Honey Bee Colony Disorder and Other Threats to Insect Pollinators*. UNEP Emerging Issues. UNEP, 2010.

Krupke, C. H., Hunt, G. J., Eitzer, B. D., Andino, G. and Given, K. "Multiple Routes of Pesticide Exposure for Honey Bees Living Near Agricultural Fields". *PLoS One* 7 (2012): 1–8.

Lang, M. and Lang, A. "CDU/CSU and SPD Present Coalition Agreement – 55% to 60% Renewables by 2035 and More", 2013. Accessed 13 August 2015. http://www.germanenergyblog.de/?p=14825.

Lautenbach, S., Seppelt, R., Liebscher, J. and Dormann, C. F. "Spatial and Temporal Trends of Global Pollination Benefit". *PLoS ONE* 7 (2012): e35954.

Le Conte, Y., Ellis, M. and Ritter, W. "Varroa Mites and Honey Bee Health: Can Varroa Explain Part of the Colony Losses?". Apidologie 41 (2010): 353–63.

Mangosing, D. A. and Gute, D. M. "Bee-Ware: Investigating Bee Colony Decline and its Ecological Effects on Human Health", 2014. Accessed 22 August 2015. http://www.academia.edu/8715722/Bee-Ware_Investigating_Bee_Colony_Decline_and_its_Ecological_Effects_on_Human_Health.

MDR. "Interview zur Sendung: Walter Haefeker. Exakt - Die Story". Mitteldeutscher Rundfunk, 2015.

Morris, C. "German Government Announces New Rules for Solar". Accessed 13 August 2015. http://energytransition.de/2015/01/german-government-announces-new-rules-for-solar/.

Morris, C. and Pehnt, M. "Energy Transition". In *The German Energiewende*, edited by D. Landgrebe and R. Bertram. Berlin: Heinrich Böll Stiftung, 2015.

Murray, T. E., Fitzpatrick, Ú., Byrne, A., Fealy, R., Brown, M. J. F. and Paxton, R. J. "Local-Scale Factors Structure Wild Bee Communities in Protected Areas". *Journal of Applied Ecology* 49 (2012): 998–1008.

Nabu Landesverband Thüringen, Buckfast Imker Sachsen-Anhalt Thüringen, Deutscher Berufs und Erwerbsimkerbund E.V. - Thüringen and Landesverband Thüriger Imker E.V. "Situation der bestäubenden Insekten". NABU Thüringen. Accessed 17 March 2016. https://thueringen.nabu.de/imperia/md/content/thueringen/positionspapiere/position_imker_und_nabu.pdf.

Nicholls, C. I. and Altieri, M. A. "Plant Biodiversity Enhances Bees and Other Insect Pollinators in Agroecosystems: A Review". *Agronomy for Sustainable Development* 33 (2012): 257–74.

oekom. "oekom corporate rating". Accessed 17 March 2016. http://www.oekom-research.com/index_en.php?content=corporate-rating.

Prosiebensat.1 Digital GmbH. *Deutschland summt*. Accessed 13 August 2015. http://www.prosieben.de/tv/greenseven/deutschland-summt.

Rosenkranz, P. *Erwiderung zur Presseerklärung des NABU vom 25.01.2011*. Accessed 13 August 2015. http://www.ag-bienenforschung.de/debimo_kritik.html.

Rucht, D. and Roose, J. "The Transformation of Environmental Activism in Berlin". Social Science Research Center Berlin, 2011. Accessed 13 August 2015. http://ecpr.eu/filestore/paperproposal/05262578-28eb-46a6-9c93-8c4da44d2d98.pdf.

Sawin, J. L. "Germany Leads Way on Renewables, Sets 45% Target by 2030". Accessed 13 August 2015. http://www.worldwatch.org/node/5430.

Schroeder, A. "Das Deutsche Bienenmonitoring". *Rheinische Bienenzeitung* 21 (2012).

Schroeder, A. "Struktur und Ergebnisse des DeBiMo". *ADIZ* 9 (2011): 8–9.

Sternenfair. "sternenfair: die 5-Sterne-Milch". Accessed 22 August 2015. http://www.sternenfair.de.

Stiftung für Mensch und Umwelt *Die Initiative - Deutschland Summt*. Accessed 13 August 2015. http://www.deutschland-summt.de/die-initiative-deutschland.html.

Stiftung für Mensch und Umwelt *Daniela Schadt - Schirmherrin von "Deutschland Summt"*. Accessed 13 August 2015. http://www.deutschland-summt.de/unsere-schirmherrin.html.

Stiftung für Mensch und Umwelt *Unser Ratgeber "Wir tun was für Bienen - Bienengarten, Insektenhotel und Stadtimkerei"*. Accessed 13 August 2015. http://www.deutschland-summt.de/unsere-schirmherrin.html.

Stiftung für Mensch und Umwelt *Umweltbildung leicht gemacht*. Accessed 13 August 2015. http://www.bienenkoffer.de.

Stiftung für Mensch und Umwelt. *Partnerstädte*. Accessed 13 August 2015. http://www.deutschland-summt.de/startseite.html.

Stiftung für Mensch und Umwelt. *Wer sind die Förderer von "Deutschland Summt!"?*. Accessed 13 August 2015. http://www.deutschland-summt.de/faq-leser/items/wer-sind-die-foerderer-von-deutschland-summt.html.

Storm, P.-C. "Environmental Laws: Introduction", 2000. Accessed 13 August 2015. http://www.iuscomp.org/gla/literature/envirmt.htm.

Sumpter, D. J. T. and Martin, S. J. "The Dynamics of Virus Epidemics in Varroa-Infested Honey Bee Colonies". *Journal of Animal Ecology* 73 (2004): 51–63.

Tirado, R., Simon, G. and Johnston, P. "A Review of Factors that Put Pollinators and Agriculture in Europe at Risk". Greenpeace Research Laboratories Technical Report (Review). Amsterdam: Greenpeace International, 2013.

Troost, C., Walter, T. and Berger, T. "Climate, Energy and Environmental Policies in Agriculture: Simulating Likely Farmer Responses in Southwest Germany". *Land Use Policy* 46 (2015): 50–64.

UNFAO. "Protecting the Pollinators". *Spotlight*, 2005. Accessed 13 August 2015. http://www.fao.org/ag/magazine/0512sp1.htm.

Vollrath, B. and Werner, A. *Wildpflanzen zur Biogasgewinnung – eine ökonomische Alternative zur Silomais*. Veitshöchheim: Bayerische Landesanstalt für Weinbau und Gartenpflege, 2012.

Vollrath, B., Kuhn, W. and Werner, A. *"Wild" statt "mono" – neue Wege für die Biogaserzeugung*. Veitshöchheim: Bayerische Landesanstalt für Weinbau und Gartenpflege, 2010.

Von der Ohe, W. and Martens, D. "Das Deutsche Bienenmonitoring, Pflanzenschutzmittel-Rückstände im Bienenbrot". *ADIZ* 10 (2011): 8–9.

Wenzel, K.-W. "Neonikotinoid-Insektizide als Verursacher des Bienensterbens". *Entomologische Zeitschrift* 2 (2015): 67–73.

Williams, G. R., Tarpy, D. R., Vanengelsdorp, D., Chauzat, M. P., Cox-Foster, D. L., Delaplane, K. S., Neumann, P., Pettis, J. S., Rogers, R. E. and Shutler, D. "Colony Collapse Disorder in Context". *Bioessays* 32 (2010): 845–46.

Williams, I. H. "Aspects of Bee Diversity and Crop Pollination in the European Union". In *Linnaean Society Symposium Series* 18, edited by A. Matheson, S.L. Buchmann, C. O'Toole, P. Westrich and I.H. Williams, 210–26. London, UK: Academic Press, 1996.

Winter, C. "Germany Reaches New Levels of Greendom, Gets 31 Percent of Its Electricity From Renewables". *Bloomberg Business*, 2014. Accessed 13 August 2015. http://www.bloomberg.com/bw/articles/2014-08-14/germany-reaches-new-levels-of-greendom-gets-31-percent-of-its-electricity-from-renewables.

38 Degrees. "Keep the Ban on Bee-Killing Pesticides. Accessed 13 August 2015. https://speakout.38degrees.org.uk/campaigns/ban-bee-killing-pesticides-for-good-937d4563-7694-41a8-a642-65e6b0e51453.

Appendix 15.1. 2014's reports of DAX30 companies that were analysed

Company Name	Type of Report
Adidas	Stand alone sustainability report (English only)
Allianz	Stand alone sustainability report
BASF	Integrated report
Bayer	Integrated report
Beiersdorf	Stand alone sustainability report
BMW	Stand alone sustainability report
Commerzbank	Stand alone sustainability report
Continental	Stand alone sustainability report
Daimler	Stand alone sustainability report
Deutsche Börse	Integrated report
Deutsche Bank	Stand alone sustainability report
Deutsche Post	Stand alone sustainability report
Deutsche Telekom	Stand alone sustainability report (online only)
E.ON	Stand alone sustainability report
Fresenius Medical Care	Annual report
Fresenius SE	Annual report
HeidelbergCement	Integrated report
Henkel	Stand alone sustainability report
Infineon Technologies	Integrated report
Deutsche Lufthansa	Stand alone sustainability report
Linde	Stand alone sustainability report
Lanxess	Stand alone sustainability report
Merck	Stand alone sustainability report
MüRück	Stand alone sustainability report
RWE	Stand alone sustainability report
SAP	Integrated report
K+S	Stand alone sustainability report
Siemens	Integrated report
ThyssenKrupp	Integrated report
Volkswagen	Stand alone sustainability report

Appendix 15.2

oekom research

BASF SE

	Weight	Rating
B.2.2.5. Major environmental controversies, fines or settlements relating to sustainable chemicals management	2.8%	D-

Research revealed controversies in recent years.

Comment: The following controversies led to a minor downgrading:

According to the European Food Safety Authority (EFSA) in May 2013, EFSA scientists identified a number of risks posed to bees by BASF's controversial insecticide fipronil, especially when used as a seed treatment. The risk assessment found a ?high acute risk to honeybees? through drifting dust. Additionally, a certain exposure risk through the nectar and pollen of some treated crops was identified. As a consequence of these findings, Fipronil was banned for outdoor applications in the EU in July 2013.
Bees play a crucial role in agriculture, as they are major pollinators for plants and crop. According to a Greenpeace report in 2013, that also criticises the bee-harming properties of biocides such as fipronil, without insect pollination, about one third of the crops for nutrition would have to be pollinated by other means to avoid significant losses. Up to 75% of the crops would suffer some decrease in productivity.
According to BASF in June 2015, fipronil - which does not belong to the controversial group of neonicotinoids - does not contribute to the observed adverse effects on bee colonies when applied correctly.

The company produces and sells pesticides which are classified as highly or extremely hazardous by the 'World Health Organization'. As proper application, adherence to respective safety measures and appropriate disposal cannot be guaranteed the substances pose a significant environmental risk. A significant amount of applied biocides do not reach the target pests but run off from fields and impact the environment in various ways. Harmful environmental impacts of pesticide use include (persistent) contamination of soil, food, water and bioaccumulation in the food chain. Additionally, the use of pesticides is linked to loss of biodiversity and elimination of key species.

ORBIT - **oekom Responsibility Benchmarking & Information Tool**

Companies Countries Controversial Weapons Alerts Archive Controversy Monitor Download Centre

Alleged decline in bee popoluation due to inseticides (BASF SE)

Current evaluation

Summary:

According to the European Food Safety Authority (EFSA) in May 2013, EFSA scientists identified a number of risks posed to bees by BASF's controversial insecticide fipronil, especially when used as a seed treatment. The risk assessment found a 'high acute risk to honeybees' through drifting dust. Additionally, a certain exposure risk through the nectar and pollen of some treated crops was identified. As a consequence of these findings, Fipronil was banned for outdoor applications in the EU in July 2013.

Bees play a crucial role in agriculture, as they are major pollinators for plants and crop. According to a Greenpeace report in 2013, that also criticises the bee-harming properties of biocides such as fipronil, without insect pollination, about one third of the crops for nutrition would have to be pollinated by other means to avoid significant losses. Up to 75% of the crops would suffer some decrease in productivity.

According to BASF in June 2015, fipronil - which does not belong to the controversial group of neonicotinoids - does not contribute to the observed adverse effects on bee colonies when applied correctly.

Case history

Last Modification: 2013-07-19

Status: ■/□

Type: Controversial Environmental Practices

Level: Company

Countermeasures: no selection

Expiry date (expected): 2017-06

Sources:

EFSA Publication

Greenpeace Report

Company Website

Other companies involved:

back

i **Glossary**

Detail view

Here, you can find background information on the case as well as details concerning the evaluation. You can access the original sources of the case by clicking on the links in the sources area (Note: In case the author of the source changes the path to the website it will no longer be possible to gain access).

History

Here, you can access the changes which have occurred over time for each case.

Legend

Controversy Score (CS)	CS
CS Business Practices	CSP
CS Business Areas	CSA
Severe Violation	■■■
Violation	■■
Minor Controversy	■
On watch (Controversy)	■/□
On watch	□
None	—

ORBIT - oekom Responsibility Benchmarking & Information Tool

Companies | Countries | Controversial Weapons | Alerts Archive | Controversy Monitor | Download Centre

High-risk pesticides (BASF SE)

Current evaluation

Summary:
The company produces and sells pesticides which are classified as highly or extremely hazardous by the 'World Health Organization'. As proper application, adherence to respective safety measures and appropriate disposal cannot be guaranteed the substances pose a significant environmental risk. A significant amount of applied biocides do not reach the target pests but run off from fields and impact the environment in various ways. Harmful environmental impacts of pesticide use include (persistent) contamination of soil, food, water and bioaccumulation in the food chain. Additionally, the use of pesticides is linked to loss of biodiversity and elimination of key species.

Last Modification: 2014-09-30

Status: ■

Type: Controversial Environmental Practices

Level: Company

Countermeasures: no selection

Expiry date (expected): 2017-03

Sources:

Media Report

NGO Press Release

Other companies involved:

Case history

back

Glossary

Detail view

Here, you can find background information on the case as well as details concerning the evaluation. You can access the original sources of the case by clicking on the links in the sources area (Note: In case the author of the source changes the path to the website it will no longer be possible to gain access).

History

Here, you can access the changes which have occurred over time for each case.

Legend

Controversy Score (CS)	CS
CS Business Practices	CSP
CS Business Areas	CSA
Severe Violation	■■■
Violation	■■
Minor Controversy	■
On watch (Controversy)	■/□
On watch	□
None	—

ORBIT - oekom Responsibility Benchmarking & Information Tool

Companies | Countries | Controversial Weapons | Alerts Archive | Controversy Monitor | Download Centre

High-risk pesticides (BASF SE)

Current evaluation

Summary:

The company produces and sells pesticides which are classified as highly or extremely hazardous by the 'World Health Organization'. As proper application, adherence to respective safety measures and appropriate disposal cannot be guaranteed in some countries, the substances may pose significant health risks for affected farmers, residents and consumers (due to residues in food). Health problems can include neurologic and endocrine (hormone) system disorders, birth defects, cancer, and other diseases. Children are especially susceptible to the harmful effects of pesticide residues. In children, exposure to certain pesticides from residues in food can cause delayed development; disruptions to the reproductive, endocrine, and immune systems; certain types of cancer; and damage to other organs. Farmworkers are also highly vulnerable to these health threats due to intensive exposure to a variety of pesticides, either from applying these chemicals or from harvesting pesticide-sprayed agricultural products.

Case history

Last Modification: 2014-09-30

Status: ■

Type: Customer and Product Responsibility (Producer)

Level: Company (Customer and Product Responsibility)

Countermeasures: no selection

Expiry date (expected): 2017-03

Sources:

NGO Press Release

Company Website

Other companies involved:

back

i **Glossary**

Detail view

Here, you can find background information on the case as well as details concerning the evaluation. You can access the original sources of the case by clicking on the links in the sources area (Note: In case the author of the source changes the path to the website it will no longer be possible to gain access).

History

Here, you can access the changes which have occurred over time for each case.

Legend

Controversy Score (CS)	CS
CS Business Practices	CSP
CS Business Areas	CSA
Severe Violation	■■■
Violation	■■
Minor Controversy	■
On watch (Controversy)	■/□
On watch	□
None	—

oekom research

Bayer AG

	Weight	Rating
B.2.2. Sustainable chemicals management	13.0%	D+
B.2.2.1. Product and substance risk management and assessment	3.1%	C

The company conducts product and substance risk assessments and provides some information on its approach (including a clear commitment to develop and sell products that do not pose risks to human and environmental safety, targets and aftersales monitoring). No or only limited information is available on further elements such as description of the assessment process (e.g. selection criteria, hazard identification and characterisation, exposure assessments, chemical mixture toxicity), follow-up measures and risk communication).

Coverage: More than 20% of relevant products.

Comment: According to Bayer, it assesses the possible health and environmental risks of products along the entire value chain (including R&D, production, use and disposal). With regard to product stewardship, Bayer states that it take into account the precautionary principle. Furthermore, the company monitors products that are already on the market. The company has set the target to conclude the assessment of the hazard potential of all substances used in quantities exceeding one metric tonne per annum by 2020. With regard to Bayer MaterialScience, the company states that it has developed a four-stage process to prioritise, characterise, manage and communicate potential risks improved with our products in line with ICCA guidelines.

	Weight	Rating
B.2.2.2. Strategy to reduce and substitute substances of concern	3.1%	D+

According to the company, it is committed to reducing and substituting substances of concern but no details are provided on a clear strategy (including a description of the substitution process (e.g. selection criteria, identification of alternatives), targets and timeline for substitutions, progress report (e.g. number of substitution activities) as well as examples of successful substitutions and eliminations).

	Weight	Rating
B.2.2.3. Strategy to enhance product life-cycle management and design	3.1%	D+

The company states that it considers environmental aspects along the entire life cycle of products. However, no information is available on a clear strategy to enhance product life-cycle management and design (e.g. scope and limits of the life-cycle assessments, targets, clear management structures/ responsibilities).

	Weight	Rating
B.2.2.4. Responsible use of nanotechnology	1.0%	B

a. Transparency on use: The company publicly reports on whether nanotechnology is used in production processes. It does not publicly report on whether it conducts research into nanotechnology or on products containing nanoparticles.

b. Product declaration: Not applicable: The company's product range does not include products processed with nanotechnology.

c. Risk management: The company implemented some measures to minimise risks at the production site (worker exposure during the research into, manufacture, or use of nanotechnology). No or only very limited information is available on measures to control nanoparticle release in the environment.

Coverage: 100% of relevant operations.

oekom research

Bayer AG

	Weight	Rating
B.2.2.5. Major environmental controversies, fines or settlements relating to sustainable chemicals management	2.6%	D-

Research revealed controversies in recent years.

Comment: The following controversy led to a minor downgrading:

According to the European Food Safety Authority (EFSA) in January 2013, EFSA scientist identified a number of risks posed to bees by three neonicotinoid insecticides. Neonicotinoids are a class of insecticides with a common mode of action that affects the central nervous system of insects, causing paralysis and death. According to EFSA, the authority was asked by the European Commission to assess the risks associated with the use of clothianidin, imidacloprid and thiamethoxam as seed treatment or as granules, with particular regard to: their acute and chronic effects on bee colony survival and development; their effects on bee larvae and bee behaviour; and the risks posed by sub-lethal doses of the three substances. The risk assessments found that these insecticides pose a 'high risks' to honey bees through the nectar and pollen of some treated crops and through drifting dust. Bees play an crucial role in agriculture, as they are major pollinators for plants and crop. According to a Greenpeace report in 2013, without insect pollination, about one third of the crops for nutrition would have to be pollinated by other means, or they would produce significantly less food. Up to 75% of the crops would suffer some decrease in productivity.
In April 2013, the EU Commission temporarily banned three widely used pesticides (including one of the company's insecticides) blamed for a sharp fall in bee numbers. Bayer stated that it considers the decision by the European Commission to be scientifically unjustified and legally flawed. According to the company, the active ingredients in question were extensively examined with regard to their impact on bee health already during the approval procedure.
According to media reports in September 2014, Canadian beekeepers have filed a class-action lawsuit against Bayer, seeking USD 400 million in damages for allegedly the devastating effects of the neonicotinoid pesticides that have been linked to the destruction of honeybee colonies. The lawsuit claims that the negligent behavior of the company directly resulted in damage or death to bee colonies and breeding stock; contamination of beeswax, honeycomb and beehives; decreases in honey production; lost profit; and increased labor and supply costs. The lawsuit is still pending.

ORBIT - oekom Responsibility Benchmarking & Information Tool

Companies | Countries | Controversial Weapons | Alerts Archive | Controversy Monitor | Download Centre

Sale of pesticides associated with massive bee death (Bayer AG)

Current evaluation

Summary:

According to the European Food Safety Authority (EFSA) in January 2013, EFSA scientist identified a number of risks posed to bees by three neonicotinoid insecticides. Neonicotinoids are a class of insecticides with a common mode of action that affects the central nervous system of insects, causing paralysis and death. According to EFSA, the authority was asked by the European Commission to assess the risks associated with the use of clothianidin, imidacloprid and thiamethoxam as seed treatment or as granules, with particular regard to: their acute and chronic effects on bee colony survival and development; their effects on bee larvae and bee behaviour; and the risks posed by sub-lethal doses of the three substances. The risk assessments found that these insecticides pose a 'high risks' to honey bees through the nectar and pollen of some treated crops and through drifting dust. Bees play an crucial role in agriculture, as they are major pollinators for plants and crop. According to a Greenpeace report in 2013, without insect pollination, about one third of the crops for nutrition would have to be pollinated by other means, or they would produce significantly less food. Up to 75% of the crops would suffer some decrease in productivity.

In April 2013, the EU Commission temporarily banned three widely used pesticides (including one of the company's insecticides) blamed for a sharp fall in bee numbers.

Case history

Last Modification: 2013-07-19

Status: ■/□

Type: Controversial Environmental Practices

Level: Company

Countermeasures: no selection

Expiry date (expected): 2016-09

Sources:

EFSA Press Release

Greenpeace Report

Media Report

Other companies involved:

Glossary

Detail view

Here, you can find background information on the case as well as details concerning the evaluation. You can access the original sources of the case by clicking on the links in the sources area (Note: In case the author of the source changes the path to the website it will no longer be possible to gain access).

History

Here, you can access the changes which have occurred over time for each case.

Legend

Controversy Score (CS)	CS
CS Business Practices	CSP
CS Business Areas	CSA
Severe Violation	■■■
Violation	■■
Minor Controversy	■
On watch (Controversy)	■/□
On watch	□
None	—

back

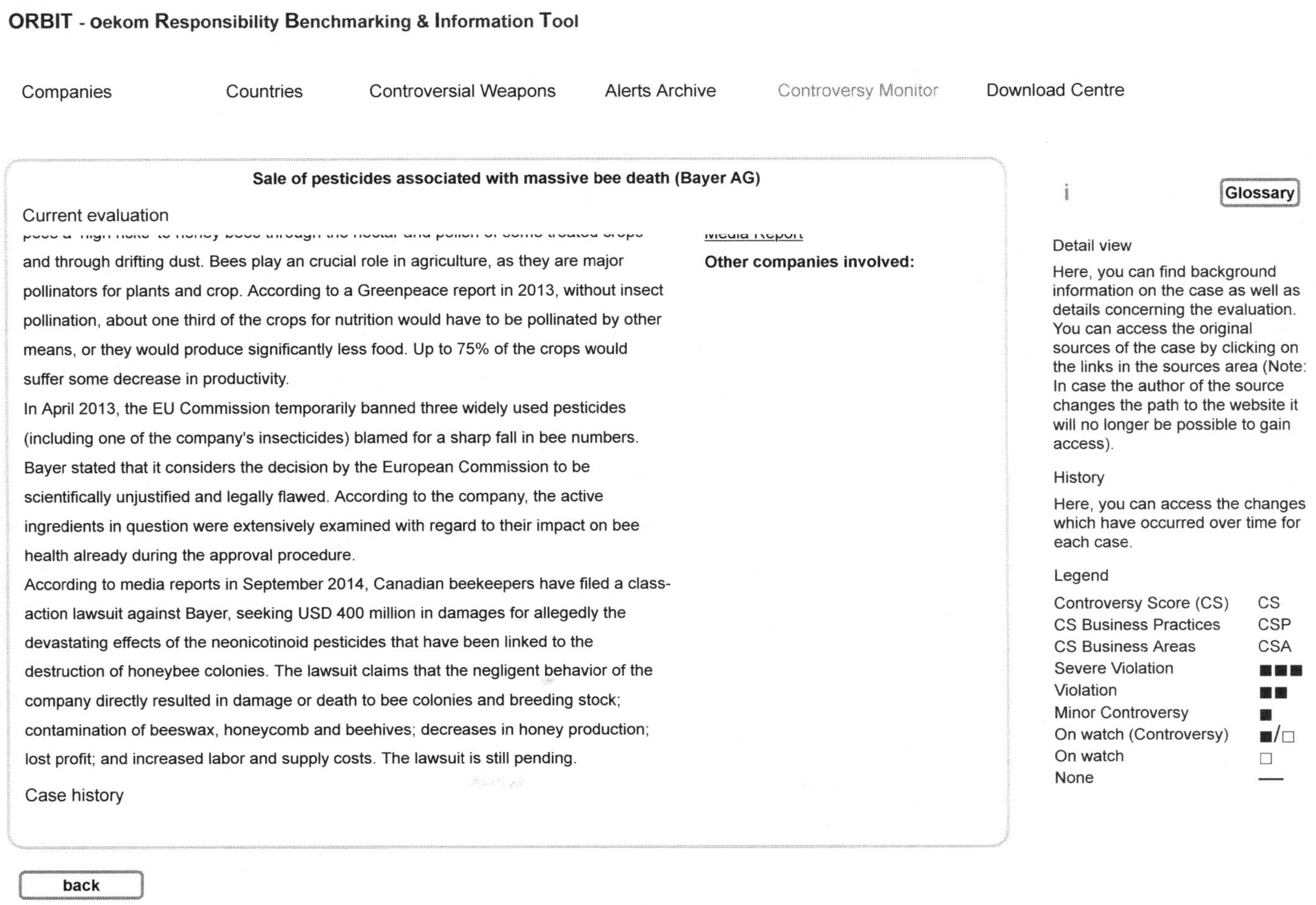

ORBIT - oekom Responsibility Benchmarking & Information Tool

Companies | Countries | Controversial Weapons | Alerts Archive | Controversy Monitor | Download Centre

Sale of pesticides associated with massive bee death (Bayer AG)

Current evaluation

and through drifting dust. Bees play an crucial role in agriculture, as they are major pollinators for plants and crop. According to a Greenpeace report in 2013, without insect pollination, about one third of the crops for nutrition would have to be pollinated by other means, or they would produce significantly less food. Up to 75% of the crops would suffer some decrease in productivity.

In April 2013, the EU Commission temporarily banned three widely used pesticides (including one of the company's insecticides) blamed for a sharp fall in bee numbers. Bayer stated that it considers the decision by the European Commission to be scientifically unjustified and legally flawed. According to the company, the active ingredients in question were extensively examined with regard to their impact on bee health already during the approval procedure.

According to media reports in September 2014, Canadian beekeepers have filed a class-action lawsuit against Bayer, seeking USD 400 million in damages for allegedly the devastating effects of the neonicotinoid pesticides that have been linked to the destruction of honeybee colonies. The lawsuit claims that the negligent behavior of the company directly resulted in damage or death to bee colonies and breeding stock; contamination of beeswax, honeycomb and beehives; decreases in honey production; lost profit; and increased labor and supply costs. The lawsuit is still pending.

Case history

Other companies involved:

back

i Glossary

Detail view

Here, you can find background information on the case as well as details concerning the evaluation. You can access the original sources of the case by clicking on the links in the sources area (Note: In case the author of the source changes the path to the website it will no longer be possible to gain access).

History

Here, you can access the changes which have occurred over time for each case.

Legend

Controversy Score (CS)	CS
CS Business Practices	CSP
CS Business Areas	CSA
Severe Violation	■■■
Violation	■■
Minor Controversy	■
On watch (Controversy)	■/□
On watch	□
None	—

oekom research

Syngenta AG

	Weight	Rating
B.2.2.5. Major environmental controversies, fines or settlements relating to sustainable chemicals management	3.1%	D-

Research revealed controversies in recent years.

Comment: The following controversies lead to a minor downgrading:

Syngenta is a major producer of thiamethoxam a controversial pesticide (neonicotinoid) linked with massive and widespread honeybee-colony death. Neonicotinoids are a class of insecticides that kill insects by attacking the central nervous system. Thiametoxam and two other neonicotinoids have been partly banned by the European Commission (and Switzerland) in response to the European Food Safety Authority's (EFSA) scientific report which identified 'high acute risks' for bees as regards exposure to dust and pollen in several crops. Recent EFSA studies suggest exposure to neonicotinoids at sub-lethal doses can harm bee health and bee colonies.

The company produces and sells pesticides which are classified as highly or extremely hazardous by the World Health Organization. As proper application, adherence to respective safety measures and appropriate disposal cannot be guaranteed the substances pose a significant environmental risk. A significant amount of applied biocides do not reach the target pests but run off from fields and impact the environment in various ways. Harmful environmental impacts of pesticide use include (persistent) contamination of soil, food, water and bioaccumulation in the food chain. Additionally, the use of pesticides is linked to loss of biodiversity and elimination of key species.

ORBIT - oekom Responsibility Benchmarking & Information Tool

Companies | Countries | Controversial Weapons | Alerts Archive | Controversy Monitor | Download Centre

High-risk pesticides (Syngenta AG)

Current evaluation

Summary:

The company produces and sells pesticides which are classified as highly or extremely hazardous by the World Health Organization. As proper application, adherence to respective safety measures and appropriate disposal cannot be guaranteed the substances pose a significant environmental risk. A significant amount of applied biocides do not reach the target pests but run off from fields and impact the environment in various ways. Harmful environmental impacts of pesticide use include (persistent) contamination of soil, food, water and bioaccumulation in the food chain. Additionally, the use of pesticides is linked to loss of biodiversity and elimination of key species.

Last Modification: 2014-09-19

Status: ■

Type: Controversial Environmental Practices

Level: Company

Countermeasures: no selection

Expiry date (expected): 2016-09

Sources:

Media Report

NGO Press Release

Company Website

Other companies involved:

Case history

back

i

Glossary

Detail view

Here, you can find background information on the case as well as details concerning the evaluation. You can access the original sources of the case by clicking on the links in the sources area (Note: In case the author of the source changes the path to the website it will no longer be possible to gain access).

History

Here, you can access the changes which have occurred over time for each case.

Legend

Controversy Score (CS)	CS
CS Business Practices	CSP
CS Business Areas	CSA
Severe Violation	■■■
Violation	■■
Minor Controversy	■
On watch (Controversy)	■/□
On watch	□
None	—

ORBIT - oekom Responsibility Benchmarking & Information Tool

Companies | Countries | Controversial Weapons | Alerts Archive | Controversy Monitor | Download Centre

High-risk pesticides in developing countries (Syngenta AG)

Current evaluation

Summary:

The company produces and sells pesticides which are classified as highly or extremely hazardous by the World Health Organization. As proper application, adherence to respective safety measures and appropriate disposal cannot be guaranteed in some countries, the substances may pose significant health risks for affected farmers, residents and consumers (due to residues in food). Health problems can include neurologic and endocrine (hormone) system disorders, birth defects, cancer, and other diseases. Children are especially susceptible to the harmful effects of pesticide residues. In children, exposure to certain pesticides from residues in food can cause delayed development; disruptions to the reproductive, endocrine, and immune systems; certain types of cancer; and damage to other organs. Farmworkers are also highly vulnerable to these health threats due to intensive exposure to a variety of pesticides, either from applying these chemicals or from harvesting pesticide-sprayed agricultural products.

Last Modification: 2014-08-13

Status: ■

Type: Customer and Product Responsibility (Producer)

Level: Company (Customer and Product Responsibility)

Countermeasures: no selection

Expiry date (expected): 2016-09

Sources:

Company Website

Other companies involved:

Case history

back

Glossary

Detail view

Here, you can find background information on the case as well as details concerning the evaluation. You can access the original sources of the case by clicking on the links in the sources area (Note: In case the author of the source changes the path to the website it will no longer be possible to gain access).

History

Here, you can access the changes which have occurred over time for each case.

Legend

Controversy Score (CS)	CS
CS Business Practices	CSP
CS Business Areas	CSA
Severe Violation	■■■
Violation	■■
Minor Controversy	■
On watch (Controversy)	■/□
On watch	□
None	—

ORBIT - oekom Responsibility Benchmarking & Information Tool

Companies | Countries | Controversial Weapons | Alerts Archive | Controversy Monitor | Download Centre

Sale of pesticides associated with massive bee death (Syngenta AG)

Current evaluation

Summary:

Syngenta is a major producer of thiamethoxam a controversial pesticide (neonicotinoid) linked with massive and widespread honeybee-colony death. Neonicotinoids are a class of insecticides that kill insects by attacking the central nervous system. Thiametoxam and two other neonicotinoids have been partly banned by the European Commission (and Switzerland) in response to the European Food Safety Authority's (EFSA) scientific report which identified 'high acute risks' for bees as regards exposure to dust and pollen in several crops. Recent EFSA studies suggest exposure to neonicotinoids at sub-lethal doses can harm bee health and bee colonies.

Last Modification: 2015-11-16

Status: ■/□

Type: Controversial Environmental Practices

Level: Company

Countermeasures: no selection

Expiry date (expected): 2017-09

Sources:

Media Report

Media Report

Company Website

Other companies involved:

Case history

back

i

Glossary

Detail view

Here, you can find background information on the case as well as details concerning the evaluation. You can access the original sources of the case by clicking on the links in the sources area (Note: In case the author of the source changes the path to the website it will no longer be possible to gain access).

History

Here, you can access the changes which have occurred over time for each case.

Legend

Controversy Score (CS)	CS
CS Business Practices	CSP
CS Business Areas	CSA
Severe Violation	■■■
Violation	■■
Minor Controversy	■
On watch (Controversy)	■/□
On watch	□
None	—

ORBIT - oekom **R**esponsibility **B**enchmarking & **I**nformation **T**ool

Companies Countries Controversial Weapons Alerts Archive Controversy Monitor Download Centre

Sale or distribution of misbranded or mislabeled pesticides (Syngenta AG)

Current evaluation

Summary:

According to media reports in May 2012, Syngenta has agreed to pay a USD 102,000 civil penalty to the US to settle a series of environmental violations related to the sale or distribution of misbranded pesticides between March and April 2011. The sale or distribution of misbranded or mislabeled pesticides can pose serious risks to human health, plant and animal life and the environment. Without proper labeling or safety instructions on packaging, users can unintentionally misapply pesticides and may not have adequate information to address needs for first aid in the event of emergency.

Last Modification: 2014-06-05

Status: Archived

Type: Customer and Product Responsibility (Producer)

Level: Company (Customer and Product Responsibility)

Countermeasures: no selection

Sources:

Media Report

Other companies involved:

Case history

back

Glossary

Detail view

Here, you can find background information on the case as well as details concerning the evaluation. You can access the original sources of the case by clicking on the links in the sources area (Note: In case the author of the source changes the path to the website it will no longer be possible to gain access).

History

Here, you can access the changes which have occurred over time for each case.

Legend

Controversy Score (CS)	CS
CS Business Practices	CSP
CS Business Areas	CSA
Severe Violation	■■■
Violation	■■
Minor Controversy	■
On watch (Controversy)	■/□
On watch	□
None	—

16

An integrated approach to bee decline

Making a bee line for the future?

Jill Atkins
University of Sheffield, UK

Barry Atkins
University of South Wales, UK

All evidence points to an ongoing and intensifying global crisis in bee populations, with local or regional declines in wild and managed pollinator populations being documented in countries across the world. As we can see from the many and varied chapters in this book, bee decline is a global issue, there is debate over the cause(s), the current and potential financial impacts are materially significant, companies and investors around the world are engaging with the problem, and society has a deep historical connection with bees which strengthens people and communities in their battle to save and restore bee populations. In this final chapter we draw together the main issues raised throughout the book and discuss some salient themes arising from the contributors' writings.

All around the world: the global phenomenon of bee loss

The statistics on bee decline and bee loss are devastating. Researchers have been recording serious declines in bee populations worldwide.[1] Yet, so many people seem unaware that there is a problem, even now. We hope fervently that this book will help to raise awareness both of global bee decline and of its potentially devastating impacts on people and planet. By way of summary, the following paragraphs provide some of the estimates on bee loss which have arisen from academic and scientific research in recent years.

Bee loss across North America has been well documented over recent decades. For the US, one survey estimated the total loss of bee colonies during the winter of 2008–2009 to be almost 29%, suggesting that between 584,000 and 771,000 colonies were lost in this period.[2] Similarly, for the winter of 2010–2011, there was an estimated total US colony loss of 29.9%.[3] US losses for 2012–2013 were even higher than previously estimated, with a survey indicating a total loss of 30.6% of US colonies over the winter. The survey also estimated bee loss in summer months and found a total loss of 25.3%. The authors of the study reported that for the entire 12-month period between April 2012 and April 2013 there was an average loss of 49.4%.[4] The latest figures produced by the US Department of Agriculture indicate a 42.1% annual loss.[5] As well as declines in honey bee colonies, there are also severe declines in wild bee populations such as bumblebees and other important pollinator species. In North America, a hefty decline of up to 96% in four bumblebee species has been recorded over the last 20 years.[6]

Across Europe, researchers have also widely documented declining bee populations for several decades. Indeed, a fall in managed honey bee colony numbers in Europe has been observed since 1965, and since 1998 there has been an increase in bee loss especially across France, Belgium, Switzerland, Germany, UK, the

1 See, for example: J.D. Evans and R.S. Schwarz, "Bees Brought to their Knees: Microbes Affecting Honey Bee Health", *Trends in Microbiology* 19(12) (2011): 614–20.

2 See D. Vanengelsdorp et al., "A Survey of Honey Bee Colony Losses in the United States, Fall 2008 to Spring 2009", *Journal of Apicultural Research* 49 (2010): 1, 7–14.

3 D. Vanengelsdorp et al., "A National Survey of Managed Honey Bee 2010–11 Winter Colony Losses in the USA: Results from the Bee Informed Partnership", *Journal of Apicultural Research* 51(5) (2012): 115–24.

4 N.A. Steinhauer et al., "A National Survey of Managed Honey Bee 2012–2013 Annual Colony Losses in the USA: Results from the Bee Informed Partnership", *Journal of Apicultural Research* 53(1) (2014): 1–18.

5 K. Kaplan, *Bee Survey: Lower Winter Losses, Higher Summer Losses, Increased Total Annual Losses* (Washington, DC: United States Department of Agriculture, 13 May 2015).

6 S.A. Cameron et al., "Patterns of Widespread Decline in North American Bumble Bees", *Proceedings of the National Academy of Sciences* 108(2) (2011): 662–67.

Netherlands, Italy, Sweden and Spain.[7] In Spain, the prevalence of "colony depopulation symptoms" among professional apiaries studied was 67.2%.[8] The intensification of agriculture in the Netherlands over the last century has been blamed for bee loss; more than half of the bee species in the Netherlands are now on their national Red List.[9] Another study provided evidence of declines in local bee diversity in both the UK and the Netherlands.[10] In Greece too, sudden deaths, tremulous movements and population declines of adult honey bees were reported in 2009 by beekeepers.[11] Eastern Europe is also no stranger to bee decline; in Poland extraordinary bee colony losses were reported for winter 2007/2008, with an average of around 15% of colonies lost.[12]

Bee loss and colony collapse disorder (CCD) are also prevalent across Africa and as we can see from Maroun's analysis in Chapter 13, South Africa is now experiencing serious problems of declining bee populations and the issue has become of political and economic significance. There is however a paucity of studies on bee decline in African countries although CCD has been reported in Egypt.[13] Across Asia, bee decline is also a serious issue. According to a UNEP report, there are 6 million bee colonies of both western and eastern species of honey bee in China and Chinese beekeepers have suffered colony loss.[14] The bumblebee trade in Japan is the focus of Chapter 4, in which Reade, Goka, Thorp, Mitsuhata and Wasbauer outline the history and development of the commercial trade in bumblebees for pollination purposes—a fascinating insight into an industry that we never knew existed before beginning the research for this book! The shift from corporate social responsibility to accountability within the Japanese business sector is an important finding of their research.

7 UNEP, *Global Honey Bee Colony Disorder and Other Threats to Insect Pollinators* (UNEP Emerging Issues, UNEP, 2010).

8 M. Higes et al., "A Preliminary Study of the Epidemiological Factors Related to Honey Bee Colony Loss in Spain", *Environmental Microbiology Reports* 2(2) (2010): 243–50.

9 J. Schepera et al., "Museum Specimens Reveal Loss of Pollen Host Plants as Key Factor Driving Wild Bee Decline in The Netherlands", *Proceedings of the National Academy of Sciences* 111(49) (2014): 17552–57.

10 J.C. Biesmeijer et al., "Parallel Declines in Pollinators and Insect-Pollinated Plants in Britain and the Netherlands", *Science* 313(5785) (2006): 351–54.

11 N. Bacandritsos et al., "Sudden Deaths and Colony Population Decline in Greek Honey Bee Colonies", *Journal of Invertebrate Pathology* 105(3) (2010): 335–40.

12 Grażyna Topolska et al., "Polish Honey Bee Colony-Loss during the Winter of 2007/2008", *Journal of Apicultural Science* 52(2) (2008): 95–104.

13 UNEP, *Global Honey Bee Colony Disorder and Other Threats to Insect Pollinators.*

14 *Ibid.*

As seen throughout this book, the potential financial ramifications of bee loss for businesses and stock markets are eye watering. As summarized by Stathers in Chapter 6, estimates of the impact of pollinator decline globally average around £130 billion per year. This is by no means a financially sustainable loss. Yet these estimates of financial loss go nowhere towards estimating the impacts on biodiversity, human health and nutrition and ultimately on the nature of life on planet Earth. Hand pollination of food crops is no longer a concept belonging to the realms of science fiction: it is already a reality in China, as we can see from discussions throughout this book.

Reasons for bee decline revisited

A whole host of reasons are given by beekeepers, scientists and other interested parties for these losses and for CCD. Throughout this book, many different culprits are identified as being responsible for CCD and especially for bee loss over winter months. As outlined in Chapter 3 by Longing and Discua, entomologists working in the US, most research has focused on the decline in managed bee populations and has found a wide variety of causes for bee loss including habitat loss, pests and disease, pesticides and nutritional deficiencies. As has been raised among the scientific community worldwide, one possible cause for CCD is the effect of neonicotinoids and their potential to (possibly) affect bees' behaviour, causing them to lose the ability to pollinate. However, the debate on whether this is a root cause of bee decline is ongoing. The controversies and legal actions involving pesticide producers are highlighted in many places throughout this book; the Canadian case covered in Chapter 8 by Clappison and Solomon is a salient illustration. There are difficulties in isolating chemical pesticides given the conflicts of interest between food producers, agrochemical companies and other interested parties. Nevertheless, scientific studies have found that chemicals used in pesticides appear to be extremely poisonous to animals. Further, scientific research has shown that chemicals in pesticides, including neonics, can cause loss of sense of direction and impair memory.[15] These are the possible effects linked to neonics which concern scien-

15 These findings are reported in the following: J.M. Bonmatin et al., "Fate of Systemic Insecticides in Fields (Imidacloprid and Fipronil) and Risks for Pollinators" (presented at the First European Conference of Apidology, Udine, 19–23 September 2004); J.M. Bonmatin et al., "Quantification of Imidacloprid Uptake in Maize Crops", *Journal of Agricultural and Food Chemistry* 53(13) (2005): 5336–41; M.E. Colin et al., "Quantitative Analysis of the Foraging Activity of Honey Bees: Relevance to the Sub-Lethal Effects Induced by Systemic Insecticides", *Archives of Environmental Contamination and Toxicology* 47 (2004): 387–95; C. Cox, "Imidacloprid, Insecticide Factsheet", *Journal of Pesticide Reform* 21(1) (2001) http://www.apiservices.com/intoxications/imidacloprid.pdf.

tists, ecologists, naturalists and the global food industry the most. A UNEP Report qualifies the findings by pointing out that laboratory studies do not necessarily replicate nature.[16] This suggests that these findings cannot be taken as definitive.

Other reasons identified to explain rising trends in bee loss figures include loss of habitat and biodiversity decline, which affect wild bees particularly badly. Adverse weather conditions, linked largely to climate change, as well as starvation in winter are identified as causal factors. Even mobile phone radiation has been raised as a possible cause of CCD. Loss of pollen host plants has been identified as a culprit, which again relates to habitat loss and commercial agricultural activities. Diseases which affect bees, carried by parasites and mites, especially *Nosema ceranae* and *Varroa destructor*, are all causes of bee death, as we saw from earlier chapters. However, these alone seem to be an insufficient explanation for current trends in bee decline and CCD. As discussed in Chapter 4 by Reade et al., the importation of "alien" bee species is a threat to native bees because of the associated introduction of foreign pests and diseases, as in the case of the Japanese bumblebee trade.

However we look at the situation and the multitude of competing and possible causes of bee decline, they all seem to have a common root cause: humans and human activity. Even threats to bees arising from diseases and natural predators/parasites seem to be exacerbated by human activity. The human race has had, in the space of a very short time (in terms of geological time and the history of the planet), an overwhelming impact on the climate, global temperatures, habitats and consequently biodiversity. There seems to be no question in our minds that bee decline is a direct result of human behaviour and activity, especially the activities of business and multinational corporations.

An integrated approach to bee decline: how can we reverse the trend?

This book has drawn together a wide range of perspectives on bee decline in order to provide an integrated approach to understanding the causes and effects and offer potential solutions to global bee decline. There are a whole array of thematic issues which arise from the accounting and accountability chapters, with similarities and differences becoming obvious between countries and continents. There are also synergies between the cultural and historical discussion of bees and humans presented in Chapter 2 and the ways in which companies are choosing to report on bee-related issues in Part III of the book. In Chapter 5, Christian provides an endearing exploration of the bumblebee from a deep ecology perspective, demonstrating the need to consider bees for their intrinsic and natural value.

16 UNEP, *Global Honey Bee Colony Disorder and Other Threats to Insect Pollinators.*

Taking an integrated approach to bee decline means that the contributors to this book have brought all aspects of human society, climate and environment into their discussions. We can see from the various chapters examining bee decline around the world that many, many industries and organizations are working incredibly hard to protect bees and other pollinators, even industries which would not be immediately obvious as interested parties. There is a worldwide crusade to protect bees.

Clearly, food security and nutrition are key areas where bees and other pollinators play a vital role. Food producers and food retailers involved in the supply chains underlying food production between plants and their final destination—the "meal on the plate"—are engaged worldwide in initiatives aimed at saving and nurturing bee populations. The studies of corporate accounts within this book provide a rich and varied source of information about the many, many projects and initiatives with which companies are involved globally to save bees, both wild bee populations and commercial bees.

Again, as we saw from the analyses of corporate bee-related disclosures, companies involved in producing luxury products such as perfumes are also extremely concerned about bee decline and are involved in many different projects internationally with the sole aim of enhancing bee populations. In a completely different field, architecture is a salient example of a profession that is involved in enhancing bee populations. Architectural companies and real estate businesses in Sweden are making huge strides to provide new urban habitats for bees, as we can see from Jonäll and Rimmel's research in Chapter 14.

In the broader cultural context of human society, there are many allusions to bee decline and related issues, as well as continual reference to bee society, honey production and bees as a representation of hard work and industry. For example, a newly published book, *Coffin Road,* brings the problems of bee decline into the heart of the fictional narrative.[17]

A unifying theme throughout the book is the crucial role that is being played by the institutional investment community in engendering change in corporate attitudes to bees through active engagement and dialogue. The responsible investment industry has grown substantially since the turn of the 21st century and institutional investors worldwide now take environmental, social and governance (ESG) issues into consideration in their investment decisions. This is because of an overwhelming perception among the global institutional investment community that ESG issues such as climate change, biodiversity loss and human rights can have financially material effects on their investment portfolios. Practitioner reports and institutional initiatives represent a principal driver of responsible investment and of the need for financial institutions to adopt a more integrated approach. To name but a few, the Principles for Responsible Investment (PRI), publications such as the Freshfields Reports (I and II), the Kay Review, and the evolution of universal

17 P. May, *Coffin Road* (Sydney: Hachette Australia, 2016).

ownership all constitute moves towards greater engagement in environmental and social issues by the global investment industry, as discussed by Herron in Chapter 7.

In other words, nature, wildlife and human suffering can hit companies' financial returns. As we can see from Part II of this book, the institutional investment community is beginning to see bee decline as one of these financially material issues. One way in which large-scale investors can address financial risks such as bee decline is to engage actively with companies in their investment portfolio. Thamotheram and Stewart, in Chapter 9, call for urgent action and engagement from the institutional investment community to tackle the bee crisis. Through a strategy of "Robust Engagement", they explain that investors, as "stewards", have the potential to stop and even reverse bee decline worldwide. Certainly, reading Chapters 6 and 7 by Stathers and Herron, respectively, we get a feeling for the power and strength of the institutional investment industry and its massive potential for transformational change. Herron, in Chapter 7, provides a list of possible questions which could be included in one-on-one meetings between institutional investors and their investee companies, in relation to bee decline. Through active strategies of engagement and dialogue with companies, today's financial institutions can represent a huge force for good. Whether "good" is pursued for financial gain and because of a "business case" does not detract from the final outcome, saving bees, which can only be good for people, planet—and financial returns. The potential power of responsible investment as a driver of change and a force for "good" cannot be overestimated. Indeed, greater and more "robust" engagement and dialogue between institutional investors and investee companies at a global level is a clear mandate arising from the research and findings contained in this book.

The chapters of this book also point to accounting and corporate reporting as an important mechanism for engendering transformation within the area of biodiversity and bee populations. We can see that companies around the world are providing (often) detailed information about their bee-related initiatives and activities. The information is currently extremely varied and differs among companies, countries and industries. The analyses in this book show that companies in countries such as Sweden and Germany are providing substantial information whereas US companies, despite the massive policy attention on pollinators, appear to be producing very little bee-related reporting. Apart from themes which come out of our analysis, there is little if any common structure or format to these corporate disclosures. However all of these disclosures are being produced in a completely voluntary reporting environment. Apart from the Global Reporting Initiative (GRI), which provides a basic framework for reporting on biodiversity and species threatened with extinction,[18] there is little guidance for companies on how they can/should disclose these types of information. We suggest that an explicit framework for bee accounting and bee accountability could assist companies not only in their reporting of bee-related information but also in identifying the areas where they

18 See Warren Maroun's discussion in Chapter 13 on this framework and on accounting for biodiversity.

need to pursue initiatives in order to halt bee decline and reduce the risks associated with the pollinator crisis more broadly.

The relatively recent development of "integrated reporting" around the world is a critical step towards potentially improving accounting for biodiversity. The increasing tendency for companies to produce "integrated reports", especially in South Africa where this is now effectively mandatory, represents a crucial development in the road towards ensuring accounting is broader and more inclusive. Integrated reporting, as discussed in Chapter 13 by Maroun, provides a basis for reporting on biodiversity and bee-related issues which have a financial impact on the company. Integrated reporting involves adopting "integrated thinking" at board level and throughout the organization. Integrated thinking means that issues such as bee decline are integrated fully into corporate strategy and business plans, as are all economic, social, environmental and governance issues which have a potentially significant financial impact on the company. In our view, integrated reporting gives companies globally a vehicle for disclosing relevant information relating to environmental issues such as biodiversity and wildlife which is far more powerful than earlier reporting frameworks. If developed appropriately, integrated reports should lead the way for companies to recognize, report and act on materially significant issues relating to the environment. Bees represent one such issue, of critical financial as well as societal importance. Through integrated reporting, we believe companies should be able to develop their strategies for tackling bee decline, as the act of producing an integrated report requires organizations to think in an integrated manner and thus should lead to changes in business behaviour.

There is a direct linkage between action and accounting, rendering disclosures such as integrated reporting emancipatory and transformational in nature. Accounting is by no means simply a calculative, technical exercise: it is an emancipatory tool which can have the potential to transform corporate activity, the environment and society.[19] Integrated reporting is about the detail, and "the devil is in the detail". It is only by drilling down into the detail, the individual items which when added together constitute a company's total material financial risk, that companies can achieve genuine, full accountability. Bees are one of many, many details which are in themselves financially significant. Any attempt at genuine accountability has to involve accounting for bees. This is the essence of what integrated reporting is about. There is a need to develop a reporting framework for bee-related disclosures. Houdet and Veldtman proffer a possible management accounting framework in Chapter 10 and further research could provide more guidance to companies attempting to integrate bees into their accounts. In Box 16.1, we provide a tentative "bee-centric framework for gently accounting for bees". This framework

19 The emancipatory potential of accounting has been recognized by accounting researchers, especially Sonja Gallhofer and Jim Haslam, in the following publications: *Accounting and Emancipation: Some Critical Interventions* (London, UK: Routledge, 2003); and S. Gallhofer et al., "Accounting as Differentiated Universal for Emancipatory Praxis", *Accounting, Auditing & Accountability Journal* 28(5) (2015): 846–74.

brings together many strands arising from the research in this book including Herron's, Stathers', and Thamotheram and Stewart's recommendations for investor engagement on bees, and the findings of research into bee accounting around the world. This tentative framework attempts to encapsulate concepts of transformative accounting and stakeholder accountability as well as corporate self-interest and a business case scenario, thereby providing what is, hopefully, an integrated approach to bee accounting and accountability.

Box 16.1 **A bee-centric framework for gently accounting for bees**

Report where, geographically, the company's activities are affected by bee decline
Report how the company's supply chain is affected by bee decline
Report the findings of studies the company has commissioned to assess the financial impact of bee decline on your business
Report the company's views on why bees are important to people and planet, and to the company's stakeholders generally: explain what the company's understanding is of stakeholder accountability
Report potential risks/impacts of bee decline on the company's operations
Report actions/initiatives taken by the company to halt bee decline and to nurture bee populations
Report educational and awareness-raising initiatives that the company has instigated among stakeholders relating to bee decline
Report decisions taken regarding the attitude of suppliers towards bee decline and their actions to ensure they nurture bee populations
Report partnerships/engagement between wildlife/nature/conservation organizations and the company which aim to address corporate impacts on bee populations
Report the outcome/impact of engagement/partnerships on bee populations
Report assessment and reflection on outcome/impact of engagement/partnerships and decisions taken about necessary changes to policy/initiatives
Report regular assessments (audit) of bee populations in areas affected by corporate operations
Report assessment of whether or not corporate initiatives/actions are assisting in halting bee decline and nurturing bee populations
Report strategy for the future development and improvement of actions/initiatives: an iterative process

From an accounting perspective, another interesting accounting-related finding from the chapters in this book is companies' preference to utilize corporate websites for bee-related disclosures rather than the published reports. Sustainability (equivalent) reporting and integrated reporting are used by some companies to discuss bee conservation initiatives they may be involved in and other bee accountability issues but generally bee accounting seems to occur on websites. This raises some questions about readership, for example. Who are these disclosures aimed at? Although we know the institutional investment community are extremely interested in the financially material impacts of bee decline, the annual report and/or integrated report is not used generally as a vehicle for bee accounting. Yet, we have seen throughout the chapters of this book that bee decline is a financial risk for companies and thus for investors which "should" be accounted for by companies that are potentially affected by the decline in populations of bees and other pollinators. Another pertinent issue relates to the different character of corporate web disclosures when compared with disclosures in published reports. As we saw from Chapter 11, companies listed on the London Stock Exchange seem to use web-based bee-related disclosures to educate "stakeholders"/readers as well as to lobby on their behalf (especially in relation to the pesticide companies). This provides insights into the ways in which corporate web disclosures differ in character, motivation and intention from disclosures in annual and other reports, where the focus is far more on financial materiality. In contrast, Swedish listed companies do appear to be using their annual and sustainability reports to discuss bee-related issues as shown in Chapter 14.

As with responsible investment, corporate disclosures, reporting and business communications generally have a massive potential for change. The act of reporting information requires there to be something to report. Also, the act of reporting feeds back to corporate activity as it can provide an imperative for the company to reflect on past practice and helps to inform practice in future. In other words, social accounting and certainly bee accounting can be transformative, or emancipatory, in nature.

Longing and Discua explore conservation efforts in depth in Chapter 3. The cultivation of large crop monocultures and its associated loss of biodiversity have been cited as a key cause of pollinator decline. Efforts by agricultural producers to increase floral diversity around crops and leave areas of undisturbed wild land between agricultural fields can only help encourage wild bees. Organic farming also seems to be associated with improved bee populations. Chapter 3 describes many initiatives around the world aimed at increasing biodiversity in agricultural areas. The authors also discuss the importance of encouraging bee populations to thrive in urban areas through the integration of roof top gardens, hives and bee hotels within the urban landscape. Chapter 14 provides examples of a wide range of initiatives employed in Swedish cities. Government and political involvement is crucial if bee populations are to be saved. As we can see from Chapter 12, Romi and Longing highlight the significant US governmental initiatives being implemented across the US to halt bee decline, since the US President established the Pollinator Health Task Force in 2014. Biehl and Macpherson show, in Chapter 15, how initiatives such as "Germany hums" can start to tackle bee decline. Indeed, the extent of the worldwide campaign on bee decline by a whole spectrum of wildlife organizations, nature lobby groups and the like is immense, as demonstrated throughout this book. The actions of lobby groups and activists are crucial to any future success in allaying bee decline. If neonics, and pesticides in general, are the root cause of bee decline, then surely the answer is to stop using them. As we saw from discussions throughout this book, the moratorium across the European Union on the use of neonicotinoids has been in place for a couple of years—why should their use be reintroduced if there is a possibility of harm to bee populations? The resonance between current concerns about neonics and concerns in the 1960s about DDT, as expounded in Rachel Carson's *Silent Spring*, is only too obvious.

Greater interconnectivity of academic disciplines can assist in bringing together vital pieces of the global jigsaw of bee decline. The "siloing" of academic disciplines is a real problem. Scientists publish in scientific academic journals. Accountants publish in esoteric accounting journals. Economists, political scientists, philosophers and entomologists all publish in their own academic journals. These journals have their own in-house writing style, are usually esoteric in expression and are seldom read by those in other fields. Indeed, it has been a huge challenge to read papers during our research for this book, from a wide range of disciplines, with difficulties in adapting to different styles of writing, difficult and diverse methods and methodological approaches, technical language and overall opaque discourse. This is, of course, not necessarily the fault of individual researchers within the academic community but rather is a direct result of the need for academics to "hit" top journals in their field as a key performance indicator: "Publish or Perish". And publish in the "right" places. It is fair to say that the higher the journal in the journal rankings, the more difficult it usually is for a layperson to read! Somehow this needs to change so that sharing of critically important findings is faster between disciplines. A systems approach, based on integrated thinking and on something akin to general systems theory is the way forward, in our view, if these academic silos are to be broken down and true interdisciplinary work is to evolve.

Although there are a whole range of reasons why bees are dying and potentially more or less significant causes of bee population decline, we feel that taking an integrated view allows us to see these in a different context. Rather than focusing on whichever cause seems to be most supported (by science) at the time, it is, instead, important to realize that it is the combination of causes, which are all generated by human activity, that is destroying bees and other pollinators. This is having, and will have, untold, unknowable consequences on the ecosystem, on other species of flora and fauna, and ultimately on all life on planet Earth. Bickering over which of the many causes is "the main one" does not seem to be an answer. Rather, each and every possible cause of bee decline needs to be addressed immediately and with urgency, whether or not it has substantial or less substantial scientific "proof". If every attempt is not made to reduce the influence of human and especially business activities on bee populations and on the lives of bees, then we are heading for untold disaster. Current attempts tend to be driven by the interests of lobby groups or businesses at the expense of "sorting the problem out".

Bee decline has certainly offered plentiful opportunities for researchers and scientists worldwide to apply for funding, set up research centres, run critically important research projects and, of course, publish voluminously in academic journals.

All this is necessary and entirely understandable as a reaction to bee decline and the very real threats of bee decline on food security, financial markets and businesses, as we have seen throughout this book. However, it may make us stop and wonder whether all this fervent research activity is leading to behavioural change quickly enough and deeply enough to stem increasing decline in bee populations and potential extinction of bee species. Yes, research is crucial for humans to understand better how their activities have led us into our current situation, but at some point shouldn't we stop researching and turn to action? As we can see from this book, there is now extensive action: bans on pesticide use among others. Is it all too little too late? Why are there still ongoing debates and controversies over what the main causes are? Surely it would be more sensible to stop all potential causes that we are aware of, immediately? There are a multitude of solutions offered throughout the chapters of this book. The best approach seems to be to continue adopting all of them!

It could be argued that this book is not about bees at all. It uses bees as an exemplar of a much greater and deeper issue: the decline of species, the threats of climate change, the failings of an anthropocentric approach, the rape of the Earth and the failure of people and planet to work together. We could have taken any species under threat, any of the 10,000 or so species on the IUCN's list of highly endangered species, and been able to show very similar phenomena and raise the same basic issues and problems.

The delicate balance between species in the natural world is crucial to survival. By altering this balance and tampering with nature, humans have produced serious deleterious effects on the planet and its species. If human society continues "business as usual" and certainly "bees' business as usual" then not only bees and other species, but also humans themselves could be heading for extinction. Drastic action is required. Is it too late to prevent further global warming and to counteract climate change? These are questions in minds worldwide. Is it too late to address the decline in bee populations and populations of other essential pollinators? Again, many people from many different communities are seeking to address this question.

In this book we have attempted to summarize the current perspectives of the scientific, financial, corporate and societal communities on bee decline. Maybe by taking a deep ecology and bee-centric perspective we are actually also embracing an anthropocentric perspective of life on Earth and bee decline. It seems increasingly likely that if bees continue to disappear then human populations will also perish. If we put bees first, they will thrive—and so will we. So, what can we do as individuals in the face of the global bee crisis? Well, you could always do what we have done—pop down to the local nursery and buy a large tray of "bee-friendly" flowering plants and put them in your garden, on your balcony or even on your windowsill. There is nothing lovelier than relaxing in a summer garden listening to the slow, gentle sound of bumblebees going about their business. In the words of Rachel Carson:

We stand now where two roads diverge. But unlike the roads in Robert Frost's familiar poem, they are not equally fair. The road we have long been traveling is deceptively easy, a smooth superhighway on which we progress with great speed, but at its end lies disaster. The other fork of the road—the one "less traveled by"—offers our last, our only chance to reach a destination that assures the preservation of our earth.[20]

20 R. Carson, *Silent Spring* (London: Penguin, 2000).

References

Bacandritsos, N., Granato, A., Budge, G., Papanastasiou, I., Roinioti, E., Caldon, M., Falcaro, C., Gallina, A. and Mutinelli, F. "Sudden Deaths and Colony Population Decline in Greek Honey Bee Colonies". *Journal of Invertebrate Pathology* 105(3) (2010): 335–40.

Biesmeijer, J.C., Roberts, S.P.M., Reemer, M., Ohlemüller, R., Edwards, M., Peeters, T., Schaffers, A.P., Potts, S.G., Kleukers, R., Thomas, C.D., Settele, J., and Kunin, W.E. "Parallel Declines in Pollinators and Insect-Pollinated Plants in Britain and the Netherlands". *Science* 313(5785) (2006): 351–54.

Bonmatin, J.M., Marchand, P.A, Charvet, R., and Colin, M.E. "Fate of Systemic Insecticides in Fields (Imidacloprid and Fipronil) and Risks for Pollinators", 1994. Presented at the First European Conference of Apidology, Udine, 19–23 September 2004.

Bonmatin J. M., Marchand P.A., Charvet, R., Moineau, I., Bengsch, E.R. and Colin, M.E. "Quantification of Imidacloprid Uptake in Maize Crops". *Journal of Agricultural and Food Chemistry* 53(13) (2005): 5336–41.

Cameron, S.A., Lozier, J.D., Strange, J.P., Koch, J.B., Cordes, N., Solter, L.F. and Griswold, T.L. "Patterns of Widespread Decline in North American Bumble Bees". *Proceedings of the National Academy of Sciences* 108(2) (2011): 662–67.

Carson, R. *Silent Spring*. London: Penguin, 2000.

Colin, M. E., Bonmatin J. M., Moineau, I., Gaimon, C., Brun, S. and Vermandere, J.P. "A Method to Quantify and Analyze the Foraging Activity of Honey Bees: Relevance to the Sub-Lethal Effects Induced by Systemic Insecticides". *Archives of Environmental Contamination and Toxicology* 47 (2004): 387–95.

Cox, C. "Imidacloprid, Insecticide Factsheet". *Journal of Pesticide Reform* 21(1) (2001) http://www.apiservices.com/intoxications/imidacloprid.pdf.

Evans, J.D. and Schwarz, R.S. "Bees Brought to their Knees: Microbes Affecting Honey Bee Health". *Trends in Microbiology* 19(12) (2011): 614–20.

Gallhofer, S. and Haslam, J. *Accounting and Emancipation: Some Critical Interventions*. London, UK: Routledge, 2003.

Gallhofer, S., Haslam, J. and Yonekura, A. "Accounting as Differentiated Universal for Emancipatory Praxis". *Accounting, Auditing & Accountability Journal* 28(5) (2015): 846–74.

Higes, M., Martín, Hernández, R., Martínez, Salvador, A., Garrido-Bailón, E., González-Porto, A.V., Meana, A., Bernal, J.L., Del Nozal, M.J. and Bernal, J. "A Preliminary Study of the Epidemiological Factors Related to Honey Bee Colony Loss in Spain". *Environmental Microbiology Reports* 2(2) (2010): 243–50.

Kaplan, K. *Bee Survey: Lower Winter Losses, Higher Summer Losses, Increased Total Annual Losses*. Washington, DC: United States Department of Agriculture, 13 May 2015.

May, P. *Coffin Road*. Sydney: Hachette Australia, 2016.

Schepera, J., Reemerb, M., van Katsa, R., Ozingac, W.A., van der Lindene, G.T., Schaminéec, J.H., Siepelf, H. and Kleijna, D. "Museum Specimens Reveal Loss of Pollen Host Plants as Key Factor Driving Wild Bee Decline in The Netherlands". *Proceedings of the National Academy of Sciences* 111(49) (2014): 17552–57.

Steinhauer, N.A., Rennich, K., Wilson, M.E., Caron, D.M., Lengerich, E.J., Pettis, J.S., Rose, R., Skinner, J.A., Tarpy, D.R., Wilkes, J.T. and VanEngelsdorp, D. "A National Survey of Managed Honey Bee 2012–2013 Annual Colony Losses in the USA: Results from the Bee Informed Partnership". *Journal of Apicultural Research* 53(1) (2014): 1–18.

Topolska, Grażyna, Anna Gajda, and Aleksandra Hartwig. "Polish Honey Bee Colony-Loss during the Winter of 2007/2008". *Journal of Apicultural Science* 52(2) (2008): 95–104.

UNEP. *Global Honey Bee Colony Disorder and Other Threats to Insect Pollinators*. UNEP Emerging Issues, UNEP, 2010.

Vanengelsdorp, D., Hayes, J., Underwood, R. M. and Pettis, J. S. "A Survey of Honey Bee Colony Losses in the United States, Fall 2008 to Spring 2009". *Journal of Apicultural Research* 49 (2010): 1, 7–14.

Vanengelsdorp, D., Caron, D., Hayes, J., Underwood, R., Henson, M., Rennich, K., Spleen, A., Andree, M., Snyder, R., Lee, K. and Roccasecca, K. "A National Survey of Managed Honey Bee 2010–11 Winter Colony Losses in the USA: Results from the Bee Informed Partnership". *Journal of Apicultural Research* 51(5) (2012): 115–24.

A final dedication

We dedicate this page to all those who have made, or are tempted to make, bee quips, puns or jokes to either of us. Let's try and get them all out of the way in one fell swoop:

> He entered the room, which was abuzz with the low hum of muted voices. His eyes rested on her, the queen bee, drawing men towards her like bees to a honey-pot. She stood proudly in her black and gold dress and stinging heels. He waggled slowly towards her, drawing in the honeyed scent. "How are you?" he droned. "Buzz off", she replied tartly. Stung, for an instant, he then regained his composure. "Darling, you seem so cold". She was surveying the room while he spoke, her antennae clearly on high alert. Dumbledore, the old butler, offered them both a cocktail. "Do you find it hot in here?" he queried. "S'warm", she replied.

About the authors

Barry Atkins has a background in script-writing and editing for BBC Radio, commercial and satellite TV, including commissions for "Weekending", "Spitting Image", *inter alia.* Barry has worked in the social care sector for a number of years and is currently training as a therapist at the University of South Wales while practising as a voluntary counsellor for a mental health charity.

Jill Atkins holds a Chair in Financial Management at Sheffield University Management School, the University of Sheffield, and is also a visiting professor at the University of the Witwatersrand, South Africa. She previously worked as a professor at Henley Business School and King's College London. Her research focuses on several areas including responsible investment, stakeholder accountability, social accounting, integrated reporting and corporate governance. Jill chairs the British Accounting & Finance Association's Special Interest Group on Corporate Governance and enjoys organizing conferences that bring together governance specialists from the academia, corporate and investor communities.

Christoph Biehl is currently working as a lecturer in Accounting for the Henley Business School in the United Kingdom. He is an active member of the Henley Centre for Governance, Accountability and Responsible Investment (GARI). His research is focused on topics in responsible investment and corporate governance. In the past Christoph has worked as Project Manager Asia for the Centre for Responsible Banking and Finance at the University of St Andrews and as consultant to the UN PRI Academic Network.

Jack Christian is a Senior Lecturer in the Accounting, Finance and Economics department at Manchester Metropolitan University Business School, UK. His main areas of interest are ethics and sustainability accounting. Prior to joining the Business School in 2007 he was a practising accountant for 30 years holding various positions including Finance Director and European Financial Controller. His interest in nature dates back to his childhood in the 1960s and in the intervening years he has taken part in many voluntary survey schemes and contributed records to bird, butterfly and other naturalist organizations.

Margaret Clappison is a member of the Chartered Professional Accountant Association in Canada. She has an undergraduate degree in Independent Studies (BIS) from the University of Waterloo, an MBA from Athabasca University and a Certificate in Executive Leadership from CGA-BC. Margaret is currently working in property management and has started a position as an academic expert at Athabasca University. She has worked in public practice in the Okanangan Valley of British Columbia and in the United Kingdom. Margaret's strong commitment to high professional standards prompted her to participate as a mentor in the CPA, CGA Ontario student-mentoring programme. Her main research interests are corporate social responsibility, corporate governance, accounting the gap between accounting standards and practices, ethics, and strategy.

Samuel Discua has a Master's degree in entomology from The Ohio State University in Columbus, Ohio and is currently working on his PhD in Plant and Soil Science from Texas Tech University in Lubbock, Texas. Samuel has conducted research on the adult emergence phenology of the emerald ash borer (*Agrilus planipennis*) and nine endemic insects of the Monahans sandhills in western Texas. Samuel's current research involves assessing the biodiversity of native bees on the Southern High Plains of Texas, evaluating local plant attractiveness to native pollinators and determining the impacts of historical land use on native bee diversity.

Koichi Goka is principal researcher, Invasive Species Research Team, Environmental Risk Research Center, National Institute for Environmental Studies, Japan. His research areas include ecological risk assessment of invasive alien species. He is actively involved in development of methods and systems for controlling invasive species, and has published on invasive bumblebees in Japan. His laboratory recently hosted a televised visit by the Emperor and Empress of Japan regarding these issues.

Abigail Herron leads responsible investment and corporate governance engagement across all asset classes and markets at Aviva Investors. She complements this work with public policy advocacy in the UK, EU and UN on a spectrum of issues from the capital markets union through to fiduciary duty, green bonds and antimicrobial resistance. Her specialisms include climate change, the built environment, sustainable finance, reputational risk and the food sector incorporating fisheries, soil depletion, palm oil and animal welfare. She focused on the impact of a loss of pollinators on investment returns when completing the Postgraduate Certificate in Sustainable Business at the University of Cambridge.

Joël R.A. Houdet holds Senior Research Fellowships at the African Centre for Technology Studies (ACTS, Nairobi, Kenya) and the Albert Luthuli Centre for Responsible Leadership (ALCRL, University of Pretoria, South Africa). Joël is based in Johannesburg, South Africa. He is an expert on corporate natural capital accounting, valuation and reporting and is involved in several high profile initiatives, including the drafting of the Natural Capital Protocol and several work streams of the Intergovernmental Platform on Biodiversity and Ecosystem Services (IPBES). Dr Houdet also works as an independent consultant at Integrated Sustainability Services, where he has acquired more than 5 years of experience in sustainability reporting, biodiversity offsetting, cost–benefit analysis, economic impact assessments and independent reviews/audits. Joël holds a PhD in Management Sciences from AgroParisTech

(France), a Master's in Practicing Accounting from Monash University (Australia) and Bachelor of Sciences from Rhodes University (South Africa).

Kristina Jonäll, PhD, is a senior lecturer at the School of Business, Economics and Law at Gothenburg University in Sweden. She received her PhD in 2009 for the thesis "Bilden av det Goda Företaget: text och siffror i VD-brev" ("Image of the Good Company: text and numbers in the CEO letters"). She is part of the research programme "Accounting for Sustainability: Communication through Integrated Reporting".

Scott Longing is an assistant professor of entomology in the Department of Plant and Soil Science at Texas Tech University. He completed graduate degrees in entomology from the University of Arkansas and Virginia Tech University. Scott conducts research on both aquatic and terrestrial insect communities across unique ecosystems of the southern US, with a focus on determining distributions, habitat associations and threats of endemic species of concern. Scott is currently directing graduate and undergraduate students on multiple projects involving pollinator ecology and conservation in the agriculturally dominant region of the US Southern High Plains.

Martina Macpherson is the Founder and Managing Partner of SI Partners Ltd, an independent consulting firm specializing in Economic, Environmental, Social and Governance (E-ESG) research, content and engagement services. Martina is a board member at the Network for Sustainable Financial Markets and an NED board member and chair at the Global Thinkers Forum. She also sits on UKSIF's analyst committee board, and on the board of Thomas Arneway Trust. Martina is a member of the Institute of Directors, and an affiliate member of ICSA. Previously, she worked at Hermes EOS and held various business, research and product development roles at MSCI, Lloyds, RBS, F&C and Deutsche Bank.

Warren Maroun is a Chartered Accountant (South Africa) and member of the Chartered Institute of Management Accountants. He joined PricewaterhouseCoopers in 2006 and the Witwatersrand in 2010 where he serves as a professor in financial reporting. He also holds a visiting managerial position at PricewaterhouseCoopers and has a PhD from King's College London.

Masahiro Mitsuhata is a researcher in the Integrated Pest Management team at ArystaLifeScience, Japan. He is a pollination specialist and technical adviser on bumblebees. He researches foraging behaviour and the effect of bumblebee pollination on crops. His work has made a significant contribution to the commercialization of Japanese native bumblebees. He is actively involved in the spread of scientific education on the usage of commercialized native bumblebees.

Carol Reade is professor of international management, San José State University. She is a regular visiting faculty at Sophia University in Tokyo, where she teaches Japanese Business and Management. Her research focuses on the interface between the global firm and the external environment. Areas of interest include culture, indigenous knowledge, societal conflict and stakeholder relations with regard to protecting the natural environment, alleviating poverty and fostering peaceful societies. She has participated in entomological field expeditions and has had the honour of having two new species named for her.

Gunnar Rimmel is full Professor of Accounting at Jönköping International Business School, Jönköping University, in Sweden. He is also affiliated with the Gothenburg Research Institute at the University of Gothenburg. His research and teaching interests include accounting communication, human resource accounting, international financial accounting and social and environmental reporting, specifically integrated reporting. During the past years his research programme "Accounting for Sustainability: communication through integrated reporting" has been externally funded by the Handelsbanken research foundations. He has also received grants from the NASDAQ OMX Nordic Foundation.

Andrea Romi, PhD, CPA, has been an Assistant Professor in the School of Accounting at the Rawls College of Business at Texas Tech University since the summer of 2012. Professor Romi's research interests include environmental and social accounting and corporate sustainability. She has worked with the Sustainability Consortium, among other leading research institutions and auditing firms, focusing efforts on sustainability issues. Andrea also has experience with both non-governmental organizations as well as regulators (e.g. CERES, AICPA, and SEC) in moving the corporate sustainability movement forward. Her research has been integrated into policy papers lobbying regulators to enhance corporate environmental and social disclosures.

Aris Solomon holds a PhD (*An Investigation into a Conceptual Framework for Environmental Reporting*) from Sheffield University, UK; a BA in Business Studies, Major Accounting from Middlesex University, UK; a BA in Business Studies, Hons, Major Accounting from Westminster University, UK; and an MA in Accounting and Finance from Essex University, UK. Aris is currently teaching at Athabasca University. He has taught at Manchester University School of Accounting, Sheffield University Management School, The Open University Business School (UK), Cardiff University Business School and Exeter University. His main research interests are corporate social responsibility, corporate governance, the integration of accounting techniques and practices into corporate social responsibility, and conceptual frameworks.

Rick Stathers has worked in responsible investment since 2000. He has been responsible for raising awareness about the materiality of sustainable development issues within the investment industry and investment processes. He was the Head of Responsible Investment at Schroders where he oversaw the development of processes to integrate sustainable development factors into the investment process as well as conceiving of the Schroders Global Climate Change fund in 2006. He has a BSc in Agriculture and Food Sciences and an MSc in Water Resource Management.

Olivia Stewart was Head of Communications for Preventable Surprises from 2015 to 2016. Prior to this, she held positions across the corporate and charitable sectors and was a founding member of a communication rights organization working across the UK, Asia and Africa. She holds a first class undergraduate degree with a focus on global politics and the environment and a Master of Science with distinction from the School of Oriental and African Studies, University of London.

Raj Thamotheram is the CEO/founder of Preventable Surprises. He is a well-recognized thought-leader on how companies and investors can adapt to put people and planet on par with profit and so deliver long-term value to their clients/customers and society. He has held senior positions in the investment management industry (USS and AXA IM) and has also worked in the NGO world (international head of advocacy at ActionAid and first manager of the Ethical Trading Initiative). He speaks and writes widely including a monthly column in *IPE* ("Long term matters"). He trained in medicine and considers this work to be an extension of his interests in public health.

Robbin Thorp is professor emeritus of entomology at University of California, Davis. A world authority on bumblebees, his research interests include bee biology, pollination ecology, foraging behaviour and management of bee populations, and systematics and ecology of bees with an emphasis on the genus *Bombus*. He is actively involved in issues related to vernal pool pollination, urban bee gardens, native bee pollination of crops and bumblebee decline.

Ruan Veldtman works in the field of applied biodiversity research specializing in ecological entomology with a PhD from the University of Pretoria. His particular interests include wild silk moth ecology, plant–insect interactions, biological control of plant invasions, pollination ecosystem services and invasive wasp management. He is also an avid supporter of interdisciplinary exchanges around entomology such as agricultural economics and sustainable agricultural production. Recently he participated in the IPBES rapid pollination assessment as a lead author in "Chapter 4: Economic Valuation of Pollinator Gains and Losses".

Marius Wasbauer is research associate at the Bohart Museum of Entomology at University of California, Davis. He is an internationally recognized authority on spider wasps and other solitary wasps. Research interests include pompilid taxonomy, morphology and behaviour. Formerly with the California Department of Food and Agriculture as senior scientist, Insect Biosystematics, he was first to identity the presence of Africanized honey bees ("killer bees") in California and participated in the subsequent state-wide eradication efforts.